I0021461

Build Your Own Programming Language

Second Edition

A programmer's guide to designing compilers, interpreters, and DSLs for modern computing problems

Clinton L. Jeffery

BIRMINGHAM—MUMBAI

Build Your Own Programming Language

Second Edition

Senior Publishing Product Manager: Denim Pinto

Acquisition Editor: Peer Reviews: Gaurav Gavas

Project Editor: Parvathy Nair

Content Development Editor: Elliot Dallow

Copy Editor: Safis Editing

Technical Editor: Aneri Patel

Proofreader: Safis Editing

Indexer: Hemangini Bari

Presentation Designer: Ajay Patule

Developer Relations Marketing Executive: Vipanshu Pareshar

First published: December 2021

Second edition: January 2024

Production reference: 3150524

Published by Packt Publishing Ltd.

Grosvenor House

11 St Paul's Square

Birmingham

B3 1RB, UK.

ISBN 978-1-80461-802-8

www.packt.com

Foreword

In the dynamic world of computer science, the creation of a programming language stands as a testament to ingenuity and a deep understanding of computational principles. *Build Your Own Programming Language* is not just a guide; it is an invitation to delve into the complexity and beauty of programming language creation.

At the helm of this voyage is Clinton L. Jeffery, a distinguished professor and Chair of the Department of Computer Science and Engineering at the New Mexico Institute of Mining and Technology. His academic journey, marked by degrees from the University of Washington and the University of Arizona, has been a path of relentless exploration in the realms of programming languages, program monitoring, and visualization, among others. His work culminates in the creation of the Unicon programming language, a testament to his expertise and vision.

This book is structured to guide the reader through the nuanced process of developing a programming language. Beginning with motivations and types of language implementations, Jeffery sets the stage for understanding the fundamental *"why"* behind language design. He intricately discusses organizing a bytecode language and differentiates between programming languages and libraries, laying a solid foundation for both novices and experienced programmers.

The detailed chapters delve into the heart of language design, parsing, and the construction of syntax trees, with practical examples and case studies like the development of Unicon and the Jzero language. Jeffery's approach is meticulous, ensuring that readers grasp the essentials of technical requirements, lexical categories, context-free grammar, and symbol tables. This comprehensive coverage ensures that readers are not just following instructions but are truly understanding the principles at play.

What makes this book exceptional is its blend of theoretical knowledge and practical application. Jeffery does not shy away from the complexities of designing graphics facilities or tackling syntax trees and symbol tables. Instead, he embraces these challenges, guiding the reader with clarity and insight. The inclusion of questions at the end of each chapter prompts critical thinking and reflection, reinforcing the overall learning experience.

As you progress through *Build Your Own Programming Language*, you will find yourself not just acquiring knowledge, but also developing a new perspective on programming languages. They are not merely tools for tasks but are expressive mediums that reflect human creativity and problem-solving skills.

Clinton L. Jeffery, with his extensive experience and pioneering work in Unicon, provides a comprehensive and enlightening guide for anyone interested in the art and science of programming language development. Whether you are a student, a professional programmer, or an enthusiast of computer science, this book is a beacon, illuminating the path to understanding and creating your own programming language.

Welcome to a journey of discovery, creativity, and technical mastery in the world of programming languages!

Imran Ahmad, PhD

Senior Data Scientist, Canadian Federal Government

Contributors

About the author

Clinton L. Jeffery is Professor and Chair of the Department of Computer Science and Engineering at the New Mexico Institute of Mining and Technology. He received his B.S. from the University of Washington, and M.S. and Ph.D. degrees from the University of Arizona, all in computer science. He has conducted research and written many books and papers on programming languages, program monitoring, debugging, graphics, virtual environments, and visualization. With colleagues, he invented the Unicon programming language, hosted at unicon.org.

Steve Wampler, Sana Algaraibeh, and Phillip Thomas provided valuable feedback and suggestions for improving this book.

About the reviewers

Steve Wampler was awarded a Ph.D. in Computer Science by the University of Arizona. He has worked as an Associate Professor of Computer Science as well as a software designer for several major telescope projects, including the Gemini 8m telescopes project and the Daniel K Inouye solar telescope. He has been a software reviewer for a number of other major telescope systems and a technical reviewer for several other programming books.

Sana Algaraibeh was awarded her Ph.D. in Computer Science from the University of Idaho. She joined the faculty at the New Mexico Institute of Mining and Technology as a Computer Science instructor in 2022. Prior to that, she worked in academia for 14+ years as a lecturer, trainer, team leader, instructional designer, and Computer Science department chair at universities in Jordan and Saudi Arabia. She teaches Internet and Web Programming, Object-Oriented Programming, Python for Data Science, Algorithms and Data Structures, and Introduction to Programming. Her area of scholarship is computer science education and compiler error messages. She is interested in developing computational solutions integrated with modern pedagogy.

Join our community on Discord

Join our community's Discord space for discussions with the authors and other readers:

https://discord.com/invite/zGVbWaxqbw

Table of Contents

Section III: Code Generation and Runtime Systems   273

Chapter 11: Preprocessors and Transpilers   275

Chapter 16: Domain Control Structures   417

Chapter 17: Garbage Collection   435

Chapter 18: Final Thoughts   457

Section IV: Appendix   469

Appendix: Unicon Essentials   471

Preface

This second edition was begun primarily at the suggestion of a first edition reader, who called me one day and explained that they were using the book for a programming language project. The project was not generating code for a bytecode interpreter or a native instruction set as covered in the first edition. Instead, they were creating a transpiler from a classic legacy programming language to a modern mainstream language. There are many such projects, because there is a lot of old code out there that is still heavily used. The Unicon translator itself started as a preprocessor and then was extended until it became in some sense, a transpiler. So, when Packt asked for a second edition, it was natural to propose a new chapter on that topic; this edition has a new *Chapter 11* and all chapters (starting from what was *Chapter 11* in the previous edition) have seen their number incremented by one. A second major facet of this second edition was requested by Packt and not my idea at all. They requested that the IDE syntax coloring chapter be extended to deal with the topic of adding syntax coloring to mainstream IDEs that I did not write and do not use, instead of its previous content on syntax coloring in the Unicon IDEs. Although this topic is outside my comfort zone, it is a valuable topic that is somewhat under-documented at present and easily deserves inclusion, so here it is. You, as the reader, can decide whether I have managed to do it any justice as an introduction to that topic.

After 60+ years of high-level language development, programming is still too difficult. The demand for software of ever-increasing size and complexity has exploded due to hardware advances, while programming languages have improved far more slowly. Creating new languages for specific purposes is one antidote for this software crisis.

This book is about building new programming languages. The topic of programming language design is introduced, although the primary emphasis is on programming language implementation. Within this heavily studied subject, the novel aspect of this book is its fusing of traditional compiler-compiler tools (Flex and Byacc) with two higher-level implementation languages. A very high-level language (Unicon) plows through a compiler's data structures and algorithms like butter, while a mainstream modern language (Java) shows how to implement the same code in a more typical production environment.

One thing I didn't really understand after my college compiler class was that the compiler is only one part of a programming language implementation. Higher-level languages, including most newer languages, may have a runtime system that dwarfs their compiler. For this reason, the second half of this book spends quality time on a variety of aspects of language runtime systems, ranging from bytecode interpreters to garbage collection.

Who this book is for

This book is for software developers interested in the idea of inventing their own language or developing a domain-specific language. Computer science students taking compiler construction courses will also find this book highly useful as a practical guide to language implementation to supplement more theoretical textbooks. Intermediate-level knowledge and experience of working with a high-level language such as Java or C++ are required in order to get the most out of this book.

What this book covers

Chapter 1, Why Build Another Programming Language?, discusses when to build a programming language, and when to instead design a function library or a class library. Many readers of this book will already know that they want to build their own programming language. Some should design a library instead.

Chapter 2, Programming Language Design, covers how to precisely define a programming language, which is important to know before trying to build a programming language. This includes the design of the lexical and syntax features of the language, as well as its semantics. Good language designs usually use as much familiar syntax as possible.

Chapter 3, Scanning Source Code, presents lexical analysis, including regular expression notation and the tools Ulex and JFlex. By the end, you will be opening source code files, reading them character by character, and reporting their contents as a stream of tokens consisting of the individual words, operators, and punctuation in the source file.

Chapter 4, Parsing, presents syntax analysis, including context-free grammars and the tools iyacc and byacc/j. You will learn how to debug problems in grammars that prevent parsing, and report syntax errors when they occur.

Chapter 5, Syntax Trees, covers syntax trees. The main by-product of the parsing process is the construction of a tree data structure that represents the source code's logical structure. The construction of tree nodes takes place in the semantic actions that execute on each grammar rule.

Chapter 6, Symbol Tables, shows you how to construct symbol tables, insert symbols into them, and use symbol tables to identify two kinds of semantic errors: undeclared and illegally redeclared variables. In order to understand variable references in executable code, each variable's scope and lifetime must be tracked. This is accomplished by means of table data structures that are auxiliary to the syntax tree.

Chapter 7, Checking Base Types, covers type checking, which is a major task required in most programming languages. Type checking can be performed at compile time or at runtime. This chapter covers the common case of static compile-time type checking for base types, also referred to as atomic or scalar types.

Chapter 8, Checking Types on Arrays, Method Calls, and Structure Accesses, shows you how to perform type checks for the arrays, parameters, and return types of method calls in the Jzero subset of Java. The more difficult parts of type checking are when multiple or composite types are involved. This is the case when functions with multiple parameter types must be checked, or when arrays, hash tables, class instances, or other composite types must be checked.

Chapter 9, Intermediate Code Generation, shows you how to generate intermediate code by looking at examples for the Jzero language. Before generating code for execution, most compilers turn the syntax tree into a list of machine-independent intermediate code instructions. Key aspects of control flow, such as the generation of labels and goto instructions, are handled at this point.

Chapter 10, Syntax Coloring in an IDE, addresses the challenge of incorporating information from syntax analysis into an IDE in order to provide syntax coloring and visual feedback about syntax errors. A programming language requires more than just a compiler or interpreter – it requires an ecosystem of tools for developers. This ecosystem can include debuggers, online help, or an integrated development environment.

Chapter 11, Preprocessors and Transpilers, gives an overview of generating output intended to be compiled or interpreted by another high-level language. Preprocessors are usually line-oriented and translate lines into very similar output, while transpilers usually translate one high-level language to a different high-level language with a full parse and significant semantic changes.

Chapter 12, Bytecode Interpreters, covers designing the instruction set and the interpreter that executes bytecode. A new domain-specific language may include high-level domain programming features that are not supported directly by mainstream CPUs. The most practical way to generate code for many languages is to generate bytecode for an abstract machine whose instruction set directly supports the domain, and then execute programs by interpreting that instruction set.

Chapter 13, Generating Bytecode, continues with code generation, taking the intermediate code from *Chapter 9, Intermediate Code Generation*, and generating bytecode from it. Translation from intermediate code to bytecode is a matter of walking through a giant linked list, translating each intermediate code instruction into one or more bytecode instructions. Typically, this is a loop to traverse the linked list, with a different chunk of code for each intermediate code instruction.

Chapter 14, Native Code Generation, provides an overview of generating native code for x86_64. Some programming languages require native code to achieve their performance requirements. Native code generation is like bytecode generation, but more complex, involving register allocation and memory addressing modes.

Chapter 15, Implementing Operators and Built-In Functions, describes how to support very high-level and domain-specific language features by adding operators and functions that are built into the language. Very high-level and domain-specific language features are often best represented by operators and functions that are built into the language, rather than library functions. Adding built-ins may simplify your language, improve its performance, or enable side effects in your language semantics that would otherwise be difficult or impossible. The examples in this chapter are drawn from Unicon, as it is much higher level than Java and implements more complex semantics in its built-ins.

Chapter 16, Domain Control Structures, covers when you need a new control structure, and provides example control structures that process text using string scanning, and render graphics regions. The generic code in previous chapters covered basic conditional and looping control structures, but domain-specific languages often have unique or customized semantics for which they introduce novel control structures. Adding new control structures is substantially more difficult than adding a new function or operator, but it is what makes domain-specific languages worth developing instead of just writing class libraries.

Chapter 17, Garbage Collection, presents a couple of methods with which you can implement garbage collection in your language. Memory management is one of the most important aspects of modern programming languages, and all the cool programming languages feature automatic memory management via garbage collection. This chapter provides a couple of options as to how you might implement garbage collection in your language, including reference counting, and mark-and-sweep garbage collection.

Chapter 18, Final Thoughts, reflects on the main topics presented in the book and gives you some food for thought. It considers what was learned from writing this book and gives you many suggestions for further reading.

Appendix, *Unicon Essentials*, describes enough of the Unicon programming language to understand those examples in this book that are in Unicon. Most examples are given side by side in Unicon and Java, but the Unicon versions are usually shorter and easier to read.

Answers, gives you some proposed answers to the revision questions placed at the end of each chapter.

To get the most out of this book

In order to understand this book, you should be an intermediate-or-better programmer in Java or a similar language; a C programmer who knows an object-oriented language will be fine.

Software/hardware covered in the book	Operating system requirements
Unicon 13.2, Uflex, and Iyacc	Windows, Linux
Java, Jflex, and Byacc/J	Windows, Linux
GNU Make	Windows, Linux

Instructions for installing and using the tools are spread out a bit to reduce the startup effort, appearing in *Chapter 3*, *Scanning Source Code*, to *Chapter 5*, *Syntax Trees*. If you are technically gifted, you may be able to get all these tools to run on macOS, but it was not used or tested during the writing of this book.

NOTE

If you are using the digital version of this book, we advise you to type the code yourself or, better yet, access the code from the book's GitHub repository (a link is available in the next section). Doing so will help you avoid any potential errors related to the copying and pasting of code.

Download the example code files

The code bundle for the book is hosted on GitHub at https://github.com/PacktPublishing/Build-Your-Own-Programming-Language-Second-Edition. We also have other code bundles from our rich catalog of books and videos available at https://github.com/PacktPublishing/. Check them out!

Code in Action

The Code in Action videos for this book can be viewed at https://bit.ly/3njc15D.

Download the color images

We also provide a PDF file that has color images of the screenshots/diagrams used in this book. You can download it here: https://packt.link/gbp/9781804618028.

Conventions used

There are a number of text conventions used throughout this book.

`CodeInText`: Indicates code words in text, database table names, folder names, filenames, file extensions, pathnames, dummy URLs, user input, and X (more commonly known as Twitter) handles. For example: "The `JSRC` macro gives the names of all the Java files to be compiled."

A block of code is set as follows:

```
public class address {
    public String region;
    public int offset;
    address(String s, int x) { region = s; offset = x; }
}
```

Any command-line input or output is written as follows:

```
j0 hello.java                    java ch9.j0 hello.java
```

Bold: Indicates a new term, an important word, or words that you see on the screen. For instance, words in menus or dialog boxes appear in the text like this. For example: "A `makefile` is like a `lex` or `yacc` specification, except instead of recognizing patterns of strings, a `makefile` specifies a graph of **build dependencies** between files".

Warnings or important notes appear like this.

Tips and tricks appear like this.

Get in touch

Feedback from our readers is always welcome.

General feedback: Email feedback@packtpub.com and mention the book's title in the subject of your message. If you have questions about any aspect of this book, please email us at questions@packtpub.com.

Errata: Although we have taken every care to ensure the accuracy of our content, mistakes do happen. If you have found a mistake in this book, we would be grateful if you reported this to us. Please visit http://www.packtpub.com/submit-errata, click **Submit Errata**, and fill in the form.

Piracy: If you come across any illegal copies of our works in any form on the internet, we would be grateful if you would provide us with the location address or website name. Please contact us at copyright@packtpub.com with a link to the material.

If you are interested in becoming an author: If there is a topic that you have expertise in and you are interested in either writing or contributing to a book, please visit http://authors.packtpub.com.

Share your thoughts

Once you've read *Build Your Own Programming Language, Second Edition*, we'd love to hear your thoughts! Scan the QR code below to go straight to the Amazon review page for this book and share your feedback.

https://packt.link/r/1804618020

Your review is important to us and the tech community and will help us make sure we're delivering excellent quality content.

Download a free PDF copy of this book

Thanks for purchasing this book!

Do you like to read on the go but are unable to carry your print books everywhere?

Is your eBook purchase not compatible with the device of your choice?

Don't worry, now with every Packt book you get a DRM-free PDF version of that book at no cost.

Read anywhere, any place, on any device. Search, copy, and paste code from your favorite technical books directly into your application.

The perks don't stop there, you can get exclusive access to discounts, newsletters, and great free content in your inbox daily

Follow these simple steps to get the benefits:

1. Scan the QR code or visit the link below

https://packt.link/free-ebook/9781804618028

2. Submit your proof of purchase
3. That's it! We'll send your free PDF and other benefits to your email directly

Section I

Programming Language Frontends

In this section, you will create a basic language design and implement the frontend of a compiler for it, including a lexical analyzer and a parser that builds a syntax tree from an input source file.

This section comprises the following chapters:

- *Chapter 1, Why Build Another Programming Language?*
- *Chapter 2, Programming Language Design*
- *Chapter 3, Scanning Source Code*
- *Chapter 4, Parsing*
- *Chapter 5, Syntax Trees*

1

Why Build Another Programming Language?

This book will show you how to build your own programming language, but first, you should ask yourself, why would I want to do this? For a few of you, the answer will be simple: because it is so much fun. However, for the rest of us, it is a lot of work to build a programming language, and we need to be sure about it before we make that kind of effort. Do you have the patience and persistence that it takes?

This chapter points out a few good reasons to build your own programming language, as well as some circumstances in which you don't *need* to build your contemplated language. After all, designing a class library for your application domain is often simpler and just as effective. However, libraries have their limitations, and sometimes, only a new language will do.

After this chapter, the rest of this book will take for granted that, having considered things carefully, you have decided to build a language. But first, we're going to consider our initial options by covering the following main topics in this chapter:

- Motivations for writing your own programming language
- Types of programming language implementations
- Organizing a bytecode language implementation
- Languages used in the examples
- The difference between programming languages and libraries
- Applicability to other software engineering tasks

- Establishing the requirements for your language
- Case study – requirements that inspired the Unicon language

Let's start by looking at motivations.

Motivations for writing your own programming language

Sure, some programming language inventors are rock stars of computer science, such as Dennis Ritchie or Guido van Rossum! Becoming a rock star in computer science was easier back in the previous century. In 1993, I heard the following report from an attendee of the second ACM History of Programming Languages Conference: *"The consensus was that the field of programming languages is dead. All the important languages have been invented already."* This was proven wildly wrong a year or two later when Java hit the scene, and perhaps a dozen times since then when important languages such as Go emerged. After a mere six decades, it would be unwise to claim our field is mature and that there's nothing new to invent that might make you famous.

In any case, celebrity is a bad reason to build a programming language. The chances of acquiring fame or fortune from your programming language invention are slim. Curiosity and a desire to know how things work are valid reasons, so long as you've got the time and inclination, but perhaps the best reason to build your own programming language is necessity.

Some folks need to build a new language, or a new implementation of an existing programming language, to target a new processor or compete with a rival company. If that's not you, then perhaps you've looked at the best languages (and compilers or interpreters) available for some domain that you are developing programs for, and they are missing some key features for what you are doing, and those missing features are causing you pain. This is the stuff Master's theses and PhD dissertations are made of. Every once in a blue moon, someone comes up with a whole new style of computing for which a new programming paradigm requires a new language.

While we are discussing your motivations for building a language, let's also talk about the different kinds of languages, how they are organized, and the examples this book will use to guide you.

Types of programming language implementations

Whatever your reasons, before you build a programming language, you should pick the best tools and technologies you can find to do the job. In our case, this book will pick them for you. First, there is a question of the implementation language, which is to say, the language that you are building your language in.

Programming language academics like to brag about writing their language in that language itself, but this is usually only a half-truth (or someone was being very impractical and showing off at the same time). There is also the question of just what kind of programming language implementation to build:

- A pure **interpreter** that executes the source code itself
- A **native compiler** and a runtime system, such as in C
- A **transpiler** that translates your language into some other high-level language
- A **bytecode compiler** with an accompanying bytecode machine, such as in Java

The first option is fun, but the resulting language is usually too slow to satisfy real-world project requirements. The second option is often optimal, but may be too labor-intensive; a good native compiler may take years of effort.

The third option is by far the easiest and probably the most fun, and I have used it before with good success. Don't discount a transpiler implementation as a kind of cheating, but do be aware that it has its problems. The first version of C++, AT&T's cfront tool, was a transpiler, but that gave way to compilers, and not just because cfront was buggy. Strangely, generating high-level code seems to make your language even more dependent on the underlying language than the other options, and languages are moving targets. Good languages have died because their underlying dependencies disappeared or broke irreparably on them. It can be the death of a thousand cuts.

For the most part, this book focuses on the fourth option; over the course of several chapters, we will build a bytecode compiler with an accompanying bytecode machine because that is a sweet spot that gives a lot of flexibility, while still offering decent performance. A chapter on transpilers and preprocessors is provided for those of you who may prefer to implement your language by generating code for another high-level language. A chapter on native code compilation is also included, for those of you who require the fastest possible execution.

The notion of a bytecode machine is very old; it was made famous by UCSD's Pascal implementation and the classic SmallTalk-80 implementation, among others. It became ubiquitous to the point of entering lay English with the promulgation of Java's JVM. Bytecode machines are abstract processors interpreted by software; they are often called **virtual machines** (as in **Java Virtual Machine**), although I will not use that terminology because it is also used to refer to software tools that implement real hardware instruction sets, such as IBM's classic platforms, or more modern tools such as **Virtual Box**.

A bytecode machine is typically quite a bit higher level than a piece of hardware, so a bytecode implementation affords much flexibility. Let's have a quick look at what it will take to get there...

Organizing a bytecode language implementation

To a large extent, the organization of this book follows the classic organization of a bytecode compiler and its corresponding virtual machine. These components are defined here, followed by a diagram to summarize them:

- A **lexical analyzer** reads in source code characters and figures out how they are grouped into a sequence of words or tokens.

- A **syntax analyzer** reads in a sequence of tokens and determines whether that sequence is legal, according to the grammar of the language. If the tokens are in a legal order, it produces a syntax tree.

- A **semantic analyzer** checks to ensure that all the names being used are legal for the operations in which they are being used. It checks their types to determine exactly what operations are being performed. All this checking makes the syntax tree heavy, laden with extra information about where variables are declared and what their types are.

- An **intermediate code generator** figures out memory locations for all the variables and all the places where a program may abruptly change execution flow, such as loops and function calls. It adds them to the syntax tree and then walks this even fatter tree, before building a list of machine-independent intermediate code instructions.

- A **final code generator** turns the list of intermediate code instructions into the actual bytecode, in a file format that will be efficient to load and execute.

In addition to the steps of this bytecode virtual machine compiler, a **bytecode interpreter** is written to load and execute programs. It is a giant loop with a switch statement in it. For very high-level programming languages, the compiler might be no big deal, and all the magic may be in the bytecode interpreter. The whole organization can be summarized by the following diagram:

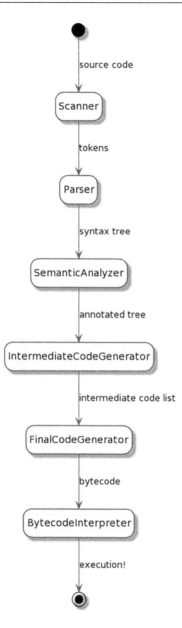

Figure 1.1: Phases and dataflow in a simple programming language

It will take a lot of code to illustrate how to build a bytecode machine implementation of a programming language. How that code is presented is important and will tell you what you need to know going in, as well as what you may learn from going through this book.

Languages used in the examples

This book provides code examples in two languages using a **parallel translations model**. The first language is **Java** because that language is ubiquitous. Hopefully, you know Java (or C++, or C#) and will be able to read the examples with intermediate proficiency. The second example language is the author's own language, **Unicon**. While reading this book, you can judge for yourself which language is better suited to building programming languages. As many examples as possible are provided in both languages, and the examples in the two languages are written as similarly as possible. Sometimes, this will be to the advantage of Java, which is a bit lower level than Unicon. There are sometimes fancier or shorter ways to write things in Unicon, but our Unicon examples will stick as close to Java as possible. The differences between Java and Unicon will be obvious, but they are somewhat lessened in importance by the compiler construction tools we will use.

This book uses modern descendants of the venerable Lex and YACC tools to generate our scanner and parser. Lex and YACC are declarative programming languages that solve some of our hard problems at a higher level than Java or Unicon. It would have been nice if a modern descendant of Lex and YACC (such as ANTLR) supported both Java and Unicon, but such is not the case. One of the very cool parts of this book is this: by choosing tools for Java and Unicon that are very compatible with the original Lex and YACC and extending them a bit, we have managed to use the same lexical and syntax specifications of our compiler in both Java and Unicon!

While Java and Unicon are our implementation languages, we need to talk about one more language: the example language we are building. It is a stand-in for whatever language you decide to build. Somewhat arbitrarily, this book introduces a language called **Jzero** for this purpose. Niklaus Wirth invented a toy language called **PL/0 (programming language zero**; the name is a riff on the language name **PL/1**) that was used in compiler construction courses. Jzero is a tiny subset of Java that serves a similar purpose. I looked *pretty hard* (that is, I googled *Jzero* and then *Jzero compiler*) to see whether someone had already posted a Jzero definition we could use and did not spot one by that name, so we will just make it up as we go along.

The Java examples in this book will be tested using Java 21; maybe other recent versions of Java will work. You can get OpenJDK from http://openjdk.org, or if you are on Linux, your operating system probably has an OpenJDK package that you can install. Additional programming language construction tools (Jflex and byacc/j) that are required for the Java examples will be introduced in subsequent chapters as they are used. The Java implementations we will support might be more constrained by which versions will run these language construction tools than anything else.

The Unicon examples in this book work with Unicon version 13.3, which can be obtained from `http://unicon.org`. To install Unicon on Windows, you must download a `.msi` file and run the installer. To install on Linux, you should follow the instructions found on the `unicon.org` site.

Having gone through the basic organization of a programming language and the implementation that this book will use, perhaps we should take another look at when a programming language is called for, and when building one can be avoided by developing a library instead.

The difference between programming languages and libraries

Unless you are in it for the "fun" or the intellectual experience, building a programming language is a lot of work that might not be necessary. If your motives are strictly utilitarian, you don't have to make a programming language when a library will do the job. Libraries are by far the most common way to extend an existing programming language to perform a new task. A **library** is a set of functions or classes that can be used together to write applications for some hardware or software technology. Many languages, including C and Java, are designed almost completely to revolve around a rich set of libraries. The language itself is very simple and general, while much of what a developer must learn to develop applications consists of how to use the various libraries.

The following is what libraries can do:

- Introduce new data types (classes) and provide public functions (an API) to manipulate them
- Provide a layer of abstraction on top of a set of hardware or operating system calls

The following is what libraries cannot do:

- Introduce new control structures and syntax in support of new application domains
- Embed/support new semantics within the existing language runtime system

Libraries do some things badly, so you might end up preferring to make a new language:

- Libraries often get larger and more complex than necessary.
- Libraries can have even steeper learning curves and poorer documentation than languages.
- Every so often, libraries have conflicts with other libraries.
- Applications that use libraries can become broken if the library changes incompatibly in a later version.

There is a natural evolutionary path from a library to a language. A reasonable approach to building a new language to support an application domain is to start by making or buying the best library available for that application domain. If the result does not meet your requirements in terms of supporting the domain and simplifying the task of writing programs for that domain, then you have a strong argument for a new language.

This book is about building your own language, not just building your own library. It turns out that learning about tools and techniques to implement programming languages is useful in many other contexts.

Applicability to other software engineering tasks

The tools and technologies you learn about from building your own programming language can be applied to a range of other software engineering tasks. For example, you can sort almost any file or network input processing task into three categories:

- Reading XML data with an XML library
- Reading JSON data with a JSON library
- Reading anything else by writing code to parse it in its native format

The technologies in this book are useful in a wide array of software engineering tasks, which is where the third of these categories is encountered. Frequently, structured data must be read in a custom file format.

For some of you, the experience of building your own programming language might be the single largest program you have written thus far. If you persist and finish it, it will teach you lots of practical software engineering skills, besides whatever you learn about compilers, interpreters, and the such. This will include working with large dynamic data structures, software testing, and debugging complex problems, among other skills.

That's enough of the inspirational motivation. Let's talk about what you should do first: figure out your requirements.

Establishing the requirements for your language

After you are sure you need a new programming language for what you are doing, take a few minutes to establish the requirements. This is open-ended. It is you defining what success for your project will look like. Wise language inventors do not create a whole new syntax from scratch. Instead, they define it in terms of a set of modifications to make to a popular existing language.

Many great programming languages (Lisp, Forth, Smalltalk, and many others) had their success significantly limited by the degree to which their syntax was unnecessarily different from mainstream languages. Still, your language requirements include what it will look like, and that includes syntax.

More importantly, you must define a set of control structures or semantics where your programming language needs to go beyond existing language(s). This will sometimes include special support for an application domain that is not well served by existing languages and their libraries. Such **domain-specific languages (DSLs)** are common enough that whole books are focused on that topic. Our goal for this book will be to focus on the nuts and bolts of building the compiler and runtime system for such a language, independent of whatever domain you may be working in.

In a normal software engineering process, requirements analysis would start with brainstorming lists of functional and non-functional requirements. Functional requirements for a programming language involve the specifics of how the end user developer will interact with it. You might not anticipate all the command-line options for your language up front, but you probably know whether interactivity is required, or whether a separate compile step is OK. The discussion of interpreters and compilers in the previous section, and this book's presentation of a compiler, might seem to make that choice for you, but Python is an example of a language that provides a fully interactive interface, even though the source code you type into Python gets compiled into bytecode and executed by a bytecode machine, rather than being interpreted directly.

Non-functional requirements are properties that your programming language must achieve that are not directly tied to the end user developer's interactions. They include things such as what operating system(s) your language must run on, how fast execution must be, or how little space the programs written in your language must run within.

The non-functional requirement regarding how fast execution must be usually determines the answer as to whether you can target a software (bytecode) machine or need to target native code. Native code is not just faster; it is also considerably more difficult to generate, and it might make your language considerably less flexible in terms of runtime system features. You might choose to target bytecode first, and then work on a native code generator afterward.

The first language I learned to program on was a BASIC interpreter in which the programs had to run within 4 KB of RAM. BASIC at the time had a low memory footprint requirement. But even in modern times, it is not uncommon to find yourself on a platform where Java won't run by default! For example, on virtual machines with configured memory limits for user processes, you may have to learn some awkward command-line options to compile or run even simple Java programs.

In addition to identifying functional and non-functional requirements, many requirements analysis approaches also define a set of use cases and ask the developer to write descriptions for them. Inventing a programming language is different from your average software engineering project, but before you are finished, you may want to go there and perform such a use case analysis. A use case is a task that someone performs using a software application. When the software application is a programming language, if you are not careful, the use cases may be too general to be useful, such as *write my application* and *run my program*. While those two might not be very useful, you might want to think about whether your programming language implementation must support program development, debugging, separate compilation and linking, integration with external languages and libraries, and so forth. Most of those topics are beyond the scope of this book, but we will consider some of them.

Since this book presents the implementation of a language called Jzero, here are some requirements for Jzero. Some of these requirements may appear arbitrary. You could certainly add your own requirements and produce your own Java dialect, but this list describes what we are aiming for in this book. If it is not clear to you where one of the following requirements came from, it either came from our source inspiration language (plzero) or previous experience teaching compiler construction:

- Jzero should be a strict subset of Java. All legal Jzero programs should be legal Java programs. This requirement allows us to check the behavior of our test programs when we are debugging our language implementation.

- Jzero should provide enough features to allow interesting computations. This includes if statements, while loops, and multiple functions, along with parameters.

- Jzero should support a few data types, including Booleans, integers, arrays, and the String type. However, it only needs to support a subset of their functionality, (as you'll see later). These types are enough to allow input and output of interesting values into a computation.

- Jzero should emit decent error messages, showing the filename and line number, including messages for attempts to use Java features not in Jzero. We will need reasonable error messages to debug the implementation.

- Jzero should run fast enough to be practical. This requirement is vague, but it implies that we won't be doing a pure interpreter. Pure interpreters that execute source code directly without any internal code generation step are a very retro thing, evocative of the 1960s and 1970s. They tend to execute unacceptably slowly by modern standards. On the other hand, you might very well decide that your language should provide the highly interactive look and feel of a pure interpreter, like Python does. Anyhow, that is not in Jzero's requirements.

- Jzero should be as simple as possible so that I can explain it. Sadly, this rules out writing a full description of a native code generator or even an implementation that targets JVM bytecode; we will provide our own simple bytecode machine.

Perhaps more requirements will emerge as we go along, but this is a start. Since we are constrained for time and space, perhaps this requirements list is more important for what it does not say, rather than for what it does say. By way of comparison, here are some of the requirements that led to the creation of the Unicon programming language.

Case study — requirements that inspired the Unicon language

This book will use the Unicon programming language, located at http://unicon.org, for a running case study. We can start with reasonable questions such as, why build Unicon, and what are its requirements? To answer the first question, we will work backward from the second one.

Unicon exists because of an earlier programming language called Icon, from the University of Arizona (http://www.cs.arizona.edu/icon/). Icon has particularly good string and list processing facilities and is used to write many scripts and utilities, as well as both programming language and natural language processing projects. Icon's fantastic built-in data types, including structure types such as lists and (hash) tables, have influenced several languages, including Python and Unicon. Icon's signature research contribution is its integration of goal-directed evaluation, including backtracking and automatic resumption of generators, into a familiar mainstream syntax. This leads us to Unicon's first requirement.

Unicon requirement #1 — preserve what people love about Icon

One of the things that people love about Icon is its expression semantics, including its **generators** and **goal-directed evaluation**. A generator is an expression that is capable of computing more than one result; several popular languages feature generators. Goal-directed evaluation is a semantic to execute code in which expressions either succeed or fail, and when they fail, generators within the expression can be resumed to try alternative results that might make the whole expression succeed. This is a big topic beyond the scope of this section, but if you want to learn more, you can check out *The Icon Programming Language, Third Edition*, by Ralph and Madge Griswold, at www.cs.arizona.edu/icon.

Icon also provides a rich set of built-in functions and data types so that many or most programs can be understood directly from the source code. Unicon's preservation goal is 100% compatibility with Icon. In the end, we achieved more like 99% compatibility.

It is a bit of a leap from *preserving the best bits* to the immortality goal of ensuring old source code will run forever, but for Unicon, we include that as part of requirement #1. We have placed a much firmer requirement on backward compatibility than most modern languages. While C is very backward compatible, C++, Java, Python, and Perl are examples of languages that have wandered away, in some cases far away, from being compatible with the programs written in them back in their glory days. In the case of Unicon, perhaps 99% of Icon programs run unmodified as Unicon programs. Unicon requirement #2 was to support programming in large-scale projects.

Unicon requirement #2 — support large-scale programs working on big data

Icon was designed for maximum programmer productivity on small-sized projects; a typical Icon program is less than 1,000 lines of code, but Icon is very high level, and you can do a *lot* of computing in a few hundred lines of code! Still, computers keep getting more capable, and modern programmers are often required to write much larger programs than Icon was designed to handle.

For this reason of scalability, Unicon adds classes and packages to Icon, much like C++ adds them to C. Unicon also improved the bytecode object file format and made numerous scalability improvements to the compiler and runtime system. It also refines Icon's existing implementation to be more scalable in many specific items, such as adopting a much more sophisticated hash function. Unicon requirement #3 is to support ubiquitous input/output capabilities at the same high level as the built-in types.

Unicon requirement #3 — high-level input/output for modern applications

Icon was designed for classic UNIX pipe-and-filter text processing of local files. Over time, more and more people wanted to use it to write programs that required more sophisticated forms of input/output, such as networking or graphics.

Arguably, despite billionfold improvements in CPU speed and memory size, the biggest difference between programming in 1970 and programming in the 2020s is that we expect modern applications to use a myriad of sophisticated forms of I/O: graphics, networking, databases, and so forth. Libraries can provide access to such I/O, but language-level support can make it easier and more intuitive.

Support for I/O is a moving target. At first, with Unicon, I/O consisted of networking facilities and GDBM and ODBC database facilities to accompany Icon's 2D graphics. Then, it grew to include various popular internet protocols and 3D graphics. The definition of what I/O capabilities are ubiquitous continues to evolve, varying by platform, but touch input and gestures or shader programming capabilities are examples of things that have become ubiquitous today, and maybe they should be added to the Unicon language as part of this requirement. The challenge posed by this requirement is increased by Unicon requirement #4.

Unicon requirement #4 – provide universally implementable system interfaces

Icon is very portable. I have run it on everything, from Amigas to Crays to IBM mainframes with EBCDIC character sets. Although the platforms have changed almost unbelievably over the years, Unicon still retains Icon's goal of maximum source code portability: code that gets written in Unicon should continue to run unmodified on all computing platforms that matter.

For a very long time, portability meant running on PCs, Macs, and UNIX workstations. But again, the set of computing platforms that matter is a moving target. These days, to meet this requirement, Unicon should be ported to support Android and iOS, if you count them as computing platforms. Whether they count might depend on whether they are open enough and used for general computing tasks, but they are certainly capable of being used as such.

All those juicy I/O facilities that were implemented for requirement #3 must be designed in such a way that they can be multi-platform portable across all major platforms.

Having given you some of Unicon's primary requirements, here is an answer to the question, why build Unicon at all? One answer is that after studying many languages, I concluded that Icon's generators and goal-directed evaluation (requirement #1) were features that I wanted when writing programs from now on. However, after allowing me to add 2D graphics to their language, Icon's inventors were no longer willing to consider further additions to meet requirements #2 and #3. Another answer is that there was a public demand for new capabilities, including volunteer partners and some financial support. Thus, Unicon was born.

Summary

In this chapter, you learned the difference between inventing a programming language and inventing a library API to support whatever kinds of computing you want to do. Several different forms of programming language implementations were considered. This first chapter allowed you to think about functional and non-functional requirements for your own language.

These requirements might be different for your language than the example requirements discussed for the Java subset Jzero and the Unicon programming language, which were both introduced.

Requirements are important because they allow you to set goals and define what success will look like. In the case of a programming language implementation, the requirements include what things will look and feel like for the programmers that use your language, as well as what hardware and software platforms it must run on. The look and feel of a programming language include answering both external questions regarding how the language implementation and the programs written in the language are invoked, as well as internal issues such as verbosity: how much the programmer must write to accomplish a given compute task.

You may be keen to get straight to the coding part. Although the classic *build-and-fix* mentality of novice programmers might work on scripts and short programs, for a piece of software as large as a programming language, we need a bit more planning first. After this chapter's coverage of the requirements, *Chapter 2, Programming Language Design*, will prepare you to construct a detailed plan for the implementation, which will occupy our attention for the remainder of this book!

Questions

1. What are the pros and cons of writing a language transpiler that generates C code, instead of a traditional compiler that generates assembler or native machine code?

2. What are the major components or phases in a traditional compiler?

3. From your experience, what are some pain points where programming is more difficult than it should be? What new programming language feature(s) address these pain points?

4. Write a set of functional requirements for a new programming language.

Join our community on Discord

Join our community's Discord space for discussions with the authors and other readers:

`https://discord.com/invite/zGVbWaxqbw`

2

Programming Language Design

Before trying to build a programming language, you need to define it. This includes the design of the features of the language that are visible on its surface, including basic rules to form words and punctuation. This also includes higher-level rules, called **syntax**, that govern the number and order of words and punctuation in larger chunks of programs, such as expressions, statements, functions, classes, packages, and programs. Language design also includes the underlying meaning, also known as **semantics**.

Programming language design often begins with you writing example code to illustrate each of the important features of your language, as well as show the variations that are possible for each construct. Writing examples with a critical eye lets you find and fix many possible inconsistencies in your initial ideas. From these examples, you can then capture the general rules that each language construct follows. Write down sentences that describe your rules as you understand them from your examples. Note that there are two kinds of rules. **Lexical rules** govern what characters must be treated together, such as words, or multi-character operators, such as ++. **Syntax rules**, on the other hand, are rules to combine multiple words or punctuation to form a larger meaning; in natural language, they are often phrases, sentences, or paragraphs, while in a programming language, they might be expressions, statements, functions, or programs.

Once you have come up with examples of everything that you want your language to do, and have written down the lexical and syntax rules, it is time to write a language design document (or language specification) to which you can refer while implementing your language. You can change things later, but it helps to have a plan to work from.

In this chapter, we're going to cover the following main topics:

- Determining the kinds of words and punctuation to provide in your language
- Specifying the control flow
- Deciding on what kinds of data to support
- An overall program structure
- Completing the Jzero language definition
- Case study – designing graphics facilities in Unicon

Let's start by identifying the basic elements that are allowed in the source code in your language.

Determining the kinds of words and punctuation to provide in your language

Programming languages have several different categories of words and punctuation. In natural language, words are categorized into **parts of speech** – nouns, verbs, adjectives, and so on. The categories that correspond to the parts of speech that you will have to invent for a programming language can be constructed by doing the following:

- Defining a set of **reserved words** or keywords
- Specifying characters in **identifiers** that name variables, functions, and constants
- Creating a format for **literal** constant values for built-in data types
- Defining single and multi-letter **operators** and punctuation marks

You should write down precise descriptions of each of these categories as part of your language design document. In some cases, you might just make lists of particular words or punctuation to use, but in other cases, you will need patterns or some other way to convey what is and is not allowed in a category.

For reserved words, a list will do for now. For names of things, a precise description must include details such as what non-letter symbols are allowed in such names. For example, in Java, names must begin with a letter and can then include letters and digits; underscores are allowed and treated as letters. In other languages, hyphens are allowed within names, so the three symbols a, -, and b make up a valid name, not a subtraction of b from a. When a precise description fails, a complete set of examples will suffice.

Constant values, also called **literals**, are a surprising and major source of complexity in lexical analyzers. Attempting to precisely describe real numbers in Java comes out something like this: Java has two different kinds of real numbers – floats and doubles – but they look the same until you get to the end, where there is an optional f (or F) or d (or D) to distinguish floats from doubles. Before that, real numbers must have either a decimal point (.), an exponent (e or E) part, or both. If there is a decimal point, there must be at least one digit on one side of the decimal or the other. If there is an exponent part, it must have an e (or E), followed by an optional minus sign and one or more digits. To make matters worse, Java has a weird hexadecimal real constant format, consisting of 0x or 0X followed by digits in hex format, with an optional mantissa consisting of a period followed by hexadecimal digits, and a mandatory power part consisting of a p (or P), followed by digits in the decimal format that multiplies the number by 2, raised to that power. If you want to write constants like 0x3.0fp8, then this IEEE-based format is for you.

Describing operators and punctuation marks is usually almost as easy as listing the reserved words. One major difference between operators and punctuation marks is that operators usually have **precedence** rules that you will need to determine. For example, in numeric processing, the multiplication operator has almost always higher precedence than the addition operator, so x + y * z will multiply y * z before it adds x to the product of y and z. In most languages, there are at least three to five levels of precedence, and many popular mainstream languages have from 13 to 20 levels of precedence that must be considered carefully.

The following diagram shows the operator precedence table for Java, from the lowest to highest precedence. We will need it for Jzero:

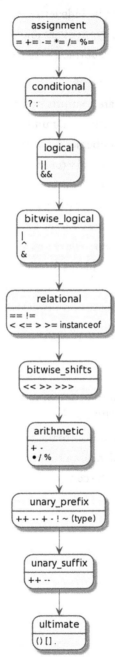

Figure 2.1: Java operator precedence

The preceding diagram shows that Java has a lot of operators, organized into 10 levels of precedence, although I might be simplifying this a bit. In your language, you might get away with fewer, but you will have to address the issue of operator precedence if you want to build a real language.

A similar issue is operator **associativity**. In many languages, most operators associate from left to right, but a few unusual ones associate from right to left. For example, the x + y + z expression is equivalent to (x + y) + z, but the x = y = 0 expression is equivalent to x = (y = 0).

The principle of least surprise applies to operator precedence and associativity, as well as to what operators you put in your language in the first place. If you define arithmetic operators and give them unusual precedence or associativity, people will reject your language out of hand. If you happen to be introducing new, possibly domain-specific data types, you have way more freedom to define operator precedence and associativity for any new operators you introduce in your language for those types.

Once you have determined what the individual words and punctuation in your language should be, you can work your way up to larger constructs. This is the transition from lexical analysis to syntax, and syntax is important because it is the level at which bits of code become large enough to specify some computation to be performed. We will look at this in more detail in later chapters, but at the design stage, you should at least think about how programmers will specify the control flow, declare data, and build entire programs. First, you must plan for the control flow.

Specifying the control flow

The **control flow** is how a program's execution proceeds from place to place within the source code. Most control flow constructs should be familiar to programmers who have been trained in mainstream programming languages. The innovations in your language design can then focus on the features that are novel or domain-specific, and that motivated you to create a new language in the first place. Make these novel things as simple and readable as possible. Envision how those new features ought to fit into the rest of the programming language.

Every language must have conditionals and loops, and almost all of them use if and while to start them. You could invent your own special syntax for an if expression, but unless you've got a good reason to, you would be shooting yourself in the foot. Here are some control flow constructs from Java that would certainly be in Jzero:

```
if (e) s;
if (e) s1 else s2;
while (e) s;
for (…) s;
```

Here are some other less common Java control flow constructs that are not in Jzero. If they were to appear in a program, what should a Jzero compiler do with them?

```
switch (e) { … }
do s while (e);
```

Since these constructs are not in Jzero, if they appear in the input source code, then by default, our compiler will print a cryptic syntax error message that doesn't explain things very well. In the next two chapters, we will make our compiler for Jzero print a nice error message about the Java features that it does not support.

Besides conditionals and loops, languages tend to have a syntax to call subroutines and return afterward. All these ubiquitous forms of control flow are abstractions of the underlying machine's capability to change the location where instructions are executing – the GOTO. If you invent a better notation for changing the location where instructions are executing, it will be a big deal.

The biggest controversy when designing many or most control flow constructs seems to be whether they are **statements**, or whether you should make them **expressions** that produce a result that can be used in a surrounding expression. I have used languages where the result of if expressions are useful – C/C++/Java even have an operator for that: the i?t:e conditional operator. I have not found a language that did something very meaningful by making a while loop an expression; the best the languages did was to have the while expressions produce a result, telling us whether the loop exited due to the test condition or an internal break.

If you are inventing a new language from scratch, one of the big questions for you is whether you should come up with some new control structure(s) to support your intended application domain. For example, suppose you want your language to provide special support for investing in the stock market. If you manage to come up with a better control structure for specifying conditions, constraints, or iterative operations within this domain, you might provide a competitive edge to those who are coding in your language for this domain. The program will have to run on an underlying von Neuman instruction set, so you will have to figure out how to map any such new control structure to instructions such as Boolean logic tests and GOTO instructions.

Whatever control flow constructs you decide to support, you will also need to design a set of data types and declarations that reflect the information that the programs in your language will manipulate.

Deciding on what kinds of data to support

There are at least three categories of data types to consider in your language design. We will describe each of these in this section. The first one is atomic, scalar primitive types, often called first-class data types. The second is composite or container types, which hold and organize collections of values. The third (which may be variants of the first or second categories) is application domain-specific types. You should formulate a plan for each of these categories.

Atomic types

Atomic types are generally built-in and immutable. As the word immutable suggests, you cannot modify existing atomic values, only combine them to compute new values. Pretty much all languages have such built-in atomic types for numbers and a few additional types. A Boolean type, null type, and maybe a string type are common atomics, but some languages have others.

You decide just how complicated to get with atomics: how many different machine representations of integers and real numbers do programs written in your language need? Some higher-level languages such as BASIC might provide a single type for all numbers, while lower-level languages such as C or C++ might provide 5 or 10 (or more) representations for different sizes and kinds of integers, and another few for real numbers. The more you add, the more flexibility and control you give to programmers who use your language, but the more difficult your implementation task will be later. In addition, the increased complexity reduces readability and makes programs harder to understand.

Similarly, it is impossible to design a single string data type that is ideal for all applications that use strings a lot. But how many string types do you want to support? One extreme is having no string type at all, only a short integer type to hold characters. Such languages consider strings to be composite types. Maybe strings are supported only by a library rather than in the language. Strings may be arrays or objects, but even such languages usually have some special lexical rules that allow string constant values to be given as double-quoted sequences of characters of some kind. Another extreme is that, given the importance of strings in many application domains, your language might want to support multiple string types for various character representations (ASCII, UTF8, and so on) with auxiliary types (character sets) and special types and control structures that support the analysis and construction of strings. Many popular languages treat strings as a special atomic type.

If you are especially clever, you may decide to support only a few built-in types for numbers and strings but make those types as flexible as possible. Once you go beyond integers, real numbers, and strings, the only types that are universal are container types, which allow you to assemble data structures.

Some of the things you must think about regarding atomic types include the following:

- How many values do they have?
- How are all those values encoded as literal constants in the source code?
- What kinds of operators or built-in functions use this type as operands or parameters?

The first question will tell you how many bytes the type will require in memory. The second and third questions tie back to determining the rules for words and punctuation in the language. The third question may also give insight into how much effort, in terms of the code generator or run-time system, will be required to implement support for the type in your language. Atomic types can be more work or less work to implement, but they are seldom as complicated as composite types, which we will discuss next.

Composite types

Composite types are types that help you allocate and access multiple values in a coordinated fashion. Languages vary enormously regarding the extent of their syntax support for composite types. Some only support arrays and structs (Java programmers: you can think of these as classes without methods) and require programmers to build all their own data structures on top of these. Many provide all higher-level composite types via libraries. However, some higher-level languages provide numerous sophisticated data structures as built-ins with syntax support.

The most ubiquitous composite type is an **array** type, where multiple values are accessed using a numerically contiguous range of integer indices. You will probably have something like an array in your language. Your main design considerations should be how the indices are given, and how changes in the size of the composite value are handled. Most popular languages use indices that start at zero. Zero-based array indexes simplify index calculations and are easier for a language inventor to implement, but they are less intuitive for new programmers. Some languages use 1-based indices or allow a programmer to specify a range of indices, starting at an arbitrary integer other than 0.

Regarding changes in size, some languages allow no changes in size at all in their array types, or they make the programmer jump through hoops to build new arrays of different sizes based on existing arrays.

Other languages are engineered to make adding values to an array a cheap and easy operation. No one design is perfect for all applications, so you just pick one and live with the consequences: do you choose to support multiple array-like data types for different purposes, or instead choose to design a very clever type that accommodates a range of common uses well?

Besides arrays, you should think about what other composite types you need. Almost all languages support a record, struct, or class type to group values of several different types together and access them by names, called fields. The more elaborate you get with this, the more complex your language implementation will be. If you need proper object orientation in your language, be prepared to pay for it in time spent writing your compiler and runtime code. Features like classes and inheritance do not come for free. Language designers are advised to keep things simple, but as a programmer, I would not want to use a programming language that did not give me this capability in some form.

You might be able to think of several other composite types that are essential for your language, which is great, especially if they will be used a lot in the programs that you care about. I will talk about one more composite type that is of great practical value: the (hash) **table** data type, also commonly called a **dictionary** type. A table type is something halfway in between an array and a record type. You index values using names, and these names are not fixed; new names can be computed while the program runs. Any modern language that omits this type is just leaving many of its prospective users out. For this reason, your language may want to include a table type. Composite types are general-purpose "glue" that's used to assemble complex data structures, but you should also consider whether some special-purpose types, either atomic or composite, belong in your language to support applications that are difficult to write in general-purpose languages.

Domain-specific types

In addition to whatever general-purpose atomic and composite types you decide to include, you should think about whether your programming language is aimed at a domain-specific niche; if so, what data types can your language include to support that domain? There is a smooth continuum between domain-specific languages that provide **domain-specific** types and control structures and general-purpose languages such as C++ and Java, which provide libraries for everything. Class libraries are powerful, but for some applications and domains, the library approach may be more complex and bug-prone than a language expressly designed to support the domain. For example, Java and C++ have string classes, but they do not support complex text-processing applications better than languages that have special-purpose types and control structures for string processing. Besides data types, your language design will need an idea of how programs are assembled and organized.

Overall program structure

When looking at the overall program structure, we need to look at how entire programs are organized and put together, as well as the important question of how much nesting is in your language. It almost seems like an afterthought, but how and where will the source code in programs begin executing? In languages based on C, execution starts from a `main()` function, while in scripting languages, the source code is executed as it is read in, so there is no need for a `main()` function to start the ball rolling.

Program structure also raises the basic question of whether a whole program must be translated and run together, or if different packages, classes, or functions can be separately compiled and then linked and/or loaded together for a program to run. A language inventor can dodge a lot of implementation complexity by either building things into the language (if it is built in, there is no need to figure out linking), requiring the whole program's source code to be presented at runtime, or by generating code for some well-known standard execution format where someone else's linker and loader will do all the hard work.

Perhaps the biggest design question relating to the overall program structure is which constructs may be nested, and what limits on nesting are present, if any. This is perhaps best illustrated by an example. Once upon a time, two obscure languages were invented around 1970 that struggled for dominance: **C** and **Pascal**.

The C language was almost flat – a program was a set of functions linked together, and only relatively small (fine-grained) things could be nested: expressions, statements, and, reluctantly, struct definitions.

In contrast, the Pascal language was fabulously more nested and recursive. Almost everything could be nested. Notably, functions could be embedded within functions, arbitrarily deep. Although C and Pascal were roughly equivalent in power, and Pascal had a bit of a head start and was by far the most popular in university courses, C eventually won. Why? There are many contributing factors that might explain why C won out over Pascal. One factor might be that nesting functions adds complexity without adding much value.

Because C won, many modern mainstream languages (I am thinking especially of C++ and Java here) started almost flat. But over time, they have added more and more nesting. Why is this? Perhaps it is natural for programming languages to add features over time until they are very complex. Niklaus Wirth saw this coming and advocated for a return to smallness and simplicity in software, but his pleas largely fell on deaf ears, and his languages support lots of nesting too.

What is the practical upshot for you, as a budding language designer? Don't over-engineer your language. Keep it as simple as possible. Don't nest things unless they need to be nested. And be prepared to pay (as a language implementor) every time you ignore this advice!

Now, it's time to draw a few programming language design examples from Jzero and Unicon. In the case of Jzero, since it is a subset of Java, the design is either a big nothingburger (we use Java's design) or it is subtractive: what do we take away from Java to make Jzero, and what will that look and feel like? Despite early efforts to keep it small, Java is a large language. If, as part of our design, we make a list of everything that is in Java that is not in Jzero, it will be a long list.

Due to the constraints of page space and programming time, Jzero must be a tiny subset of Java. However, ideally, any legal Java program that is input to Jzero would not fail embarrassingly – it would either compile and run correctly, or it would print a useful explanatory message that conveys what Java feature(s) are being used that Jzero does not support. So that you can easily understand the rest of this book, as well as to help keep your expectations to a manageable size, the next section will cover additional details regarding what is in Jzero and what is not.

Completing the Jzero language definition

In the previous chapter, we listed the requirements for the language that will be implemented in this book, and the previous section elaborated on some of its design considerations. For reference purposes, this section will describe additional details regarding the Jzero language. If you find any discrepancies between this section and our Jzero compiler, then they are bugs. Programming language designers use more precise formal tools to define various aspects of a language; notations to describe lexical and syntax rules will be presented in the next two chapters. This section will describe the language in layman's terms.

A Jzero program consists of a single class in a single file. This class may consist of multiple methods and variables, but all of them are **static**. A Jzero program starts by executing a static method called main(), which is required. The kinds of statements that are allowed in Jzero are assignment statements, if statements, while statements, and the invocation of void methods. The kinds of expressions that are allowed in a Jzero program include arithmetic, relational, and Boolean logic operators, as well as the invocation of non-void methods.

The Jzero language supports the bool, char, int, and long atomic types. The int and long types are equivalent to 64-bit integer data types.

Jzero also supports arrays. Jzero supports the `String`, `InputStream`, and `PrintStream` class types as built-ins, along with subsets of their usual functionality. Jzero's `String` type supports the concatenation operator and the `charAt()`, `equals()`, `length()`, and `substring(b,e)` methods. The `String` class's `valueOf()` static method is also supported. Jzero's `InputStream` type supports the `read()` and `close()` methods, while Jzero's `PrintStream` type supports the `print()`, `println()`, and `close()` methods.

With that, we have defined the minimal features necessary to write basic computations in a toy language resembling Java. It is not intended to be a real language. However, you are encouraged to extend the Jzero language with additional features that we didn't have room for in this book, such as floating-point types and user-defined classes with non-static class variables. Now, let's see what we can observe about language design by looking at one aspect of the Unicon language.

Case study – designing graphics facilities in Unicon

Unicon's 2D and 3D graphics are built-in and non-trivial in size. The design of Unicon's graphics facilities is a real-world example that illustrates some of the trade-offs in programming language design. Most programming languages don't feature built-in graphics (or any built-in input/output), instead relegating all input/output to libraries. The C language, for example, performs input/output via libraries, and Unicon's graphics facilities are built on top of C language APIs. When it comes to libraries, many languages emulate the lower-level language they are implemented in (such as C or Java) and attempt to provide an exact 1:1 translation of the APIs of the implementation language. When higher-level languages are implemented on top of lower-level languages, this approach provides full access to the underlying API, at the cost of lowering the language level when using those facilities.

This wasn't an option for Unicon; Unicon's design emphasizes ease of programming and portability, both of which preclude providing a 1:1 mapping of complex C graphics libraries such as Xlib or OpenGL. Instead, Unicon's graphics were added via two separate large additions to the language: first 2D, and then 3D. We will consider the design issues pertaining to 2D and 3D graphics separately in the following sections. The next section describes Unicon's 2D graphics facilities.

Language support for 2D graphics

Unicon's 2D graphics facility was the last major feature to be introduced to the Icon language before it was frozen. The public motivation to add graphics to Icon was to support the rapid experimentation and development of software visualization tools. I did not mention to my Ph.D. advisor that I also wanted to be able to use these graphics capabilities to write video games more easily.

Icon was almost frozen at the time its graphics facilities were created. The design of the graphics facilities minimized the changes to the language syntax because a large change would have been rejected. The only surface changes were the addition of 19 keywords denoting special values in the graphics system. Keywords in Icon and Unicon look like variable names with an ampersand preceding them.

All but one of the keywords are devoted to simplifying the processing of *input* mouse and keyboard events. The primary keyword addition is &window. All graphics functions use this window unless another window value is supplied as an optional first argument.

The requirement mentioned in the previous chapter, to preserve what people love about Icon, extends to new features added to the language. For this reason, the graphics facility's design is consistent with Icon's existing input and output features. Icon's input and output facilities include a file type and built-in functions and operators that perform input and output.

For Unicon, a single new type ("window") was introduced as a subtype (and extension) of Icon's file data type. A window is a single, simple thing for a Unicon programmer to create and draw on. All the existing (text) input/output operations on files were made to work on windows, and then graphics output capabilities were added.

Graphics output capabilities in Unicon are comprised of a set of 40 or so built-in functions to draw different graphics primitives. The design decisions for these functions involved selecting a minimal set of non-overlapping graphics features, and designing their parameters and return values to be as simple and flexible as possible. These design goals are achieved by extensive parameter defaults, and designing functions to handle an arbitrary number of graphics primitives, by accepting an arbitrary number of parameters, wherever that is possible.

Control structures and program organization are major factors when designing language features. When writing graphics programs in most languages, a programmer is immediately taught (and forced) to give up the control flow to the library and organize their programs as a set of **callback** functions. These are functions that are called when various events occur. In Unicon, the program is allowed to retain control, and instead, the language runtime system checks for graphics events every so often, handles common tasks such as repainting the window's contents from the backing store, and queues other events for later processing at the Unicon language level, when the application control flow requests it.

After several years, 3D graphics hardware support became ubiquitous. The next section describes the design issues surrounding the addition of 3D graphics in Unicon.

Adding support for 3D graphics

2D graphics were added to Icon (and Unicon) as an extension of the file data type and supported normal file operations such as open, close, read, and write. The associated window in which individual pixels and other graphics primitives could be manipulated was a bonus. Similarly, 3D graphics were added as an extension of 2D graphics. The 3D windows support camera viewing primitives in a 3D space, but they support the same attributes, such as color and fonts in the same notation as the 2D facilities, with appropriate extensions. The 3D windows also provide the same input capabilities as 2D windows, along with additional graphics output primitives.

A 2D window's canvas is a 2D array of pixels that can be read and written, but a 3D window's canvas includes a **display list** that is redrawn for each frame. In Unicon, the display list can be manipulated directly to cause various animation effects, such as changing the size or position of individual 3D objects. The display list is central to both **level of detail** (**LOD**) management and 3D object selection. A control structure was added to mark and name sections of the display list, which can then be enabled/disabled or selected for user input.

This discussion of the design of Unicon's graphics facilities is necessarily incomplete due to space limitations. Initially, in the 2D facilities, the design was intentionally minimalist. Although the result was successful, you can argue that Unicon's graphics facilities should do more. It might be possible, for example, to invent new control structures that simplify graphics output operations even further. In any case, this design discussion has given you an idea of some of the issues that may come up when adding support for a new domain to an existing language.

Summary

This chapter presented some of the issues involved in language design. The skills you acquired from this chapter include those surrounding lexical design, including creating literal constant notations for data types; syntax design, including operators and control structures; and program organization, including deciding how and where to start execution.

The reason you should spend some time on your design is that you will need a good idea of what your programming language will be capable of, in order to implement it. If you defer design decisions until you need to implement them, mistakes will cost you more at that point. Designing your language includes what data types it supports, ways to declare variables and introduce values, control structures, and the syntax needed to support code at different levels of granularity, from individual instructions to whole programs. Once you have finished or think you have finished, it is time to code, beginning with a function for reading the source code, which is the focus of the next chapter.

Questions

1. Some programming languages do not have any reserved words at all, but most popular mainstream languages have several dozen. What are the advantages and disadvantages of adding more reserved words to a language?

2. The lexical rules for literal constants are often the largest and most complex rules in a programming language's lexical specification. Give examples of how even something as simple as integer literals can become quite a challenge to the language implementor.

3. Semicolons are often used to either terminate statements or separate adjacent statements from each other. In many popular mainstream languages, the single most common syntax error is a missing semicolon. Describe one or more ways that semicolons can be made unnecessary in a programming language's syntax.

4. Many programming languages define a program as starting from a function named `main()`. Java is unusual in that, although execution starts from `main()`, every class may have its own `main()` procedure, which is another way to start the program. Is there any value to this odd program organization?

5. Most languages feature automatic, pre-opened files for standard input, standard output, and error messages. On modern computers, however, these pre-opened files may have no meaningful mapping, and a program is more likely to utilize the pre-opened standard network, database, or graphics window resources. Should we add pre-opened networks, databases, or windows to programming languages? Explain whether this proposition is practical and why.

Join our community on Discord

Join our community's Discord space for discussions with the authors and other readers:

`https://discord.com/invite/zGVbWaxqbw`

3

Scanning Source Code

The first step in any programming language is reading the individual characters of the input source code and figuring out which characters are grouped together. In a natural language, this would include looking at the adjacent sequences of letters to identify the words. In a programming language, clusters of characters form variable names, reserved words, or sometimes operators or punctuation marks that are several characters long. This chapter will teach you how to read source code and identify the words and punctuation from the raw characters using **pattern matching**.

In this chapter, we're going to cover the following main topics:

- Lexemes, lexical categories, and tokens
- Regular expressions
- Using UFlex and JFlex
- Writing a scanner for Jzero
- Regular expressions are not always enough

First, let's look at the several kinds of words that appear in program source code. A natural language reader must distinguish the nouns from the verbs and adjectives to understand what a sentence means. In the same way, your programming language must categorize each entity in the source code to determine how it is to be interpreted.

Technical requirements

This chapter will take you through some real technical content. You can download this book's examples from our GitHub repository: `https://github.com/PacktPublishing/Build-Your-Own-Programming-Language-Second-Editon/tree/master/ch3`. The Code in Action video for the chapter can be found here: `https://bit.ly/3Fnn2c2`.

To follow along, you will need to install some tools and download the examples. Let's start by looking at how to install **UFlex** and **JFlex**. UFlex comes with Unicon and requires no separate installation.

For JFlex, download `jflex-1.9.1.tar.gz` (or newer) from `http://jflex.de/download.html`. Depending on your version of the `tar(1)` program, you may have to first decompress it with gunzip, converting the file from a `.tar.gz` file into a `.tar` file. You can get gunzip from places such as `www.gzip.org/` or `gnuwin32.sourceforge.net/packages/gzip.htm`.

After that, you can then extract the files from the `.tar` file with `tar`. It will extract itself into a subdirectory under the directory where you run `tar`. Modern versions of `tar` may include gunzip capabilities built in. For example, you will see a subdirectory named `jflex-1.9.1`. On Windows, wherever you extract JFlex, if you do not move your JFlex installation into `C:\JFLEX`, you will need to set a `JFLEX_HOME` environment variable to where you install it, and you will also want to put your `JFLEX\bin` directory in your `PATH`. On Linux, you can add your `JFLEX/bin` directory to your `PATH` or create a symbolic link to the `JFLEX\bin\jflex` script.

If you unpacked JFlex in `/home/myname/jflex-1.9.1`, you can make a symbolic link from `/usr/bin/jflex` to the untarred `/home/myname/jflex-1.9.1/bin/jflex` script:

```
sudo ln -s /home/myname/jflex-1.9.1/bin/jflex /usr/bin/jflex
```

Previously, we mentioned that the examples in this book will be delivered in both Unicon and Java in a **parallel translation** model. There is not enough horizontal space on a printed page to show the code side by side. Instead, the Unicon example will be given first, followed by the corresponding Java code. Usually, the Unicon code constitutes good executable **pseudocode** from which the Java implementation is derived. Once you have UFlex and/or JFlex installed and ready to go, it is time to discuss what we are doing. Then, we will talk about how to use UFlex and JFlex to generate the code for the lexical analyzer, also called a scanner.

Lexemes, lexical categories, and tokens

Programming languages read characters and group adjacent characters together when they are part of the same entity in the language. This can be a multi-character name or reserved word, a constant value, or an operator.

A **lexeme** is a string of adjacent characters that form a single entity. Most punctuation marks are lexemes unto themselves, in addition to separating what came before from what comes after them. In reasonable languages, whitespace characters such as spaces and tabs are ignored other than to separate lexemes.

Almost all languages also have a way of including comments in the source code, and comments are typically treated the same as whitespace: they can be the boundary that separates two lexemes, but they are discarded and not considered further.

Each lexeme has a **lexical category**. In natural languages, lexical categories are called parts of speech: nouns, verbs, and adjectives are examples. In a programming language implementation, the lexical category is generally represented by an integer code and used in parsing. Variable names are another lexical category. Constants are at least one category; in most languages, there are several different categories for different constant data types. Most reserved words get their own category because they are allowed in distinct places in the syntax; in a lot of grammars, all reserved words are given their own category, even though some of them are syntactically interchangeable. Similarly, operators usually get at least one category per precedence level, and often, each operator will be given its own category. A typical programming language has between 50 and 100 different lexical categories, which is a lot more than the number of parts of speech most of us can name for natural languages.

The bundle of information that a programming language gathers for each lexeme that it reads in the source code is called a **token**. Tokens are typically represented by a struct (pointer) or an object. The fields in the token include the following:

- The lexeme (a string)
- The category (an integer)
- Filename (a string name of the file in which the lexeme occurred)
- Line number (an integer for the line within that file where the lexeme occurred)
- Possibly other data about the lexeme (column number, binary representation, etc.)

When reading books about programming languages, you may find that some authors will use the word "token" in various ways to mean the string (lexeme), the integer category, or the struct/object, depending on context. With the vocabulary of lexemes, categories, and tokens in hand, it is time to look at the notation that is used to associate sets of lexemes with their corresponding categories. Patterns in this notation are called regular expressions.

Regular expressions

Regular expression (RE) notation is the most widely used way to describe patterns of symbols within files. They are formulated from very simple rules that are easy to understand. The set of symbols over which a set of regular expressions is written is called the **alphabet**. Our "alphabet" in this section will not be the colloquial A through Z of English but, instead, is closer to what is called the ASCII set.

In some sets of input symbols, regular expressions are patterns that describe sets of strings using the members of the input symbol set and a few regular espression operators. Since they are a notation for sets, terms such as **member, union,** or **intersection** apply when talking about the sets of strings that regular expressions can match. We will look at the rules for building regular expressions in this section, followed by examples.

Regular expression rules

Over the years, many different tools have used regular expressions, featuring many non-standard extensions to the notation. This book will show only those regular expression operators that are needed for the examples given. This is a superset of the minimum set of operators that theory books say are required for regular expressions, while it avoids the overkill of seldom-used operators found in some tools' regular expression implementations. The rules of regular expressions that we will consider are as follows. After the first rule, the rest are all about chaining regular expressions together into larger regular expressions that match more complicated patterns:

Any symbol, such as a from the alphabet, is a regular expression that matches that symbol. The escape symbol, the backslash (\), turns an RE operator into a regular expression that just matches that operator symbol.

Parentheses may be placed around a regular expression, (r), so that it matches the same thing as r. This is used to force operator **precedence** of the regular expression operators inside the parentheses so that they are applied before operators outside the parentheses.

When two regular expressions, re1 and re2, are adjacent, the resulting pattern, re1 re2, matches an instance of the left regular expression, followed by an instance of the right regular expression. This is called **concatenation** and it is sneaky because it is an invisible or implicit operator. An arbitrary string enclosed in double quotes is that sequence of characters, concatenated. Regular expression operators do not apply inside double quotes, and the usual escape sequences such as \n can be used.

Any two regular expressions, re1 and re2, can have a vertical bar placed between them to create a regular expression, re1 | re2, that matches a member of either re1 or re2. This is called **alternation** because it allows either alternative. Square brackets are a special shorthand for regular expressions composed of lots of vertical bar operators separating individual character symbols: [abcd] is equivalent to (a|b|c|d), either a or b or c or d. The shorthand also has shorthand: the [a-d] regular expression is an even shorter equivalent of (a|b|c|d), while the [^abcd] regular expression means any one character that is neither a nor b nor c nor d. A useful shorthand for the shorthand of the shorthand is the period character, or dot (.). The period, or dot character, ., is equivalent to [^\n] and matches any one character except a newline.

Any regular expression, re, can be followed by an asterisk, or **star** operator. The re* regular expression matches zero or more occurrences of the re regular expression. Similarly, any regular expression can be followed by a plus sign. The re+ regular expression matches one or more occurrences of the re regular expression. Lastly, any regular expression can be followed by a question mark. The re? regular expression matches zero or one occurrence of the re regular expression.

These rules do not say anything about whitespace in regular expressions, or comments. Programming languages have these things, but they are not part of regular expression notation! If you need a space character as part of the pattern you are matching, sure, you can escape one, or put it in double quotes or square brackets. But if you see a comment or a space that is not escaped in a regular expression, it is a bug. If you want to insert whitespace into a regular expression just to make it more pretty, you can't. If you need to write a comment to explain what a regular expression is doing, you are probably making your regular expression too complicated; regular expressions are supposed to be self-documenting. If yours are not, you should consider re-writing them.

Despite my argument of keeping things simple, the five simple rules for forming regular expressions can be combined in various ways to form powerful patterns that match very interesting sets of strings. Before we dive into the lexical analyzer generator tools that use them, we'll look at some additional examples that will give you a feel for some of the kinds of patterns that can be described by regular expressions.

Regular expression examples

Regular expressions are easy once you have written a few of them. Here are some that could conceivably be used in your scanner:

The regular expression while is a concatenation of five regular expressions, one for each letter: w, h, i, l, and e. It matches the "while" string, without the double quotes.

The regular expression "+"|"-"|"*"|"/" matches a string of length one that is either a plus, a minus, an asterisk, or a slash. Double quotes are used to ensure that none of these punctuation marks are interpreted as a regular expression operator. You could specify the same pattern as [+\-*/]. Regular expression operators such as * do not apply inside square brackets. Inside square brackets punctuation marks are just treated as symbols, with the exception of characters such as minus or caret that have special interpretations inside square brackets, which must be escaped with a backslash.

The `[0-9]*\.[0-9]*` regular expression matches zero or more digits, followed by a period, followed by zero or more digits. The dot is escaped because, otherwise, it would mean any character other than a newline. Although this pattern looks like a good effort at matching real numbers, it allows the dot to be there without any digits on either side! You will have to do better than this. It is pretty cumbersome, I admit, to say `([0-9]+\.[0-9]*|[0-9]*\.[0-9]+)`, but at least you know that token will be a number of some kind.

The `"\""[^"]*"\""` regular expression matches a double quote character, followed by zero or more occurrences of any character that is not a double quote character, followed by a double quote character. This is a typical newbie attempt at a regular expression for string constants. One thing that is wrong with it is that it allows newlines in the middle of the string, which most programming languages do not allow. Another problem with it is that it has no way to put a double quote character inside a string constant. Most programming languages will provide an escape mechanism that allows this. Once you start allowing escaped characters, you must be very specific about them. To just allow escaped double quotes, you might write `"\""([^"\\\n]|\\")*"\""`. A more general version for a language such as C might look closer to `"\""([^\\\n]|\\([abfnrtv\\?0]|[0-7][0-7][0-7]|x[0-9a-fA-F][0-9a-fA-F]))*"\""`.

These examples show that regular expressions range from trivial to gigantic. Regular expressions are something of a write-only notation – much harder to read than to write. Sometimes, if you get your regular expression wrong, it may be easier to rewrite it from scratch than to try and debug it. Having looked at several examples of regular expressions, it is time to learn about the tools that use regular expression notation to generate scanners for reading source code – namely, UFlex and JFlex.

Using UFlex and JFlex

Writing a scanner by hand is an interesting task for a programmer who wants to know exactly how everything works, but it will slow down the development of your language and make it more difficult to maintain the code afterward.

Good news, everyone! A family of tools that originated as part of UNIX, known as lex, takes regular expressions and generates a scanner function for you. Lex-compatible tools are available for the most popular programming languages. For C/C++, the most widely used lex-compatible tool is **Flex**, hosted at `https://github.com/westes/flex/`. For Unicon, we use **UFlex**, while for Java, you can use **JFlex**. These tools have various custom extensions, but to the extent that they are compatible with UNIX lex, we can present them together as one language for writing scanners. This book's examples have been crafted carefully so that we can even use the same input file for both the Unicon and Java implementation!

The input files for lex are often called (lex) specifications. Lex specifications use the extension .l and consist of several sections, separated by %%. This book refers generically to lex specifications, meaning the input file provided to either UFlex or JFlex, and, for the most part, those files would also be valid input for C Flex.

There are required sections in a lex specification: a **header section** followed by a **regular expression section**, and an optional **helper functions section**. JFlex adds an **imports section** to the front because Java needs imports and needs separate places to insert code fragments before the class and inside the class definition. The lex header section and the regular expression section are the sections you need to know about right now. We will start by looking at the header section.

Header section

Most Flex tools have options you can enable in the header section; they vary, and we will only cover them if we use them. You can also include bits of host language code there, such as variable declarations. However, the main purpose of the header section is to define named macros for patterns that may appear multiple times in the regular expression section. In lex, these named macros are on single lines of the following form:

```
name        regex
```

On a macro line, name is a sequence of letters, and then there are one or more spaces, and then there is a regular expression. Later, in the regular expressions section, these macros may be substituted into a regular expression by surrounding the name with curly braces – for example, {name}. The most common error that newbies make with lex macros is to try and insert a comment after the regular expression, so don't do that. The lex language does not support comments on these lines and will try to interpret what you write as part of the regular expression.

In a kind of epic tragedy, JFlex breaks compatibility and requires an equals sign after the name, so its macros are like this:

```
name=regex
```

This incompatibility with UNIX lex is egregious enough that we elected not to use macros in this book. While writing this book, we extended UFlex to handle macros with either syntax. If you add some macros, then the code here can be shortened a little. Without macros, your header section will be almost empty, so let's look at the next part of the lex specification: the regular expressions section.

Regular expressions section

The primary section of a lex specification is the regular expression section. Each regular expression is given on a line by itself, followed by some whitespace, followed by a **semantic action** consisting of some host language code (in our case, Unicon or Java) to execute when that regular expression has been matched. Note that although each regular expression rule starts on a new line, if the semantic action uses curly braces to enclose a statement block in the usual way, it can span multiple lines of source code and lex will not start looking for the next regular expression until it finds the matching closing curly brace.

The most common mistake made by novices in the regular expression section is that they try to insert spaces or comments in the regular expression to improve readability. Don't do that; inserting a space into the middle of the regular expression cuts off the regular expression at the space, and the rest of the regular expression is interpreted as being host language code. You can get some cryptic error messages when you do this.

When you run UNIX lex, which is a C tool, it generates a function called yylex() that returns an integer category for each lexeme; global variables are set with other useful bits of information. An integer called yychar holds the category; a string called yytext holds the characters that were matched for the lexeme; and yyleng tells us how many characters were matched. Lex tools vary in their compatibility with this public interface and some tools will compute more for you automatically. For example, JFlex must generate the scanner within a class and provide yytext() using a member function. Programming languages will certainly want more details, such as what line number the token came from. Now, it is time to work our way through examples that get us there.

Writing a simple source code scanner

This example lets you check whether you can run UFlex and JFlex. It is short enough that you can try it out either by downloading the code from the book's GitHub site or by typing in the examples as we go along. The example helps to establish the extent to which the use of UFlex and JFlex is similar. The example scanner just recognizes names and numbers and whitespace; the lex specification will be placed in a file named nnws.l. The first thing a programming language tool must do when reading source code is identify the category of each lexeme and return what category was found. This example returns a 1 for a name and a 2 for a number. Whitespace is discarded. Anything else is an error.

The body of nnws.l is given in this section. This specification will work as input for both UFlex and JFlex. After you download it or type it in, the book will show you how to build it for both Unicon and Java— you can do either or both. Since the semantic actions for UFlex are Unicon code and for JFlex they are Java code, this requires some restraint.

A semantic action will be legal in both Java and Unicon only if we limit the semantic action code to the syntax that is common to the two languages, such as method calls and return expressions. If you start inserting if statements or assignments and language-specific syntax, your lex specification will become specific to one host language, such as Unicon or Java.

Even this short example contains some ideas we will need later. The first two lines are for JFlex and are ignored by UFlex. The initial %% ends an empty JFlex import section. The second line is a JFlex option in the header section. By default, JFlex's yylex() function returns an object of the Yytoken type; the %int option tells it to instead return a type of integer like C Flex and UFlex. The third line, which starts with %%, transitions us into the regular expressions section. On the fourth line, the [a-zA-Z]+ regular expression matches one or more lowercase or uppercase letters; it matches as many adjacent letters as it can find and returns a 1. As a by-product, the characters matched will be stored in the yytext variable. On the fifth line, the [0-9]+ regular expression matches as many digits as it can find and returns a 2. On the sixth line, whitespace is matched by the [\t\r\n]+ regular expression, and nothing is returned; the scanner keeps on going into the input file looking for its next lexeme by matching some other regular expression. You probably know the other whitespace besides the actual space character inside the square brackets, but \t is a tab character, \r is a carriage return character, and \n is a newline. The dot (.) on the seventh line will match any character other than the newline, so it will catch any source code that was not allowed in any of the previous patterns and report an error in that case. Errors are reported using a function named lexErr() for reporting lexical errors, in an object named simple. We will need additional error reporting functions for later phases of our compiler:

```
%%
%int
%%
[a-zA-Z]+  { return 1; }
[0-9]+     { return 2; }
[ \t\r\n]+ {  }
    .          { simple.lexErr("unrecognized character"); }
```

This specification will be called from a main() function, once for each word in the input. Each time it is called, it will match the current input against all the regular expressions (four, in this case) and select whichever regular expression will match the most characters at the current location. If two or more regular expressions tie for the longest match, whichever one appears first in the specification file wins.

Various lex tools can provide a default main() function, but for full control, you should write your own. Writing our own main() function also allows the sample example to demonstrate how to call yylex() from a separate file. You will need to be able to do that to hook your scanner up to the parser in the next chapter.

The main() function varies by language. Unicon has a C++-style program organization model where main() starts outside of any object, while Java places main() functions inside of classes, but otherwise, the Unicon and Java code have many similarities.

The Unicon implementation of the main() function can be put in any filename with Unicon's .icn extension; let's call this one simple.icn. This file contains a main() procedure and a **singleton** class called simple that is only needed because in nnws.l, we called a lexical error helper function in a Java-compatible way – that is, simple.lexErr(). The main() procedure initializes the simple class by replacing the class constructor function with a single instance returned by that function. main() then opens the input file from a name given in the first command-line argument. The lexical analyzer is informed of what file to read by yyin. The code then calls yylex() in a loop until the scanner has finished:

```
procedure main(argv)
   simple := simple()
   yyin := open(argv[1])
   while i := yylex() do
      write(yytext, ": ", i)
end
class simple()
   method lexErr(s)
      stop(s, ": ", yytext)
   end
end
```

The corresponding Java main() function must be put in a class, and the filename must be the class name with a .java extension appended. We'll call this one simple.java. It opens a file by creating a FileReader object and attaches it to the lexical analyzer by passing the FileReader object as a parameter when it creates a lexical analyzer Yylex object. Because FileReader can fail, we have to declare that main() throws an exception. After constructing the Yylex object, main() then calls yylex() over and over again until the input is exhausted, as denoted by the Yylex.YYEOF sentinel value returned from yylex().

Despite being a bit longer, main() is doing the same thing as in the Unicon version. Compared to Unicon's simple class, the Java version has an extra proxy method, yytext(), so that other functions in the simple class or the rest of the compiler can access the most recent lexeme string without having a reference to the simple class's Yylex object:

```
import java.io.FileReader;
public class simple {
    static Yylex lex;
    public static void main(String argv[]) throws Exception {
        lex = new Yylex(new FileReader(argv[0]));
        int i;
        while ((i=lex.yylex()) != Yylex.YYEOF)
            System.out.println("token "+ i +": "+ yytext());
    }
    public static String yytext() {
        return lex.yytext();
    }
    public static void lexErr(String s) {
        System.err.println(s + ": " + yytext());
        System.exit(1);
    }
}
```

We've now covered all the code for your first scanner: a lexical specification in a .l file from which a yylex() function is produced, plus a main() function in a .icn or a .java, that calls yylex() and checks its operation. This simple scanner is intended mainly to show you how the plumbing is all wired together. Whether you downloaded it or typed it in, you should now be ready to try it out in Unicon, Java, or both. To ensure that the plumbing works as intended, we had better run it and find out.

Running your scanner

Let's run this example on the following (trivial) input file, named dorrie.in:

```
Dorrie is 1 fine puppy
```

Before you can run this program, you must compile it. UFlex and JFlex write out Unicon and Java code that is called from the rest of your programming language, which is written either in Unicon or Java. If you are wondering what the compilation looks like, it is shown in the following diagram. In Unicon, the two source files are compiled and linked together into an executable file named simple. In Java, the two files are compiled into separate .class files; you run Java on the simple.class file where the main() method lives, and it loads others as needed:

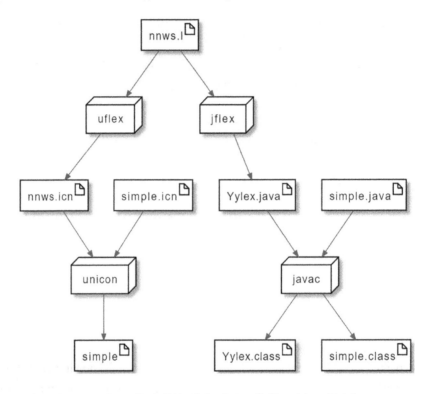

Figure 3.1: nnws.l *used to build both the Unicon (left) and Java (right) programs*

You can compile and run the program in either Unicon or Java by using the left column or the right column, as shown here:

```
uflex nnws.l                jflex nnws.l
unicon simple nnws          javac simple.java Yylex.java
simple dorrie.in            java simple dorrie.in
```

From either implementation, the output that you should see is the five lines shown here:

```
token 1: Dorrie
token 1: is
```

```
token 2: 1
token 1: fine
token 1: puppy
```

So far, all the example does is categorize groups of input characters using a regular expression to identify what kind of lexeme has been found. For the rest of the compiler to work, we will need more information about that lexeme, which we will store in a token.

Tokens and lexical attributes

In addition to identifying what integer category each lexeme belongs to, the rest of the programming language implementation (in our case, the compiler) requires the scanner to allocate an object that holds all the associated information about the lexeme. This object is called a **token**.

A token holds a group of named fields, called **lexical attributes**. The pieces of information that must be recorded pertaining to a given lexeme will depend on the language and implementation. Tokens will normally track the integer category, the string lexeme, and what line number the token came from. In a real compiler, tokens usually contain additional information about the lexeme. This is likely to include the filename and the column within the line where the lexeme occurred. For some tokens (literal constants), a compiler or interpreter may find it useful to store the actual binary value represented by that literal.

You might be wondering why you should store what column a token came from on a line. Given the lexeme text itself, you can usually see it easily enough by just looking at the line of source code, and most compilers only give line numbers when they report errors, not column numbers. In truth, not all programming language implementations store column numbers in their lexical attributes. The ones that do, however, can disambiguate errors when the same token appears more than once on a line: is the error at the first closing parenthesis, or the third? You can leave it to the human to guess, or you can record the extra details. Whether you elect to store column information or not might also depend on whether your lexical analyzer will be used in an IDE that jumps the cursor to the offending token when an error occurs. If that is among your requirements, you will need column information in order to implement that feature.

Expanding our example to construct tokens

Normally, a new token instance is allocated for each call to yylex(). In lex, tokens are transmitted to the parser by placing a pointer to the new instance in a global variable named yylval each time yylex() is called. As a transition toward a real programming language scanner, we will extend the example given previously so that it allocates these token objects.

The most elegant and portable way of doing that is to insert a function called `scan()` into the semantic actions; the `scan()` function allocates the token objects and then (usually) returns its parameter, which is the integer category code in the previous example.

A lex specification to do this can be found in the nnws-tok.l file. Fascinatingly, in JFlex, a carriage return character is neither part of a newline, nor part of the anything-but-newline dot operator, so if you use JFlex, you must account for carriage returns explicitly. In this example, they are optional in front of newlines:

```
%%
%int
%%
[a-zA-Z]+  { return simple2.scan(1); }
[0-9]+     { return simple2.scan(2); }
[ \t]+     { }
\r?\n      { simple2.increment_lineno(); }
.          { simple2.lexErr("unrecognized character"); }
```

The revised `main()` procedure in Unicon is shown in the following file, named `simple2.icn`. The `scan()` function depends on a global variable called `yylineno` that is set from `main()` and updated in `yylex()` every time a newline is matched. As per the previous example, the `simple2` class is a singleton class that is here so that the lex specification can work unchanged for both Unicon and Java. The representation of tokens is defined by a Unicon record type, which is like a `struct` in C/C++ or a class with no methods. So far, it only contains the integer category code, the lexeme string itself, and what line number it came from:

```
global yylineno, yylval
procedure main(argv)
    simple2 := simple2()
    yyin := open(argv[1]) | stop("usage: simple2 filename")
    yylineno := 1
    while i := yylex() do
        write("token ", i, " (line ",yylval.lineno, "): ", yytext)
end
class simple2()
    method lexErr(s)
        stop(s, ": line ", yylineno, ": ", yytext)
    end
    method scan(cat)
        yylval := token(cat, yytext, yylineno)
```

```
        return cat
    end
    method increment_yylineno()
        yylineno +:= 1
    end
end
record token(cat, text, lineno)
```

The corresponding Java main() in the simple2.java file looks like this:

```java
import java.io.FileReader;
public class simple2 {
    static Yylex lex;
    public static int yylineno;
    public static token yylval;
    public static void main(String argv[]) throws Exception {
        lex = new Yylex(new FileReader(argv[0]));
        yylineno = 1;
        int i;
        while ((i=lex.yylex()) != Yylex.YYEOF)
            System.out.println("token "+ i +
                    " (line " +yylval.lineno + "): "+ yytext());
    }
    public static String yytext() {
        return lex.yytext();
    }
    public static void lexErr(String s) {
        System.err.println(s + ": line " + yylineno +
            ": " + yytext());
        System.exit(1);
    }
    public static int scan(int cat) {
        yylval = new token(cat, yytext, yylineno);
        return cat;
    }
    public static void increment_lineno() {
        yylineno++;
    }
}
```

Another Java file is required for the `simple2` example. The `token.java` file contains our representation of the class token. This class token will be expanded in the next section:

```
public class token {
    public int cat;
    public String text;
    public int lineno;
    public token(int c, String s, int l) {
        cat = c; text = s; lineno = l;
    }
}
```

The following input file, `dorrie2.in`, has been extended to multiple lines and has a period added so that we can see the line number when unrecognized characters are reported:

```
Dorrie
is 1
fine puppy.
```

You can run the program in either Unicon or Java, as follows:

```
uflex nnws-tok.l                    jflex nnws-tok.l
                                    javac token.java
unicon simple2 nnws-tok             javac simple2.java Yylex.java
simple2 dorrie2.in                  java simple2 dorrie2.in
```

From either implementation, the output that you should see is as follows:

```
token 1 (line 1): Dorrie
token 1 (line 2): is
token 2 (line 2): 1
token 1 (line 3): fine
token 1 (line 3): puppy
unrecognized character: line 3: .
```

The output from this example includes line numbers, and the input file includes an unrecognized character so that we can see that the error message includes a line number as well.

Writing a scanner for Jzero

This section presents a larger example: a scanner for Jzero, our subset of the Java language. From here on out, the examples are large enough that most readers will want to download the code from the book's GitHub site if they want to run it. This example extends the previous simple2 example to a realistic language size and adds column information, as well as additional lexical attributes for literal constants. The big change is the introduction of many regular expressions for more complex patterns than what we've seen previously. The entire Java language is recognized, but a significant fraction of Java categories cause executions to terminate with an error so that our grammar in the next chapter, along with the rest of the compiler, does not have to consider them.

The Jzero flex specification

Compared to the previous examples, a real programming language lex specification will have a lot more, and more complicated, regular expressions. The following file is called javalex.l and it will be presented in several pieces.

The beginning of javalex.l includes the header and the regular expressions for comments and whitespace. These regular expressions match and consume characters from the source code without returning integer code for them; they are invisible to the rest of the compiler. As a subset of Java, Jzero includes both C-style comments bounded by /* and */ as well as C++-style comments starting with // that go to the end of the line. The regular expression for C comments is a whopper; if your language has any patterns like this, it is very easy and common to get them wrong. It reads as follows: start with a /* and then eat chunks of non-asterisk characters or asterisks so long as they don't end the comment, and finish when you find an asterisk(s) followed by a slash:

```
%%
%int
%%
"/*"([^*]|"*"+[^/*])*"*"+"/"  { j0.comment(); }
"//".*\r?\n                   { j0.comment(); }
[ \t\r\f]+                    { j0.whitespace(); }
\n                           { j0.newline(); }
```

The next part of javalex.l contains the reserved words, whose regular expressions are trivial. Since these words are common in semantic actions, use double quotes to emphasize that they are just the characters themselves and that you are not accidentally looking at some semantic action code.

Many of the integer category codes here are accessed from the parser class, specified in a separate file. In the remaining chapters of this book, the integer codes are specified by the parser. The lexical analyzer must use the parser's codes for these two phases of the compiler to communicate successfully.

You might be wondering, why use a separate integer category code for each reserved word? You only need a separate category code for each unique role in the syntax. Reserved words that can be used in the same places may use the same integer category code. If you do so, your grammar will be shorter, but you defer their differences to later in semantic analysis and make your grammar a bit vague. An example of this would be true and false; they are syntactically the same kind of thing, so they are both returned as a BOOLLIT. We might find other reserved words, such as the names of types, where we could assign them the same category code. This is a design decision to consider. When in doubt, play it safe and be un-vague by giving each reserved word its own integer:

```
"break"              { return j0.scan(parser.BREAK); }
"double"             { return j0.scan(parser.DOUBLE); }
"else"               { return j0.scan(parser.ELSE); }
"false"              { return j0.scan(parser.BOOLLIT); }
"for"                { return j0.scan(parser.FOR); }
"if"                 { return j0.scan(parser.IF); }
"int"                { return j0.scan(parser.INT); }
"null"               { return j0.scan(parser.NULLVAL); }
"return"             { return j0.scan(parser.RETURN); }
"string"             { return j0.scan(parser.STRING); }
"true"               { return j0.scan(parser.BOOLLIT); }
"bool"               { return j0.scan(parser.BOOL); }
"void"               { return j0.scan(parser.VOID); }
"while"              { return j0.scan(parser.WHILE); }
"class"              { return j0.scan(parser.CLASS); }
"static"             { return j0.scan(parser.STATIC); }
"public"             { return j0.scan(parser.PUBLIC); }
```

The third part of javalex.1 consists of the operators and punctuation marks. The regular expressions are quoted to indicate that they just mean the characters themselves. As with reserved words, in some cases, operators can be lumped together into a shared category code if they have the same operator precedence and associativity. This would make the grammar shorter at the expense of vagueness.

Another wrinkle compared to reserved words is that many operators and punctuation marks are only a single character. In that case, it is shorter and more readable to use their ASCII code as their integer category code, so we do. The `j0.ord(s)` function provides a way to do this that runs on both Unicon and Java. For multi-character operators, a `parser` constant is defined, as per the reserved words:

```
"("               { return j0.scan(j0.ord("("));  }
")"               { return j0.scan(j0.ord(")"));  }
"["               { return j0.scan(j0.ord("["));  }
"]"               { return j0.scan(j0.ord("]"));  }
"{"               { return j0.scan(j0.ord("{"));  }
"}"               { return j0.scan(j0.ord("}"));  }
";"               { return j0.scan(j0.ord(";"));  }
":"               { return j0.scan(j0.ord(":"));  }
"!"               { return j0.scan(j0.ord("!"));  }
"*"               { return j0.scan(j0.ord("*"));  }
"/"               { return j0.scan(j0.ord("/"));  }
"%"               { return j0.scan(j0.ord("%"));  }
"+"               { return j0.scan(j0.ord("+"));  }
"-"               { return j0.scan(j0.ord("-"));  }
"<"               { return j0.scan(j0.ord("<"));  }
"<="              { return j0.scan(parser.LESSTHANOREQUAL);}
">"               { return j0.scan(j0.ord(">"));  }
">="              { return j0.scan(parser.GREATERTHANOREQUAL);}
"=="              { return j0.scan(parser.ISEQUALTO);  }
"!="              { return j0.scan(parser.NOTEQUALTO);  }
"&&"              { return j0.scan(parser.LOGICALAND);  }
"||"              { return j0.scan(parser.LOGICALOR);  }
"="               { return j0.scan(j0.ord("="));  }
"+="              { return j0.scan(parser.INCREMENT);  }
"-="              { return j0.scan(parser.DECREMENT);  }
","               { return j0.scan(j0.ord(","));  }
"."               { return j0.scan(j0.ord("."));  }
```

The fourth and final part of `javalex.l` contains the more difficult regular expressions. The rule for variable names, whose integer category is IDENTIFIER, must come after all the reserved words. The regular expressions for the reserved words override the far more general identifier regular expression, but only because lex's semantics break ties by picking whichever regular expression comes first in the lex specification.

If it will make your code more readable, you can have as many regular expressions as you want, all returning the same integer category. This example uses multiple regular expressions for real numbers, which are numbers with either a decimal point, a scientific notation, or both. After the last regular expression, a catch-all pattern is used to generate a lexical error if some binary or other strange characters appear in the source code:

```
[a-zA-Z_][a-zA-Z0-9_]*{ return j0.scan(parser.IDENTIFIER);}
[0-9]+                 { return j0.scan(parser.INTLIT); }
[0-9]+"."[0-9]*([eE][+-]?[0-9]+)? { return j0.scan (parser.DOUBLELIT);}
[0-9]*"."[0-9]+([eE][+-]?[0-9]+)? { return j0.scan (parser.DOUBLELIT);}
([0-9]+)([eE][+-]?([0-9]+))  {return j0.scan (parser.DOUBLELIT);}
\"([^\"])|(\\.)*\"     { return j0.scan(parser.STRINGLIT); }
.                      { j0.lexErr("unrecognized character");}
```

Although it has been split into four portions for presentation here, the javalex.l file is not very long, at around 58 lines of code. Since it works for both Unicon and Java, this is a lot of bang for your coding buck. The supporting Unicon and Java code is non-trivial, but we are letting lex (UFlex and JFlex) do most of the work here.

Unicon Jzero code

The Unicon implementation of the Jzero scanner resides in a file named j0.icn. Unicon has a pre-processor and normally introduces defined symbolic constants via $include files. To use the same lex specification in Unicon and Java, this Unicon scanner creates a parser object whose fields, such as parser.WHILE, contain the integer category code:

```
global yylineno, yycolno, yylval
procedure main(argv)
   j0 := j0()
   parser := parser(257,258,259,260,261,262,263,264,265,
                    266, 267,268,269,270,273,274,275,276,
                    277,278,280,298,300,301,302,303,304,
                    306,307,256)
   yyin := open(argv[1]) | stop("usage: simple2 filename")
   yylineno := yycolno := 1
   while i := yylex() do
```

```
        write("token ", i, ":",yylval.lineno, " ", yytext)
    end
```

The second part of j0.icn consists of the j0 class. Compared to the simple2 class from the previous simple2.icn example, additional methods have been added for the semantic actions to call when various whitespace and comments are encountered. This allows the current column number to be calculated in a global variable called yycolno:

```
class j0()
    method lexErr(s)
        stop(s, ": ", yytext)
    end
    method scan(cat)
        yylval := token(cat, yytext, yylineno, yycolno)
        yycolno +:= *yytext
        return cat
    end
    method whitespace()
        yycolno +:= *yytext
    end
    method newline()
        yylineno +:= 1; yycolno := 1
    end
    method comment()
        yytext ? {
            while tab(find("\n")+1) do newline()
            yycolno +:= *tab(0)
        }
    end
    method ord(s)
        return proc("ord",0)(s[1])
    end
end
```

In the third part of j0.icn, the token type has been promoted from a record to a class, because now it has added complexity in its constructor, as well as a method for processing string escape characters and computing the binary representation of string literal constants. In Unicon, the constructor code comes at the end of the method in an initially section.

The deEscape() method discards leading and trailing double-quote characters and then processes a string literal character by character using Unicon string scanning. Inside the string scanning control structure, s ? { … }, the s string is examined from left to right. The move(1) function grabs the next character from the string and moves the scanning position forward by 1. A longer explanation of string scanning is given in the *Appendix, Unicon Essentials*.

In the deEscape() method, normal characters are copied over from the sin input string to the sout output string. Escape characters cause one or more characters that follow to be interpreted differently. The Jzero subset only handles tabs and newlines; Java has a lot more escapes that you could add. There is something funny about turning a backslash followed by a "t" into a tab character, but every compiler that you have ever used has had to do something like that:

```
class token(cat, text, lineno, colno, ival, dval, sval)
    method deEscape(sin)
        local sout := ""
        sin := sin[2:-1]
        sin ? {
            while c := move(1) do {
                if c == "\\" then {
                    if not (c := move(1)) then
                        j0.lexErr("malformed string literal")
                    else case c of {
                        "t":{ sout ||:= "\t" }
                        "n":{ sout ||:= "\n" }
                    }
                }
                else sout ||:= c
            }
        }
        return sout
    end
initially
    case cat of {
        parser.INTLIT:    { ival := integer(text) }
        parser.DOUBLELIT: { dval := real(text) }
        parser.STRINGLIT: { sval := deEscape(text) }
    }
```

```
end
record parser(BREAK,PUBLIC,DOUBLE,ELSE,FOR,IF,INT,RETURN,VOID,
              WHILE,IDENTIFIER,CLASSNAME,CLASS,STATIC,STRING,
              BOOL,INTLIT,DOUBLELIT,STRINGLIT,BOOLLIT,
              NULLVAL,LESSTHANOREQUAL,GREATERTHANOREQUAL,
              ISEQUALTO,NOTEQUALTO,LOGICALAND,LOGICALOR,
              INCREMENT,DECREMENT,YYERRCODE)
```

The singleton parser record here looks rather silly to an experienced Unicon programmer who can just use $define for all these token category names and skip introducing a parser type. If you are a Unicon programmer, just remind yourself that this is for Java compatibility – specifically byacc/j compatibility.

Java Jzero code

The Java implementation of the Jzero scanner includes a main class in the j0.java file. It resembles the simple2.java example. It is presented here in four parts. The first part includes the main() function and should be familiar, other than the addition of extra variables such as the yycolno variable, which tracks the current column number:

```java
import java.io.FileReader;
public class j0 {
    static Yylex lex;
    public static int yylineno, yycolno;
    public static token yylval;
    public static void main(String argv[]) throws Exception {
        lex = new Yylex(new FileReader(argv[0]));
        yylineno = yycolno = 1;
        int i;
        while ((i=lex.yylex()) != Yylex.YYEOF) {
            System.out.println("token " + i + ":" + yylineno + " " +
                yytext());
        }
    }
}
```

The j0 class continues with several helper functions that were seen in previous examples:

```java
    public static String yytext() {
        return lex.yytext();
    }
}
```

```
    public static void lexErr(String s) {
        System.err.println(s + ": line " + yylineno +
                                ": " + yytext());
        System.exit(1);
    }
    public static int scan(int cat) {
        last_token = yylval =
            new token(cat, yytext(), yylineno, yycolno);
        yycolno += yytext().length();
        return cat;
    }
    public static void whitespace() {
        yycolno += yytext().length();
    }
    public short ord(String s) {return(short)(s.charAt(0));}
```

The j0 class's function for handling newline characters in the source code increments the line number and sets the column back to 1. The comment-handling method goes character by character through the comment to keep the line number and column number correct:

```
    public static void newline() {
        yylineno++; yycolno = 1;
    }
    public static void comment() {
        int i, len;
        String s = yytext();
        len = s.length();
        for(i=0; i<len; i++)
            if (s.charAt(i) == '\n') {
                yylineno++; yycolno=1;
            }
            else yycolno++;
    }
}
```

There is a supporting module named parser.java. It provides a set of named constants, similar to an enumerated type, but it declares the constants directly as short integers so that they're compatible with the iyacc parser, which will be discussed in the next chapter. The integers that are chosen start above 256 because that's where iyacc starts them so that they don't conflict with integer codes of single-byte lexemes that we produce via calls to j0.ord():

```
public class parser {
public final static short BREAK=257;
public final static short PUBLIC=258;
public final static short DOUBLE=259;
public final static short ELSE=260;
public final static short FOR=261;
public final static short IF=262;
public final static short INT=263;
public final static short RETURN=264;
public final static short VOID=265;
public final static short WHILE=266;
public final static short IDENTIFIER=267;
public final static short CLASSNAME=268;
public final static short CLASS=269;
public final static short STATIC=270;
public final static short STRING=273;
public final static short BOOL=274;
public final static short INTLIT=275;
public final static short DOUBLELIT=276;
public final static short STRINGLIT=277;
public final static short BOOLLIT=278;
public final static short NULLVAL=280;
public final static short LESSTHANOREQUAL=298;
public final static short GREATERTHANOREQUAL=300;
public final static short ISEQUALTO=301;
public final static short NOTEQUALTO=302;
public final static short LOGICALAND=303;
public final static short LOGICALOR=304;
public final static short INCREMENT=306;
public final static short DECREMENT=307;
public final static short YYERRCODE=256;
}
```

There is also a supporting module named token.java that contains the token class. It has grown to include a column number, and for literal constants, their binary representation is stored in ival, sval, and dval for integers, strings, and doubles, respectively. The deEscape() method, which is used to construct the binary representation of string literals, was discussed in the Unicon implementation of this class. Once again, the algorithm goes character by character and just copies the character unless it is a backslash, in which case it grabs the following character and interprets it differently. You can see the efficacy of the Java String class by comparing this code with the Unicon version:

```java
public class token {
    public int cat;
    public String text;
    public int lineno, colno, ival;
    String sval;
    double dval;
    private String deEscape(String sin) {
        String sout = "";
        sin = String.substring(sin,1,sin.length()-1);
        int i = 0;
        while (sin.length() > 0) {
            char c = sin.charAt(0);
            if (c == '\\') {
                sin = sin.substring(1);
                if (sin.length() < 1)
                    j0.lexErr("malformed string literal");
                else {
                    c = sin.charAt(0);
                    switch(c) {
                    case 't': sout = sout + "\t"; break;
                    case 'n': sout = sout + "\n"; break;
                    default: j0.lexErr("unrecognized escape");
                    }
                }
            }
            else sout = sout + c;
            sin = sin.substring(1);
        }
```

```
        return sout;
    }
    public token(int c, String s, int ln, int col) {
        cat = c; text = s; lineno = ln; colno = col;
        switch (cat) {
        case parser.INTLIT:
            ival = Integer.parseInt(s);
            break;
        case parser.DOUBLELIT:
            dval = Double.parseDouble(s);
            break;
        case parser.STRINGLIT:
            sval = deEscape(s);
            break;
        }
    }
}
```

The token constructor performs the same four assignments – that is, initializing the token fields for all tokens. It then uses a `switch` statement with branches for three categories of tokens. For literal constant values only, there is an extra lexical attribute that must be initialized. Using Java's built-in `Integer.parseInt()` and `Double.parseDouble()` to convert the lexeme is a simplification for Jzero – a real Java compiler would have to do some more work here. The `sval` string is constructed by the `deEscape()` method because no built-in converter in Java takes a Java source code string and builds the actual string value for you. There are third-party libraries that you can find, but for Jzero purposes, it is simpler to provide our own.

Running the Jzero scanner

You can run the program in either Unicon or Java, as follows. This time, let's run the program on the following sample input file, named `hello.java`:

```
public class hello {
    public static void main(String argv[]) {
        System.out.println("hello, jzero!");
    }
}
```

Remember that, to your scanner, this hello.java program is just a sequence of lexemes. The commands to compile and run the Jzero scanner are similar to those in earlier examples, with more Java files creeping in:

```
uflex javalex.l                    jflex javalex.l
unicon j0 javalex                  javac j0.java Yylex.java
                                   javac token.java parser.java
j0 hello.java                      java j0 hello.java
```

From either implementation, the output that you should see should look like this:

```
token 258:1 public
token 269:1 class
token 267:1 hello
token 123:1 {
token 258:2 public
token 270:2 static
token 265:2 void
token 267:2 main
token 40:2 (
token 267:2 String
token 267:2 argv
token 91:2 [
token 93:2 ]
token 41:2 )
token 123:2 {
token 267:3 System
token 46:3 .
token 267:3 out
token 46:3 .
token 267:3 println
token 40:3 (
token 277:3 "hello, jzero!"
token 41:3 )
token 59:3 ;
token 125:4 }
token 125:5 }
```

The Jzero scanner will make a lot more sense in the next chapter when its output provides the parser's input. Before we move on, though, we should remind you that regular expressions can't do everything a programming language lexical analyzer might need. Sometimes, you must go beyond the lex scanning model. The next section is a real-world example of that.

Regular expressions are not always enough

If you take a theory of computation course, you'll probably be treated to proof that regular expressions cannot match some common patterns that occur in programming languages, particularly patterns that **nest** instances of the same pattern inside themselves. This section shows that regular expressions are not always enough in other aspects.

If regular expressions are not always able to handle every lexical analysis task in your language, what do you do? A lexical analyzer written by hand can handle weird cases that a lexical analyzer generated from regular expressions can't handle, perhaps at the cost of an extra day, week, or month of your time. However, in almost all real programming languages, regular expressions can get you close enough to where you only need a few extra tricks to produce the finished scanner. Here is a small real-world example.

Unicon and Go are examples of languages that provide **semicolon insertion**. The language defines lexical rules under which semicolons are inserted so that programmers don't have to worry about them for the most part. You may have noticed that the Unicon code examples tend to contain very few semicolons. Unfortunately, these semicolon insertion rules are not something that can be described with a regular expression.

In the case of the Go language, you can almost do it by remembering the previously returned token and doing some checks in the semantic action for a newline character; that newline can return as a semicolon if the checks are satisfied. But in Unicon, you must scan further forward and read the next token after the newline to decide whether a semicolon ought to be inserted! This allows Unicon semicolon insertion to be more precise and create fewer problems than in the Go language. As an example, in Go, you cannot format your code in classic C style:

```
func main()
{
    ...
}
```

Instead, you must write the curly brace on the function header line:

```
func main() {
    ...
}
```

To avoid this laughable limitation, the lexical analyzer must provide one token of look ahead. It will have to read the first token on the next line to decide whether a semicolon should be inserted at a newline.

It would be very un-Javalike to implement semicolon insertion in our Jzero scanner. But if we were going to do that, we could do it the Go way, or we could do it the Unicon way. We will show you a subset of the Go way. For your reference, the Go definition of semicolon insertion semantics can be found at https://golang.org/ref/spec#Semicolons.

This example illustrates rule #1 from the Go semicolon insertion semantics. OK, so you see a newline – do you insert a semicolon? Let's just remember the last token we saw, and if it is an identifier, a literal, a break, continue, return, ++, --,),], or }, then the newline itself should return a new dummy semicolon token. You can modify the newline() method so that it returns a Boolean true if a semicolon is to be inserted.

This defeats our strategy of using a common lex specification for both Unicon and Java. We need to write a conditional in the lex specification to say whether to return a semicolon or not, but the syntax is different in the two languages. In Unicon, our lex specification would have an if statement that might look like the following line:

```
\n        { if j0.newline() then return j0.semicolon() }
```

However, in Java, it would require parentheses and it would not say the then reserved word:

```
\n        { if (j0.newline()) return j0.semicolon(); }
```

The Unicon version of the modified j0 main module with semicolon insertion code has been provided in this book's GitHub repository, in the j0go.icn file. It is j0.icn with a new global variable called last_token, a modification of the scan() and newline() methods, and the addition of a method called semicolon() that constructs an artificial token. Here are the changed methods. Checking whether the last token category is one of several that triggers a semicolon shows off Unicon's generators. The !")]}" expression is a clever way of writing ")"|"]"|"}", which will be fed one at a time into ord() until all three are tried:

```
    method scan(cat)
        last_token := yylval := token(cat, yytext, yylineno)
        return cat
    end
    method newline()
        yylineno +:= 1
        if (\last_token).cat ===
                ( parser.IDENTIFIER|parser.INTLIT|
                  parser.DOUBLELIT|parser.STRINGLIT|
                  parser.BREAK|parser.RETURN|
                  parser.INCREMENT|parser.DECREMENT|
                  ord(!")]}") ) then return
    end
    method semicolon()
        yytext := ";"
        yylineno -:= 1
        return scan(parser.SEMICOLON)
    end
```

There are two fascinating things here. One is that a given element of source code – a newline character, which is just whitespace in most languages – will sometimes return an integer code (for an inserted semicolon) and sometimes not. That is why we introduced an if statement into the lex specification semantic actions for newlines. The other fascinating thing is the artificial token produced by the semicolon() method. It produces output that's indistinguishable from if the programmer had typed a semicolon themselves into the source code input of your programming language.

The Java implementation is provided in this book's GitHub repository, in the j0go.java file. Here are the key parts of it. The Java implementation behaves the same as the Unicon version in j0go. icn, with a new global variable called last_token, a modification of the scan() and newline() methods, and the addition of the semicolon() method, which constructs an artificial token. However, it is a bit longer. In the newline() method within the following block, a Java switch statement is being used to check if the last token's category triggers a semicolon insertion:

```
    public static int scan(int cat) {
        last_token = yylval =
            new token(cat, yytext(), yylineno);
        return cat;
    }
```

```
    public static boolean newline() {
        yylineno++;
        if (last_token != null)
            switch(last_token.cat) {
                case parser.IDENTIFIER: case parser.INTLIT:
                case parser.DOUBLELIT: case parser.STRINGLIT:
                case parser.BREAK: case parser.RETURN:
                case parser.INCREMENT: case parser.DECREMENT:
                case ')': case ']': case '}':
                    return true;
            }
        return false;
    }
    public int semicolon() {
        yytext = ";";
        yylineno--;
        return scan(parser.SEMICOLON);
    }
```

The full Go semicolon insertion semantics are a bit more involved but inserting a semicolon when the scanner has seen the regular expression for a newline is rather easy. If you want to learn how Unicon does better semicolon insertion, check out the Unicon Implementation Compendium at `http://www.unicon.org/book/ib.pdf`.

Summary

In this chapter, you learned about the crucial technical skills and tools used in programming languages when they are reading the characters of program source code. Thanks to these skills, the rest of your programming language compiler or interpreter has a much smaller sequence of words/tokens to deal with, instead of the enormous number of characters that were in the source file.

We covered a lot of ground in this chapter. You learned about what happens as input characters are read in: they are analyzed and grouped into lexemes. Lexemes are either discarded (in the case of comments and whitespace) or categorized for subsequent parsing purposes.

Besides categorizing lexemes, you learned how to make tokens from them. A token is an object instance that is created for each lexeme when it is categorized. The token is a record of that lexeme, its category, and where it came from.

The lexemes' categories are the main input of the parsing algorithm described in the next chapter. During parsing, the tokens will eventually be inserted as leaves into an important data structure called a syntax tree.

You are now ready to start stringing together the words into phrases in your source code. The next chapter will cover parsing, which checks whether the phrases make sense according to the grammar of the language.

Questions

1. Write a regular expression to match dates in dd/mm/yyyy format. Is it possible to write this regular expression so that it only allows legal dates?

2. Explain the difference between the return value that `yylex()` returns to the caller, the lexeme that `yylex()` leaves in yytext, and the token value that `yylex()` leaves in yylval.

3. Not all the `yylex()` regular expressions return an integer category after they match. When a regular expression does not return a value, what happens?

4. Lexical analysis has to deal with ambiguity, and it is entirely possible to write several regular expressions that all can match at a given point in the input. Describe Flex's tie-breaking rules for when more than one regular expression can match at the same place.

Join our community on Discord

Join our community's Discord space for discussions with the authors and other readers:

`https://discord.com/invite/zGVbWaxqbw`

4

Parsing

In this chapter, you will learn how to take individual words and punctuation, **lexemes**, and group them into larger programming constructs, such as expressions, statements, functions, classes, and packages. This task is called **parsing**. The code module is called a **parser**. You will make a parser by specifying syntax rules using grammars, and then using a parser generator tool that takes your language grammar and generates a parser for you. We will also look at writing useful syntax error messages.

This chapter covers the following main topics:

- Syntax analysis
- Context-free grammars
- Using iyacc and BYACC/J
- Writing a parser for Jzero
- Improving syntax error messages

We will review the technical requirements for this chapter, and then it will be time to refine your ideas of syntax and syntax analysis.

Technical requirements

In this chapter, you will need the following tools:

- Iyacc, a parser generator for Unicon. Iyacc comes with all recent Unicon builds, so you should already have it.

- BYACC/J, a parser generator for Java. BYACC/J is descended from Berkeley YACC, an open source YACC implementation. You can get the BYACC/J source code from SourceForge.net. Instructions are given below.

You can download this chapter's code from our GitHub repository: `https://github.com/PacktPublishing/Build-Your-Own-Programming-Language-Second-Edition/tree/master/ch4`.

The Code in Action video for the chapter can be found here: `https://bit.ly/3ClVCSf`.

At the time of writing, the BYACC/J source distribution on GitHub contains a command-line option that we will use that is not in the binary distributions or other source `.zip` distributions on that site. To obtain a copy of BYACC/J, first clone the Git via a command such as the following:

```
git clone https://git.code.sf.net/p/byaccj/git byaccj-git
```

Change into the newly created `byaccj-git` directory and go into its `src/` directory. Type make to build a `yacc` executable, and note its name, such as `yacc.linux`. Rename it `yacc`, or else modify all references to `yacc` in this book to whatever your BYACC/J is named. Add the BYACC/J `src/` directory containing this `yacc` executable to your path.

For both iyacc and BYACC/J, you will want to verify that they have been added to your path by opening a new Command Prompt or Terminal window and trying the `iyacc` and `yacc` commands. Note that you may already have a different program on your computer named yacc! In this case, we recommend renaming the BYACC/J executable that you install for this book as `byaccj` or `byaccj.exe` instead of `yacc` or `yacc.exe`. If you do this, everywhere in this book that it says to use `yacc`, you should type `byaccj` instead. To use this book successfully, you will have to keep your yaccs straight! You have been warned!

There is one additional technical requirement for this chapter. You must set your `CLASSPATH` environment variable. If you are working with the examples in this chapter in `C:\users\Alfrede Newmann\ch4`, you may need to set `CLASSPATH` to point at the `Alfrede Newmann` directory above the ch4 directory. On Windows, it is best to set this once and for all in the `Control Panel` or `Settings`, but you can set it manually if you have to with a command such as this one:

```
set CLASSPATH=".;c:\users\Alfrede Newmann"
```

Adding the directory above on `CLASSPATH` can be achieved with `..` on Linux, while you must supply the full path of the parent directory on Windows. On Linux, this is best set in `~/.bashrc` or similar, but on the command line, it looks like this:

```
export CLASSPATH=.:..
```

In the Linux-like shell that comes with MSYS2 Mingw64, I found I had to write:

```
export CLASSPATH=/c/users/clint/books/byop12
```

Before we get into the nuts and bolts of yacc, let's look at the bigger picture of what we are trying to accomplish by parsing, which is to analyze the syntax of the source code for a program written in the programming language that we are building.

Syntax analysis

As a programmer, you are probably already familiar with **syntax error messages** and the general idea of syntax, which is to understand what kinds of words or lexemes must appear, in what order, for a given communication to be well formed, which is to say grammatically correct, in a language. Most human languages are picky about this, while a few are very flexible about word order. Most programming languages are far simpler and more restrictive than natural human languages about what constitutes a legal input.

The input for syntax analysis consists of the output of the previous chapter on **lexical analysis**. Communication, such as a message or a program, is broken down into a sequence of component words and punctuation. This could be an array or list of token objects, although for parsing, all the algorithm requires is the sequence of integer codes returned from calls to yylex( ), one after another. It is the job of syntax analysis to determine whether the communication, in a given language such as English or Java, is correct or not. The result of syntax analysis is a simple Boolean true or false. In practice, in order to interpret or translate the message, more is needed than a Boolean value that tells us whether its syntax is correct. In the next chapter, you will learn how to build a syntax tree that forms the basis for the subsequent translation of a program into code. But first, we must check the syntax, so let's look at how programming language syntax is specified, in a format that is called **context-free grammar** notation.

Context-free grammars

In this section, we will define a notation used by programming language inventors to describe the syntax of their language. You will be able to use what you learn in this section to supply syntax rules as input to the parser generators used in the next section. Let's begin by understanding what context-free grammars are.

Context-free grammars are the most widely used notation for describing the syntax allowed in a programming language in terms of patterns of lexemes. They are formulated from very simple rules that are easy to understand. Context-free grammars are built from the following components:

- **Terminal symbols:** A set of input symbols are called terminal symbols. Terminal symbols in a grammar are read in from a scanner such as the one we produced in the last chapter. Although they are referred to as symbols, terminal symbols correspond to an entire word, operator, or punctuation mark; a terminal symbol identifies the category of a lexeme. As you saw in the previous chapter, these symbols' categories are represented by integer codes that are usually given mnemonic names such as IDENTIFIER, INTCONST, or WHILE. In our grammars, we will also use character literal notation for the more trivial terminal symbols; a single character inside apostrophes is just a terminal symbol that consists of that character itself. For example, ';' is the terminal symbol that consists of just a semi-colon and literally denotes the integer 59, which is the ASCII code for a semi-colon.

- **Non-terminal symbols:** Unlike regular expressions, context-free grammar rules utilize a second set of symbols called non-terminal symbols. Non-terminal symbols refer to sequences of symbols that make sense together, such as noun phrases or sentences (in natural languages), function or class definitions, or entire programs (in programming languages). One special non-terminal symbol is designated as the **start symbol** of the entire grammar. In a programming language grammar, the start symbol denotes an entire well-formed source file.

- **Production rules:** A set of rules called production rules explains how to form non-terminal symbols from smaller words and component phrases. Each production rule specifies how one non-terminal symbol can be constructed from a sequence of zero or more terminal or non-terminal symbols. Because the production rules control what terminal and non-terminal symbols are used, it is common to give the grammar by just listing all of its production rules.

Now it is time to look in more detail at the rules for building context-free grammars, followed by examples.

Writing context-free grammar rules

Production rules, also called context-free grammar rules, are patterns that describe legal sequences of lexemes using terminal symbols and additional non-terminal symbols that represent other sequences of zero or more symbols. In this book, we will use yacc notation for writing context-free grammars.

Each production rule consists of a single non-terminal symbol, followed by a colon, followed by zero or more terminal and non-terminal symbols, ending with a semi-colon, as shown in the following notation:

```
X : symbols ;
```

There is only one symbol to the left of the colon and, by definition, it is non-terminal because the meaning of the grammar rule is as follows: a non-terminal X can be constructed from a sequence of terminals and non-terminals that appear on the right-hand side of a production rule.

A context-free grammar can have as many such rules as desired, including many rules that build the same non-terminal with different combinations of symbols on the right-hand side. In fact, giving multiple rules for the same non-terminal is so common that it has its own shorthand consisting of a vertical bar. You can see an example of a vertical bar in the following code:

```
X : symbols | other_symbols ;
```

When the vertical bar (read as *or*) is used in a grammar, it indicates that there are multiple ways to build a non-terminal X. Using the vertical bar is optional because you could write the same rules as separate statements of the non-terminal, colon, right-hand side, and semicolon. For example, here are three different ways to build an X:

```
X : A | B | C ;
```

This line is equivalent to the following three lines:

```
X : A ;
X : B ;
X : C ;
```

The preceding two cases describe the same three production rules. The vertical bar is just a shorthand notation for writing multiple production rules.

So, what does a production rule mean, anyhow? It can be read and used either forward or backward. If you start from the start symbol and replace a non-terminal with one of its production rules' right-hand sides (called a **derivation step**), you work your way down from the top. If you repeat this process and eventually get to a sequence of terminal symbols with no non-terminals remaining, you have generated a legal instance of that grammar.

On the other hand, checking the syntax of a program written in a programming language starts from the other end. Executed on some input, the scanner from the last chapter will produce a sequence of terminal symbols. Given a sequence of terminal symbols, can you find within it the right side of a production rule, and replace it with its non-terminal? If you can do that repeatedly and make your way back to the start symbol, you have proved that the input source program is legal according to the grammar. This is called **parsing**.

Now it is time to look at some simple grammar examples. Some of the most intuitive grammars that we can suggest come from natural (human) languages. Other simple examples show how context-free grammars apply to programming language syntax.

 Recursion: Are you on top of your **recursion**? In math and computer science, recursion is when something is defined in terms of a simpler version of itself; see `https://en.wikipedia.org/wiki/Recursion` if you need a refresher. You will need that concept to build your programming language syntax. In context-free grammars, a non-terminal X is often used on the right side of a production rule that builds an X. This is a form of recursion. The one logical rule that you must learn when you use recursion is this: there must be another grammar rule (a **basis case**) that is not recursive. Otherwise, the recursion never ends, and your grammar doesn't make sense.

Writing rules for programming constructs

Context-free grammars are easy once you have written a few of them. You should start with the simplest rules you can think of and work your way up one tiny bit at a time. The simplest values in a language are its literal constants. Suppose we have two kinds of values in our language, Booleans and integers:

```
literal : INTLIT | BOOLLIT ;
```

The preceding production rule says that there are two kinds of literal values: integers and Booleans. Some language constructs, such as addition, may be defined only for certain types, while other constructs, such as assignment, are defined for all types. It is often best to feature a common syntax for all types, and then ensure that types are correct later, during semantic analysis. We will cover that in *Chapter 7, Checking Base Types*, and *Chapter 8, Checking Types on Arrays, Method Calls, and Structure Accesses*. Now consider a grammar rule that allows either variables or literal constants:

```
simple_expr : IDENTIFIER | literal ;
```

As you saw in *Chapter 3, Scanning Source Code,* IDENTIFIER denotes a name. The preceding production rule says that both variables and literals are allowed in simple expressions. Complex expressions are constructed by applying operators or functions to simple expressions:

```
expr : expr '+' expr | expr '-' expr | simple_expr ;
```

The preceding three production rules present a common design question. The first two rules are recursive, multiple times over. They are also ambiguous. When multiple operators are chained together, the grammar does not specify which operator is applied first.

Ambiguity: When a grammar can accept the same string in two or more different ways, the grammar is **ambiguous**. In the preceding example, $1 + 2 - 3$ could be parsed by applying the production rule for the plus sign first, and then the subtraction, or vice versa. Ambiguity can sometimes force you to rewrite your grammar so there is only one way to parse the input.

There are a lot more operators in real languages, and there is the issue of operator precedence to consider. You can look at these topics in the *Writing a parser for Jzero* section in this chapter. For now, let's briefly explore larger language structures, such as statements. A simple representation of an assignment statement is given here:

```
statement : IDENTIFIER '=' expr ';' ;
```

This version of assignment allows only a name on the left side of the equals sign. The right side can take any expression. There are several other fundamental kinds of statements found in many languages. Of these, consider the two most common ones, the IF statement and the WHILE statement:

```
statement : IF '(' expr ')' statement ;
statement : WHILE '(' expr ')' statement ;
```

These statements contain other (sub)statements. Grammars build larger constructs from smaller ones using recursive rules such as this. IF and WHILE statements have almost identical syntax. After an identifying terminal symbol reserved word, they both precede a statement with a conditional expression in parentheses. Now consider an example production rule for a sequence of one or more statements:

```
statements : statements statement | statement ;
```

Multiple statements can be accepted by repeated application of the first rule in this grammar. Good language designers write recursive rules all the time in order to repeat a construct. In the case of languages such as Java, semi-colons do not appear in this grammar rule as a statement separator, but they appear as terminators at the ends of various grammar rules, like the previous rules for assignment statements.

In this section, you saw that grammar rules for a programming language use reserved words and punctuation marks as building blocks. Larger expressions and statements are composed of smaller ones using *recursion*. Now it is time to learn about some tools that use context-free grammar notation to generate parsers for reading source code, namely iyacc and BYACC/J.

Using iyacc and BYACC/J

The name **yacc** stands for **yet another compiler-compiler**. This category of tools takes a context-free grammar as input and generates a parser from it. Yacc-compatible tools are available for most popular programming languages.

In this book, for Unicon, we use **iyacc** (short for **Icon yacc**) and for Java, you can use **BYACC/J** (short for **Berkeley YACC extended for Java**). They are highly compatible with UNIX yacc and we can present them together as one language for writing parsers. In the rest of this chapter, we will just say yacc when we mean both iyacc and BYACC/J. Complete compatibility required a bit of a Kobayashi Maru solution, mostly when it came to the semantic actions, which are written in native Unicon and Java, respectively.

 Kobayashi Maru: A **Kobayashi Maru** scenario is a no-win situation where the best answer is to change the rules of the game. In this case, I modified **iyacc** and **BYACC/J** a bit so that our no-win situation was winnable. For more information, see en.wikipedia.org/wiki/Kobayashi_Maru.

Yacc files are often called (yacc) specifications. They use the extension .y and consist of several sections, separated by %%. This book refers generically to yacc specifications, meaning the input file provided to either iyacc or BYACC/J and, for the most part, those files would also be valid input for C yacc.

There are required sections in a yacc specification: a **header section** followed by a **context-free grammar section**, and an optional **helper functions section**. The yacc header section and the context-free grammar section are the sections you need to know about for this book. In the following section, you will learn how to declare your terminal symbols in the yacc header section. Some versions of yacc require these declarations.

Declaring symbols in the header section

Most yacc tools have options that you can enable in the header section; they vary, and we will only cover them if we use them. You can also include bits of host language code there, such as variable declarations, inside %{ ... %} blocks. The main purpose of the header section is to declare the terminal and non-terminal symbols in the grammar. In the context-free grammar section, these symbols are used in production rules.

Whether a symbol is terminal or non-terminal can be inferred from how the symbol is used in a grammar, but unless they are ASCII codes, you must declare all your terminal symbols anyhow. Terminal symbols are declared in the header section using a line beginning with %token, followed by as many terminal symbol names as you want, separated by spaces. Non-terminals may be declared by a similar %nonterm line. Non-terminal declarations are not mandatory, perhaps because non-terminals can be inferred from their presence on the left-hand side of one or more grammar rules. Among other things, yacc uses terminal symbol declarations to generate a file that assigns integer constants to those names, for use in your scanner.

Advanced yacc declarations

There are other declarations that can be placed in the yacc header section beyond those used in this book. If you don't want to place your starting non-terminal at the top of your grammar, you can put it anywhere and then identify it explicitly in the header via the %start declaration for some non-terminal symbol. Also, instead of just declaring tokens with %token, you can use %left, %right, and %nonassoc to specify operator precedence and associativity in increasing order.

Now that we have learned about the header section, let's have a look at the context-free grammar section.

Putting together the yacc context-free grammar section

The primary section of a yacc specification is the context-free grammar section. Each production rule of the context-free grammar is given, followed by an optional **semantic action** consisting of some host language code (in our case, Unicon or Java) to execute when that production rule has been matched. The yacc syntax is typically like the following example:

```
X : symbols { semantic action code } ;
```

It is also legal to place semantic actions before or in between symbols in addition to the end of the rule, but if you do that, you are really declaring a new non-terminal with an empty production rule that just contains that semantic action. We will not do that in this book, as it is a frequent source of bugs.

Yacc is less picky about whitespace than lex was. The following example shows three equivalent ways to format production rules with different whitespace. Which you prefer depends on what you think is best for readability:

```
A : B | C;


A : B |
    C ;


A : B
  | C
  ;
```

Although each production rule starts on a new line, it can span multiple lines and is terminated by one of the following: a semi-colon, a vertical bar indicating another production rule for the same non-terminal, a %% indicating the start of the helper functions section, or an end-of-file. Like in lex, if the semantic action uses curly braces to enclose a statement block in the usual way, the semantic action can span multiple lines of source code. Yacc will not start looking for the next production rule until it finds the matching closing curly brace to finish the semantic action, and then goes on to find one of the terminators listed earlier, such as a semi-colon or vertical bar that ends the production rule.

A common mistake that newbies make in the context-free grammar section is trying to insert comments in the production rules to improve readability. Don't do that; you can get some very cryptic error messages when you do this.

When you run classic UNIX yacc, which is a C tool, it generates a function called `yyparse()` that returns whether the input sequence of terminal symbols returned from `yylex()` was legal according to the grammar. Global variables may be set with other useful bits of information. You can use such global variables to store anything you want, such as the root of your syntax tree. Before we progress to some larger examples, first, let's look at how yacc parsers work. You will need to know this to debug your parser when things do not go according to plan.

Understanding yacc parsers

The algorithm of the parser generated by yacc is called **LALR(1)**. It comes from a family of parsing algorithms invented by Donald Knuth of Stanford and made practical by Frank DeRemer of UC Santa Cruz and others.

If you are interested in the theory, you should check out the Wikipedia page for the LALR parser at https://en.wikipedia.org/wiki/LALR_parser or consult a serious compiler construction book, such as Douglas Thain's *Introduction to Compilers and Language Design*, from https://www3.nd.edu/~dthain/compilerbook/.

For our purposes, you need to know that the generated algorithm consists of a long while loop. In each iteration of the loop, the parser takes one tiny step forward. The algorithm uses a stack of integers to keep track of what it has seen. Each element on the parse stack is an integer code called a **parse state** that encodes the terminal and non-terminal symbols seen up to that point. The parse state on top of the stack and the **current input** symbol, which is an integer terminal symbol obtained from the yylex() function, are the two pieces of information used to decide what to do at each step. For no intrinsic reason, it is common to visualize this like a horizontal piece of string, with a string of beads on the right being slid left onto a stack that is depicted horizontally. *Figure 4.1* illustrates the yacc parse stack on the left and its input on the right.

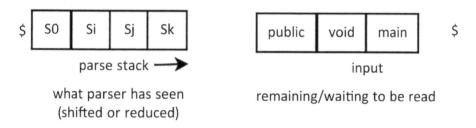

Figure 4.1: yacc's parse stack and its input

The dollar sign on the left denotes the bottom of the stack, while the dollar sign on the right denotes end-of-file. Yacc generates two big tables, computed from the grammar, called the **action table** and the **goto table**. These tables encode what to do at each step. The primary table is an action table that looks up the parse state and current input and returns one of the following possibilities:

- The top elements on the stack contain a production rule that can be used to get us (eventually) back to the starting non-terminal. This is called a **reduce**.
- The algorithm needs to look at the next input symbol. Place the current input onto the parse stack, and read the next one using yylex(). This is called a **shift**.
- If neither a shift nor a reduce will work, a syntax error is reported by calling the yyerror() function that you must write.

- If we are looking at the starting non-terminal on the top of the stack and there is no more input pending, you win! yyparse() returns the code that says there were no errors.

The following example shows the yacc parsing algorithm in pseudocode form. In this code, there are several key variables and operations, described here:

- parsestk is the parse stack, an array of integer finite automaton parse states.
- index top tracks the subscript of the top of the parse stack.
- current is the current input symbol.
- shift_n means to move the input from the right to the left, pushing parse state n onto the stack and moving current to the next input symbol.
- reduce_m means to apply production rule m by popping the number of parse states equal to the right side of production rule m and pushing the new parse state corresponding to the non-terminal on the left side of production rule m. The goto table tells what the new parse state is that the reduce is to push.

Here is the parsing algorithm in pseudocode form:

```
repeat:
    x = action_table[parsestk[top], current]
    if x == shift_n then {
        push(state_n, parsestk)
        current = next
        }
    else if x == reduce_m then {
        pop(parsestk) |m| times
        push(goto_table[parsestk[top],m], parsestk)
        }
    else if x == accept then return 0 // no errors
    else { yyerror("syntax error") }
```

This pseudocode is a direct embodiment of the preceding bulleted list. A large percentage of the world's programming languages perform their syntax analysis with this method. You may find it interesting to compare this pseudocode with the generated .icn or .java file output by iyacc or BYACC/J. Importantly, because this parsing algorithm is just a loop with a couple of table lookups, it runs quite fast.

The point of the yacc tool is to just supply the context-free grammar and get a parser without having to worry about how it works; yacc is thus a special-purpose **declarative** language for creating parsers. The algorithm works, and you don't have to know a lot about it, but if you change a grammar or use the yacc tool to create a parser for a new language that you have invented, you might have to know about these shifts and reduce operations in order to debug your context-free grammar if your parser isn't doing what you want. The most common way that a yacc programmer encounters this is when you run yacc and it reports conflicts that you may need to fix.

Fixing conflicts in yacc parsers

Earlier in this chapter, in the section titled *Writing rules for programming constructs*, you learned that grammars can be ambiguous. When a grammar is ambiguous, yacc will have more than one possible action that it can encode for a given (parse state or current input) lookup in the action table. Yacc reports this as a problem but produces a parser anyhow, and in that case, the generated parser will use only one of the possible interpretations of the ambiguity. There are two kinds of conflicts that yacc reports:

- A **shift/reduce** is when one production rule says it can shift the current input at this point, but another production rule says it is all finished and ready to reduce. In this case, yacc will only shift and you are in trouble if you need it to reduce.

- A **reduce/reduce** is even worse. Different production rules are saying they want to reduce at this point. Which one is correct? Yacc will arbitrarily pick whichever one appears earlier in your .y file, which is correct 50% of the time.

For shift/reduce conflicts, the default rule is usually correct. I have seen production language grammars with literally hundreds of shift/reduce conflicts that are ignored with seemingly no ill effects – they are asymptomatic. But once in a blue moon, and I have seen it in real life, the default on a shift/reduce conflict is not what the language needs.

For reduce/reduce conflicts, the default rule is almost surely wrong at least part of the time. Part of your grammar will never be used. In any reduce/reduce situation, or if you determine a shift/reduce conflict is a problem, you will need to modify your grammar to eliminate the conflict. Modifying your grammar to avoid conflicts is beyond the scope of this book, but it usually involves refactoring to eliminate redundant bits of grammar or creating new non-terminals and making production rules pickier. Now we will explore what happens when the parser encounters an error.

Syntax error recovery

Syntax error recovery is when your parser continues after reporting a syntax error. If recovery is successful, the compiler can go on to find the rest of the errors, if any. In the days of batch processing, it was important to recover well and show as many distinct errors from a compilation as possible. However, error recovery is known for its spectacular failures! Compilers tend to give numerous **cascading error messages** after the first one because the attempt to recover and continue parsing is often based on wild guesses as to whether tokens were missing, extra tokens were present, or the wrong token was used unintentionally... there are just too many possibilities. For this reason, we will stick to minimal error recovery in this book.

A yacc parser tries to recover if extra production rules are added to the grammar that depict likely locations of errors using a special token named error where a syntax error is expected. When an actual syntax error occurs, the shift/reduce parser throws away parse states from its parse stack and tokens from its input, until it finds a state that has a rule that allows it to proceed forward on an error. In the Jzero language, we might have a rule that throws away a syntax error within statements and discards tokens until it sees a semi-colon. There might be one or two higher-level locations in the grammar where an error token skips to the end of a function body or a declaration, and that is it.

Although we are only just touching on the topic, if your programming language becomes famous and popular, you should probably eventually learn how to recover from at least the simplest and most common errors. Since errors are inevitable, besides recovering and continuing parsing, you need to think about reporting error messages. Error reporting is covered in the *Improving syntax error messages* section at the end of this chapter. Now let's put together some working parsers using the scanners developed in the previous chapter.

Putting together a toy example

This example lets you check whether you have installed and can run iyacc and BYACC/J. The example parser just parses sequences of alternating names and numbers. The filename ns.y ("name sequence") will be used for the yacc specification. The code generated by yacc from this specification will use two helper functions: the yylex() method from the lexer class and yyerror() from the yyerror class.

The yylex() and yyerror() methods are placed in different classes and source files instead of the helper functions section of the .y file because they will be different in Unicon and Java. Another reason is that yylex() and yyerror() may be generated by separate tools.

The yylex() function is commonly generated by uflex and jflex from the preceding chapter, and yyerror() can be generated by the Merr tool described later in this chapter. Unfortunately, Java cannot utilize these static methods without placing these classes and methods inside a package. The package is named ch4 because this chapter's code is in a directory named ch4, and Java requires package names and directory names to match. Thanks to packages, some of the code from *Chapter 3, Scanning Source Code*, must be altered slightly, and you also can look forward to tricky CLASSPATH issues and cryptic error messages.

The current version of BYACC/J on GitHub has command-line options for static imports that are required. This allows the ns.y file to work unmodified as input for both Unicon and Java projects.

In the following ns.y example, there is no semantic action code (this chapter focuses solely on syntax analysis – the next chapter deals with semantic actions extensively):

```
%token NAME NUMBER
%%
sequence : pair sequence | ;
pair : NAME NUMBER ;
```

From this specification, yacc will produce a function, yyparse(). It executes the LALR parsing algorithm with a net effect described as follows:

1. yyparse() is called from a main() function.
2. yyparse() calls yylex() to get a terminal symbol.
3. yyparse() matches each terminal symbol returned from yylex() against all possible parses using all possible combinations of production rules.
4. Parsing eventually selects whichever production rule is correct at the current location and executes its semantic action (if any).

Steps 2–4 repeat until the entire input is parsed or a syntax error is found. The yylex() function is generated from the following lex specification:

```
package ch4;
%%
%int
%%
[a-zA-Z]+    { return Parser.NAME; }
[0-9]+       { return Parser.NUMBER; }
[ \t\n]+     { }
.            { lexer.lexErr("unrecognized character"); }
```

This is the nnws.l file from the previous chapter, modified in order to be used with this yacc-generated parser. For one thing, in Java, it must be made a part of the ch4 package. For another thing, it must return the integers that yacc uses for NAME and NUMBER. As you may recall from the previous chapter, the Java-compatible way to access those integers by name is through a Parser object that contains them. The BYACC/J tool generates this Parser object automatically for Java. For Unicon, iyacc's traditional -d option generates macro definitions in an include file (for ns.y, it would be in ns_tab.icn) à la classic UNIX C yacc. For this book, iyacc was extended with a command-line option, -dd, that instead generates a Java-compatible Parser object that contains the names and their values.

The main() function necessarily varies by language. By the time you add the yacc yyparse() module into the program, things start to get complicated. For this reason, the previous chapter's main() functionality is tweaked to pull out lexical analyzer initialization and lexical error handling in separate files. We will discuss the main() function first. After initialization, main() calls yyparse() to check the syntax of the source code. Here is the Unicon version of the main module, in the trivial.icn file:

```
procedure main(argv)
    yyin := open(argv[1]) | stop("usage: trivial file")
    lexer := lexer()
    Parser := Parser()
    if yyparse() = 0 then write("no errors")
end
procedure yyerror(s)
    stop(s)
end
class lexer()
    method lexErr(s)
        stop("lexical error: ", s)
    end
end
```

This Unicon implementation of main() opens the input file from a name given in the first command-line argument. The lexical analyzer is informed what file to read from via an assignment to the yyin variable. Lexical analyzer and Parser objects are initialized; they are here just for the Java compatibility of our flex specification. The code then calls yyparse() to parse the input file. The following Java code in the trivial.java file contains a main() function that corresponds with the previous Unicon example:

```
package ch4;
public class trivial {
    static ch4.j0p par;
    public static void main(String argv[]) throws Exception
    {
        ch4.lexer.init(argv[0]);
        par = new ch4.Parser();
        int i = par.yyparse();
        if (i == 0)
            System.out.println("no errors");
    }
}
```

This main module is shorter than the simple class in the previous chapter. All it does is initialize
lexical analysis, initialize the parser, and then call yyparse() to see if the input is legal. In order to
call the yylex() function from yyparse() without a reference to the Yylex object and without a
circular reference back to the main trivial class, the Yylex object and its initialization have been
pulled out into a wrapper class named lexer. The following lexer.java file contains that code:

```
package ch4;
import java.io.FileReader;
public class lexer {
    public static Yylex yylexer;
    public static void init(String s) throws Exception {
        yylexer = new Yylex(new FileReader(s));
    }
    public static int YYEOF() { return Yylex.YYEOF; }
    public static int yylex() {
        int rv = 0;
        try {
          rv = yylexer.yylex();
        } catch(java.io.IOException ioException) {
          rv = -1;
        }
        return rv;
    }
    public static String yytext() {
        return yylexer.yytext();
    }
```

```
    public static void lexErr(String s) {
        System.err.println(s);
        System.exit(1);
    }
}
```

The init() method instantiates a Yylex object for later use by a static method, yylex(), which is callable from yyparse(). The yylex() here is just a proxy that turns around and calls yylexer.yylex().

There is one more piece to the puzzle: yyparse() calls a function named yyerror() when it encounters a syntax error. The yyerror.java file contains a yyerror class that has a yyerror() static method, shown here:

```
package ch4;
public class yyerror {
    public static void yyerror(String s) {
        System.err.println(s);
        System.exit(1);
    }
}
```

This version of the yyerror() function just calls println() and exits, but we can modify it as needed. Although you might be willing to do this just for the sake of sharing a yacc specification file across both Unicon and Java, it will also pay off when we improve our syntax error messages in the next section.

Now it is time to run our toy program and see what it does. Run it with the following input file, dorrie3.in:

```
Dorrie 1 Clint 0
```

You can build and run the program in either Unicon or Java as follows. The sequence of commands to execute under Unicon looks like this:

```
uflex nnws.l
iyacc -dd ns.y
unicon trivial nnws ns ns_tab
trivial dorrie3.in
```

The sequence of commands to execute under Java is as follows:

```
jflex nnws.l
yacc -Jpackage=ch4 -Jyylex=ch4.lexer.yylex \
                    -Jyyerror=ch4.yyerror.yyerror ns.y
javac trivial.java Yylex.java Parser.java lexer.java \
        yyerror.java ParserVal.java
java ch4.trivial dorrie3.in
```

From either implementation, the output that you should see is as follows:

```
no errors
```

So far, all the example does is categorize groups of input characters using a regular expression to identify what kind of lexeme has been found. For the rest of the compiler to work, we will need more information about that lexeme, which we will store in a token.

In this section, you learned how to integrate a yacc-generated parser with a lex-generated scanner from the previous chapter. The same lex and yacc specifications were used for Unicon and Java, after some slight tweaks to iyacc and BYACC/J. The main challenges encountered were in integrating these declarative languages into Java, which involved writing and importing two static methods from helper classes. Happily, we were able to make these tools work on a toy example. Now it is time to use them on an actual programming language.

Writing a parser for Jzero

The next example is a parser for Jzero, our subset of the Java language. This extends the previous chapter's Jzero example. The big change is the introduction of many context-free grammar rules for more complex syntax constructs than have been seen up to this point. If you wrote a new language not based on an existing one, you would have to come up with a context-free grammar from scratch. For Jzero, this is not the case. The grammar we use for Jzero was adapted from a Java dialect named Godiva. To work from a real Java grammar, you can look at `https://docs.oracle.com/javase/specs/`.

The Jzero lex specification

The Jzero lex specification is as given in the previous chapter, with a one-line package declaration added to the top. The parser must be generated before the scanner is compiled. This is because yacc turns `j0gram.y` into a parser class whose constant values are referenced from the scanner. Because the static import of `yylex()` entails using packages, you must add the following line to the top of `javalex.l` from the previous chapter:

```
package ch4;
```

In order to be compatible with the previous chapter's javalex.1, the module called lexer in the trivial parser earlier in this chapter is called j0 in the Jzero parser.

With the understanding of this slight change to the Jzero lexical specification in order to call it from the parser, let's move on to the next section to learn about the Jzero yacc specification.

The Jzero yacc specification

Compared with the previous examples, a real(ish) programming language yacc specification has a lot more (and more complicated) production rules. The following file is called j0gram.y and it is presented in several parts.

The first section of j0gram.y includes the header and declarations of terminal symbols. These declarations are the source of the symbolic constants in the parser class used in the previous chapter. It is not enough for the names to match in the scanner and parser; the integer codes must be identical for the two tools to talk. The scanner must return the parser's integer codes for its terminal symbols. Per the preceding description of the yacc header section, declarations of terminal symbols are made by giving their name on a line beginning with %token. Jzero declares approximately 27 symbols for reserved words, different kinds of literal constants, and multi-character operators:

```
%token BREAK DOUBLE ELSE FOR IF INT RETURN VOID WHILE
%token IDENTIFIER CLASSNAME CLASS STRING BOOL
%token INTLIT DOUBLELIT STRINGLIT BOOLLIT NULLVAL
%token LESSTHANOREQUAL GREATERTHANOREQUAL
%token ISEQUALTO NOTEQUALTO LOGICALAND LOGICALOR
%token INCREMENT DECREMENT PUBLIC STATIC
%%
```

After the %% are the production rules for the context-free grammar of the language we are specifying. By default, the non-terminal on the first rule listed is the starting non-terminal, which, in Jzero, denotes one whole source file, module, or compilation unit. In Jzero, this is just one class; this is a severe simplification of Java where there are usually several declarations such as imports before the class in each source file.

A class declaration consists of the word class followed by an identifier giving the class name, followed by a body. The identifier that follows the word class becomes the name of a type:

```
ClassDecl:      PUBLIC CLASS IDENTIFIER ClassBody ;
```

A class body is a sequence of declarations for fields, methods, and constructors. Notice how the production rules for ClassBody allow for zero or more occurrences of declarations within the curly braces: one rule requires a list of one or more ClassBodyDecls, while a second rule explicitly allows the unusual but legal case of an empty class:

```
ClassBody:      '{' ClassBodyDecls '}' | '{' '}' ;
ClassBodyDecls: ClassBodyDecl | ClassBodyDecls ClassBodyDecl;
ClassBodyDecl:  FieldDecl | MethodDecl | ConstructorDecl ;
```

Field declarations consist of a type followed by a comma-separated list of variables. Some language implementations make the lexical analyzer report a different integer category code for an identifier once it has become a type name instead of a variable name; Jzero does not:

```
FieldDecl:      Type VarDecls ';' ;
Type:           INT | DOUBLE | BOOL | STRING | Name ;
Name:           IDENTIFIER | QualifiedName ;
QualifiedName:  Name '.' IDENTIFIER ;
VarDecls:       VarDeclarator | VarDecls ',' VarDeclarator;
VarDeclarator:  IDENTIFIER | VarDeclarator '[' ']' ;
```

The next part of j0gram.y consists of the syntax rules for the other two kinds of things that can be declared within a class, which use function syntax: methods and constructors. To begin with, they have slightly different headers followed by a block of statements:

```
MethodDecl: MethodHeader Block ;
ConstructorDecl: ConstructorDeclarator Block ;
```

Method headers have a return type, but otherwise, methods and constructors have a similar syntax:

```
MethodHeader: PUBLIC STATIC MethodReturnVal MethodDeclarator ;
MethodReturnVal: Type | VOID ;
```

A method's name (or in the case of a constructor, the class name) is followed by a parenthesized list of parameters:

```
MethodDeclarator: IDENTIFIER '(' FormalParmListOpt ')' ;
```

A parameter list is zero or more parameters. Non-terminal `FormalParmListOpt` has two production rules: either there is a (non-empty) `FormalParmList` or there isn't. The suffix `Opt` in the name is intended to indicate that the construct is optional. The empty production after the vertical bar is called an **epsilon rule**:

```
FormalParmListOpt: FormalParmList | ;
```

A formal parameter list is a comma-separated list where each formal parameter consists of a type and a variable name:

```
FormalParmList: FormalParm | FormalParmList ',' FormalParm;
FormalParm: Type VarDeclarator ;
```

The next part of `j0gram.y` contains the **statement grammar**. A **statement** is a chunk of code that does not provide a value for use by the surrounding code. Jzero has several kinds of statements. A `Block` (such as the body of a method) is a statement consisting of a sequence of (sub)statements enclosed in curly braces, {}:

```
Block: '{' BlockStmtsOpt '}' ;
```

Since a `Block` may contain zero substatements, a non-terminal with an epsilon rule is used:

```
BlockStmtsOpt:    BlockStmts | ;
```

Having dispensed with the optional case, `BlockStmts` are chained together using recursion:

```
BlockStmts:       BlockStmt | BlockStmts BlockStmt ;
```

The kinds of statements allowed within a `Block` include variable declarations and ordinary executable statements:

```
BlockStmt:        LocalVarDeclStmt | Stmt ;
```

Local variable declarations consist of a type followed by a comma-separated list of variable names ending with a semi-colon. Non-terminal `VarDecls` was presented where it was previously used in class variable declarations:

```
LocalVarDeclStmt: LocalVarDecl ';' ;
LocalVarDecl:     Type VarDecls ;
```

There are many kinds of ordinary executable statements, including expressions, break and return statements, `if` statements, and `while` and `for` loops:

```
Stmt:    Block | ';' | ExprStmt | BreakStmt | ReturnStmt
```

```
        | IfThenStmt | IfThenElseStmt | IfThenElseIfStmt
        | WhileStmt | ForStmt ;
```

Most expressions produce a value that must be used in a surrounding expression. Three kinds of expressions can be turned into a statement by following them with a semi-colon:

```
ExprStmt:  StmtExpr ';' ;
StmtExpr:  Assignment | MethodCall | InstantiationExpr ;
```

Several forms of if statements are provided, allowing for chains of else statements. If they seem excessive, it is because the Jzero subset of Java generally requires bodies of conditional and loop constructs to use curly braces, avoiding a common source of bugs:

```
IfThenStmt:       IF '(' Expr ')' Block ;
IfThenElseStmt:   IF '(' Expr ')' Block ELSE Block ;
IfThenElseIfStmt: IF '(' Expr ')' Block ElseIfSequence
        | IF '(' Expr ')' Block ElseIfSequence ELSE Block ;
ElseIfSequence:   ElseIfStmt | ElseIfSequence ElseIfStmt ;
ElseIfStmt:       ELSE IfThenStmt ;
```

WHILE loops have a simple syntax similar to IF statements:

```
WhileStmt:        WHILE '(' Expr ')' Block ;
```

FOR loops, on the other hand, are quite involved:

```
ForStmt: FOR '(' ForInit ';' ExprOpt ';' ForUpdate ')' Block ;
ForInit:          StmtExprList | LocalVarDecl | ;
ExprOpt:          Expr | ;
ForUpdate:        StmtExprList | ;
StmtExprList:     StmtExpr | StmtExprList ',' StmtExpr ;
```

The BREAK and RETURN statements are very simple, the only difference in their syntax being that RETURN can have an optional expression after it. VOID methods return without this expression, while non-VOID methods must include it; this must be checked during semantic analysis:

```
BreakStmt:        BREAK ';' ;
ReturnStmt:       RETURN ExprOpt ';' ;
```

The next part of j0gram.y contains the **expression grammar**. An **expression** is a chunk of code that computes a value, typically for use in a surrounding expression. This expression grammar uses one non-terminal symbol per level of operator precedence.

For example, the way that multiplication is forced to be higher precedence than addition is that all multiplications are performed on a `MulExpr` non-terminal and then `MulExpr` instances are chained together using plus (or minus) operators in the `AddExpr` production rules:

```
Primary: Literal | '(' Expr ')' | FieldAccess | MethodCall;
Literal:  INTLIT | DOUBLELIT | BOOLLIT | STRINGLIT | NULLVAL;
InstantiationExpr: Name '(' ArgListOpt ')' ;
ArgListOpt:  ArgList | ;
ArgList: Expr | ArgList ',' Expr ;
FieldAccess: Primary '.' IDENTIFIER ;
MethodCall: Name '(' ArgListOpt ')'
    | Name '{' ArgListOpt '}'
    | Primary '.' IDENTIFIER '(' ArgListOpt ')'
    | Primary '.' IDENTIFIER '{' ArgListOpt '}' ;
PostFixExpr: Primary | Name ;
UnaryExpr:  '-' UnaryExpr | '!' UnaryExpr | PostFixExpr ;
MulExpr: UnaryExpr | MulExpr '*' UnaryExpr
    | MulExpr '/' UnaryExpr | MulExpr '%' UnaryExpr ;
AddExpr: MulExpr | AddExpr '+' MulExpr | AddExpr '-' MulExpr ;
RelOp: LESSTHANOREQUAL | GREATERTHANOREQUAL | '<' | '>' ;
RelExpr: AddExpr | RelExpr RelOp AddExpr ;
EqExpr: RelExpr | EqExpr ISEQUALTO RelExpr | EqExpr
NOTEQUALTO RelExpr ;
CondAndExpr: EqExpr | CondAndExpr LOGICALAND EqExpr ;
CondOrExpr: CondAndExpr | CondOrExpr LOGICALOR CondAndExpr;
Expr: CondOrExpr | Assignment ;
Assignment: LeftHandSide AssignOp Expr ;
LeftHandSide: Name | FieldAccess ;
AssignOp: '=' | INCREMENT | DECREMENT ;
```

Although it is split into five portions for presentation here, the j0gram.y file is not very long: around 120 lines of code. Since it works for both Unicon and Java, this is a lot of bang for your coding buck. The supporting Unicon and Java code is non-trivial, but we are letting yacc (iyacc and BYACC/J) do most of the work here. The j0gram.y file will get longer in the next chapter when we extend the parser to build syntax trees.

Now it is time to look at the supporting Unicon Jzero code that invokes and works with the Jzero yacc grammar.

Unicon Jzero code

The Unicon implementation of the Jzero parser uses almost the same organization as in the previous chapter, starting in a file named j0.icn. Instead of calling yylex() in a loop, in a yacc-based program, the main() procedure calls yyparse(), which calls yylex() every time it does a shift operation.

As was mentioned in the last chapter, the Unicon scanner uses a Parser object whose fields, such as parser.WHILE, contain the integer category codes. The Parser object is no longer in j0.icn; it is now generated by iyacc in a j0gram.icn file that is enormous and will not be shown here:

```
global yylineno, yycolno, yylval, parser
procedure main(argv)
   j0 := j0()
   parser := Parser()
   yyin := open(argv[1]) | stop("usage: j0 filename")
   yylineno := yycolno := 1
   if yyparse()=0 then
      write("no errors, ", j0.count, " tokens parsed")
end
```

The second part of j0.icn consists of the j0 class. See the explanations in *Chapter 3, Scanning Source Code*, in the *Unicon Jzero code* section:

```
class j0(count)
   method lexErr(s)
      stop(s, ": ", yytext)
   end
   method scan(cat)
      yylval := token(cat, yytext, yylineno, yycolno)
      yycolno +:= *yytext
      count +:= 1
      return cat
   end
   method whitespace()
      yycolno +:= *yytext
   end
   method newline()
      yylineno +:= 1; yycolno := 1
```

```
        end
    method comment()
        yytext ? {
            while tab(find("\n")+1) do newline()
            yycolno +:= *tab(0)
        }
    end
    method ord(s)
        return proc("ord",0)(s[1])
    end
initially
    count := 0
end
```

In the third part of j0.icn, the token type with its deEscape() method has been preserved from the previous chapter:

```
class token(cat, text, lineno, olon, ival, dval, sval)
    method deEscape(sin)
        local sout := ""
        sin := sin[2:-1]
        sin ? {
            while c := move(1) do {
                if c == "\\" then {
                    if not (c := move(1)) then
                        j0.lexErr("malformed string literal")
                    else case c of {
                        "t":{ sout ||:= "\t" }
                        "n":{ sout ||:= "\n" }
                        }
                    }
                }
                else sout ||:= c
            }
        }
        return sout
    end
initially
```

```
      case cat of {
        parser.INTLIT:    { ival := integer(text) }
        parser.DOUBLELIT: { dval := real(text) }
        parser.STRINGLIT: { sval := deEscape(text) }
      }
  end
```

You might notice that the Unicon Jzero code got a bit shorter in this chapter compared with the last, thanks to yacc doing some of the work for us. Now let's look at the corresponding code in Java.

Java Jzero parser code

The Java implementation of the Jzero parser includes a main class in the j0.java file. It resembles the same file in the previous chapter, except its main() function calls yyparse():

```
package ch4;
import java.io.FileReader;
public class j0 {
    public static Yylex yylexer;
    public static parser par;
    public static int yylineno, yycolno, count;
    public static void main(String argv[]) throws Exception
    {
        init(argv[0]);
      par = new parser();
      yylineno = yycolno = 1;
      count = 0;
      int I = par.yyparse();
      if (i == 0) {
          System.out.println""no errors,"" + j0.count +
                          " tokens parsed");
      }
    }
    public static void init(String s) throws Exception {
      yylexer = new Yylex(new FileReader(s));
    }
    // rest of j0.java methods are the same as in Chapter 3.
  }
```

To run the program, you will also have to compile the module named `Parser.java`, which is generated by yacc from our input `j0gram.y` file. That module provides the `yyparse()` function along with a set of named constants declared directly as short integers. While this book lists `j0gram.y` instead of the `Parser.java` file that is generated from it, you can run yacc and look at its output yourself.

There is also a supporting module named `token.java` that contains the token class. It is identical to that presented in the previous chapter, so we will not duplicate it here.

If you like to plan ahead, it may interest you to know that the instances of the class token contain exactly the information that you need in the leaves of the syntax tree that you will build in the next chapter. There are different ways that a person could wire up this lexical information into the tree leaves. We will deal with that in *Chapter 5, Syntax Trees*.

Running the Jzero parser

You can run the program in either Unicon or Java as follows. This time, let's run the program on the following sample input file, named `hello.java`:

```
public class hello {
    public static void main(String argv[]) {
        System.out.println("hello, jzero!");
    }
}
```

Remember, to your parser, this `hello.java` program is a sequence of lexemes that must be checked to see if it follows the grammar of the Jzero language that we gave earlier. The commands to compile and run the Jzero parser resemble earlier examples, with more files creeping in. The Unicon commands look like the following example:

```
uflex javalex.l
iyacc -dd j0gram.y
unicon j0 javalex j0gram j0gram_tab yyerror
j0 hello.java
```

The machine-generated code output by uflex for `javalex.l` contains a single function that is large enough to cause earlier versions of Unicon's code generator (`icont`) to fail with its own parse stack overflow! I had to modify the `icont yacc` grammar to use a larger stack to run this example. You need a new or recent version of Unicon to compile the lexical analyzers produced by uflex from large lexical specifications.

In the next to last line in the preceding list of commands, compiling the j0 executable with a single invocation to perform compilation plus linking is a lazy presentation choice on Unicon. On Java, there is enough of a circular dependency between the lexical analyzer (which uses parser integer constants) and the parser (which calls yylex()) that you will find it necessary to continually resort to the big inhale model of compilation. While this is a sad state of affairs, if that's what it takes for Java to smoothly combine jflex and BYACC/J, let's just relax and enjoy it.

> **Big inhale model**: All serious programming languages, especially object-oriented ones, allow modules to be compiled separately and, in fact, encourage modules to be small, such that a build consists of many tiny module compilations. When some code is changed, only a small portion of the whole program needs to be recompiled. Unfortunately, many programming language features – in this case, classes that use each other's static members – can cause you to need to compile several or many modules at once under Java (highly ironic for a language that eschews linking). Sometimes you can tease out a sequence of single compilations that will work in Java, and sometimes not. When you must submit many or all the Java source files on the command line at once, behavior that would be unwise for a C/C++ programmer becomes routine and necessary for a Java programmer. Don't sweat it. That's what fast CPUs, multiple cores, and overengineered IDEs are for.

The Java commands to build and run the j0 parser are as follows:

```
jflex javalex.l
yacc -Jclass=parser -Jpackage=ch4 -Jyylex=ch4.j0.yylex\
    -Jyyerror=ch4.yyerror.yyerror j0gram.y
javac parser.java Yylex.java j0.java parserVal.java \
      token.java yyerror.java
java ch4.j0 hello.java
```

From either the Unicon or the Java implementation, you should see the output like this:

```
no errors, 26 tokens parsed
```

Not a very interesting output. The Jzero parser will become a lot more useful in the next chapter when you learn how to construct a data structure that is a record of the complete syntactic structure of the input source program. That data structure is the fundamental skeleton upon which any interpreter or compiler implementation of a programming language is based. In the meantime, what if we give an input file that is missing some required punctuation, or uses some Java constructs that are not in Jzero? We expect an error message.

The following example input file named `helloerror.java` serves to motivate our next section:

```
public class hello {
    public static void main(String argv[]) {
        System.out.println("hello, jzero!")
    }
}
```

Can you see the error? It is the oldest and most common syntax error of all. A semi-colon is missing at the end of the `println()` statement.

Based on the parser written so far, running `j0 helloerror.java` prints the following yacc default error message and exits:

```
syntax error
```

While no errors was uninteresting, saying `syntax  error` when there is a problem is not user-friendly at all. It is time to consider syntax error reporting and recovery.

Improving syntax error messages

Earlier, we saw a bit about the yacc syntax error reporting mechanism. Yacc just calls a function named yyerror(s). Very rarely, this function can be called for an internal error such as a parse stack overflow, but usually, when it is called, it is passed the string "parse error" or "syntax error" as its parameter. Neither is adequate for helping programmers find and fix their errors in the real world. If you write a function called yyerror() yourself, you can produce a better error message. The key is to have extra information available that the programmer can use. Usually, that extra information will have to be placed in a global or public static variable in order for yyerror() to access it. Let's look at how to write a better yyerror() function in Unicon, and then in Java.

Adding detail to Unicon syntax error messages

In the *Putting together a toy example* section earlier in this chapter, you saw a Unicon implementation of yyerror(s) that just consisted of calling stop(s). It is easy to do better than this, especially if we have global variables such as yylineno available. In Unicon, your yyerror() function might look like the following:

```
procedure yyerror(s)
    write(&errout, "line ", yylineno, " column ", yycolno,
                  ", lexeme \"", yytext, "\": ", s)
end
```

This prints the line and column numbers, as well as the current lexeme at the time that the syntax error was discovered. Because yylineno, yycolno, and yytext are global variables, it is no problem to access them from the yyerror() helper procedure. The main thing that you might want to do even better than this is figure out how to produce a message that's more helpful than just saying parse error.

Adding detail to Java syntax error messages

The corresponding Java yyerror() function is given below. In BYACC/J, you could place this method in the helper functions section of j0gram.y, where it will be included within the Parser class where it is called from. Unfortunately, if you do this, you give up Unicon/Java portability in the yacc specification file. So instead, we place the yyerror() function in its own class and its own file. This example shows the degree of pathos inflicted by Java's semi-pure object-oriented model, where everything must be in a class, even when it is inane to do so:

```
public class yyerror {
    public static void yyerror(String s) {
        System.err.println("line "+ j0.yylineno +
                    " column "+ j0.yycolno +
                    ", lexeme \"" + j0.yytext()+ "\": " + s);
    }
}
```

As we saw earlier in this chapter, using this yyerror() from another file from within a parser class generated by BYACC/J requires an import static declaration for which we added -Jyylex=… and -Jyyerror=… command-line options to BYACC/J.

With either the Unicon or the Java implementation, when you link this yyerror() into your j0 parser and run j0 helloerror.java, you should see output that looks like the following:

```
line 4 column 1, lexeme "end": parse error
```

Until recently, this was as good as many production compilers such as GCC managed to do. For an expert programmer, it is enough. Looking before and after the point of failure, an expert will see a missing semi-colon. But for a novice or an intermediate programmer having a bad day, even the line number, column, and token at which an error is discovered are not enough. Good programming language tools must be able to deliver better error messages.

Using Merr to generate better syntax error messages

How do we write a better message that clearly indicates a parse error? The parsing algorithm was looking at two integers when it realized there was an error: a **parse state** and a **current input symbol**. If you can map those two integers to a set of better error messages, you win. Unfortunately, it is not trivial to figure out what the integer parse states mean. You can do it by painful trial and error, but every time you change the grammar, those numbers change.

A tool was created just to solve this problem, called **Merr** (for **Meta error**). Merr lives at http://unicon.org/merr. It takes as input the name of your compiler, a makefile for building it, and a meta.err specification file that contains a list of error fragments and their corresponding error messages. In order to generate yyerror(), Merr builds your compiler and runs it in a mode that causes it to print out the parse state and current input token on each of the fragment errors. It then writes out a yyerror() that contains a table showing, for each parse state and error fragment, what the associated error message is. A sample meta.err file for a few errors, including the missing semi-colon error shown earlier, is as follows:

```
public {
::: class expected
public class {
::: missing class name
public class h public
::: { expected
public class h{public static void m(S a[]){S.o.p("h")}}
::: semi-colon expected
```

You invoke the Merr tool by telling it the name of the compiler you are building; it uses this name as a target argument when it calls make to build your compiler. Various command-line options let you specify what yacc version you have and other important details. The following command lines invoke merr on Unicon (left) or Java (right):

```
merr -u j0                          merr -j j0.class
```

This command grinds for a while. Merr rebuilds your compiler with a modified yyerror() function to report the parse state and input token at the time of each error. Merr then runs your compiler on each of the error fragments and records what parse states they die in. Finally, merr writes out a yyerror() containing a table mapping parse states to error messages.

As you saw in both the Unicon and Java cases, writing an error message that includes line numbers or the current input symbol when a syntax error is found is easy. On the other hand, saying something more helpful about it can be challenging.

Summary

In this chapter, you learned about the crucial technical skills and tools used in programming languages when they are parsing the sequence of lexemes from the program source code to check its organization and structure.

You learned how to write context-free grammars and how to use the iyacc and BYACC/J tools to take your context-free grammar and generate a parser for it.

When input fails to follow the rules, an error reporting function, yyerror(), is called. You learned some basics about this error-handling mechanism.

You learned how to call a generated parser from a main() function. The parser that yacc generates is called via the yyparse() function.

You are now ready to learn how to build the syntax tree data structure that reflects the structure of the input source code. The next chapter will cover the construction of syntax trees in detail.

Questions

1. What does it really mean to say a grammar symbol is terminal? Is it dying or something?
2. YACC parsers are called shift/reduce parsers. What exactly is a shift? What is a reduce?
3. Does the semantic action code in a YACC grammar execute when the parser performs a shift, a reduce, or both?
4. How does syntax analysis make use of the lexical analysis described in the previous chapter?

Join our community on Discord

Join our community's Discord space for discussions with the authors and other readers:

`https://discord.com/invite/zGVbWaxqbw`

5

Syntax Trees

The preceding two chapters covered lexical and syntax analysis, the first two phases of a compiler. Next, we will need to perform semantic analysis and code generation, but those phases will need some information to work from. The parser we constructed in the last chapter can detect and report syntax errors, which is a big, important job. When there is no syntax error, you need to build a data structure during parsing that represents the whole program logically. This data structure is based on how the different tokens and larger pieces of the program are grouped together. A **syntax tree** is a tree data structure that records the branching structure of the grammar rules used by the parsing algorithm to check the syntax of an input source file. A branch occurs whenever two or more symbols are grouped together on the right-hand side of a grammar rule to build a non-terminal symbol. This chapter will show you how to build syntax trees, which are the central data structures for your programming language implementation. A syntax tree is what the parser produces and passes along to the semantic analysis and code generation phases.

This chapter covers the following main topics:

- Learning about trees
- Creating leaves from terminal symbols
- Building internal nodes from production rules
- Forming syntax trees for the Jzero language
- Debugging and testing your syntax tree

We will learn about tree data structures and how to build them. But first, let's learn about some new tools that will make building your language easier for the rest of this book.

Technical requirements

There are two tools for you to install for this chapter, as follows:

- **Dot** is part of a package called **Graphviz** that can be downloaded from `http://graphviz.org`. After successfully installing Graphviz, you should have an executable named dot (or `dot.exe`) on your path. We will use dot to check our work after we finish building the syntax tree. Dot will generate a graphic image of the tree for us, but first, we will have to build the tree. For this reason, we will cover the dot tool and our use of it at the end of the chapter.

- **GNU's Not Unix (GNU) Make** is a tool to help manage large programming projects that supports both Unicon and Java. It is available for Windows from `http://gnuwin32.sourceforge.net/packages/make.htm`. Most programmers probably get it along with their C/C++ compiler or with a development suite such as MSYS2 or Cygwin. On Linux, you typically get Make from a C development suite, although it is often also a separate package you can install.

- You can download this book's examples from our GitHub repository: `https://github.com/PacktPublishing/Build-Your-Own-Programming-Language-Second-Edition/tree/master/ch5`.

The Code in Action video for the chapter can be found here: `https://bit.ly/3DgRcgC`.

While our use of dot can wait until we build our syntax trees, our use of Make will be pervasive. For this reason, before we dive into the main topics of this chapter, let's explore the basics of how to use GNU Make and why you need it for developing your language.

Using GNU Make

Command lines are growing longer and longer, and you will get very tired of typing the commands required to build a programming language. We are already using Unicon, Java, uflex, jflex, iyacc, and BYACC/J. Few tools for building large programs are multi-platform and multi-language enough for this toolset. We will use the ultimate multi-platform, multi-language software build tool: GNU Make.

Once the Make program is installed on your path, you can store the build rules for Unicon or Java, or both, in a file named a makefile (or Makefile), and then just run Make whenever you have changed the code and need to rebuild. A full treatment of Make is beyond the scope of this book, but here are the key points.

A makefile is like a lex or yacc specification, except instead of recognizing patterns of strings, a makefile specifies a graph of **build dependencies** between files. For each file, the makefile contains the source files it depends on as well as a list of one or more command lines needed to build that file. The makefile header just consists of macros defined by NAME= strings that are used in later lines by writing $(NAME) to replace a name with its definition. The rest of the makefile lines are dependencies written in the following format:

```
file: source_file(s)
    build rule
```

In the first line, file is an output file you want to build, also called a target. The first line specifies that the target depends on current versions of the source file(s). These files are required to make the target. build rule is the command line that you execute to make that output file from those source file(s).

Don't forget the tab! The **make** program supports multiple lines of build rules. Each build rule line must start with a tab. The most common newbie mistake in writing a makefile is that the build rule line(s) must begin with an **American Standard Code for Information Interchange (ASCII) Ctrl-I**, also known as a tab character. Some text editors will totally blow this, replacing the tab character with a sequence of space characters. If your build rule lines don't start with a tab, make will probably give you some confusing error message. If that is happening to you, switch to a real code editor and don't forget the tab.

The following example makefile will build both Unicon and Java if you just say make. If you run make unicon or make java, then it only builds one or the other. Added to the commands from the last chapter is a new module (tree.icn or tree.java) for this chapter. The makefile is presented in two halves, for the Unicon and then the Java build, respectively.

The target named all specifies what to build if Make is invoked without an argument saying what to build. The rest of the first half of the makefile is concerned with building the Unicon implementation of our compiler. The U macros (and IYU for iyacc ucode) list the Unicon modules that are separately compiled into a machine code format called **ucode**. The strange %.u:%.icn dependency is called a **suffix rule**. It says that all .u files are built from .icn files by running unicon -c on the .icn file. The executable named j0 is built from the ucode files by running unicon on all the .u files to link them together.

The `javalex.icn` and `j0gram.icn` files are built using uflex and iyacc, respectively. Let's look at the first half of our `makefile` for this chapter, as follows:

```
all: unicon java
LYU=javalex.u j0gram.u j0gram_tab.u
U=j0.u token.u tree.u serial.u yyerror.u $(LYU)
unicon: j0
%.u : %.icn
    unicon -c $<
j0: $(U)
    unicon $(U)
javalex.icn: javalex.l
    uflex javalex.l
j0gram.icn j0gram_tab.icn: j0gram.y
    iyacc -dd j0gram.y
```

The Java build rules occupy the second half of our `makefile`. The `JSRC` macro gives the names of all the Java files to be compiled. `BYSRC` macros for BYACC/J-generated sources, `BYJOPTS` for BYAC-C/J options, and `IMP` and `BYJIMPS` for BYACC/J static imports serve to shorten later lines in the `makefile` so that they fit within this book's formatting constraints. We are sticking carefully to a `makefile` that will run on both Windows and Linux. As a reminder, the Java rules of our `makefile` depend on a `CLASSPATH` environment variable, and the syntax for that varies with your operating system and its Command Prompt (or shell) syntax. On Windows, you might say the following:

```
set CLASSPATH=".;c:\users\username\byopl"
```

Here, `username` is your username, while on Linux, you might instead say the following:

```
export CLASSPATH=..
```

In any case, here is the second half of our `makefile`:

```
BYSRC=parser.java parserVal.java Yylex.java
JSRC=j0.java tree.java token.java yyerror.java serial.java $(BYSRC)
BYJOPTS= -Jclass=parser -Jpackage=ch5
BYJIMPS= -Jyylex=ch5.j0.yylex -Jyyerror=ch5.yyerror.yyerror
java: j0.class

j: java
    java ch5.j0 hello.java
    dot -Tpng hello.java.dot >hello.png

j0.class: $(JSRC)
    javac $(JSRC)
parser.java parserVal.java: j0gram.y
    yacc $(BYJOPTS) $(BYJIMPS) j0gram.y
Yylex.java: javalex.l
    jflex javalex.l
```

In addition to the rules for compiling the Java code, the Java part of the `makefile` has an artificial target, `make j`, that runs the compiler and invokes the dot program to generate a **Portable Network Graphic (PNG)** image of your syntax tree. We will look at the dot program and the images of tree data structures that it generates at the end of this chapter, after we have shown you how to construct syntax trees during parsing.

If you find `makefiles` strange and scary-looking, don't worry—you are in good company. This is a red pill/blue pill moment. You can close your eyes and just type `make` at the command line. Alternatively, you can dig in and take ownership of this universal multi-language software development build tool. If you want to read more about make, you might want to check out *GNU Make: A Program for Directing Compilation*, by Stallman and McGrath, or one of the other fine books on make. Now, it's time to get on with syntax trees, but first, you must know what a tree is and how to define a tree data type for use in a programming language.

Learning about trees

Mathematically, a **tree** is a kind of **graph** structure; it consists of **nodes** and **edges** that connect those nodes. All the nodes in a tree are connected. A single node at the top is called the **root**. Tree nodes can have zero or more children, and at most one parent. A tree node with zero children is called a **leaf**; most trees have a lot of leaves. A tree node that is not a leaf has one or more children and is called an **internal node**. *Figure 5.1* shows an example tree with a root, two additional internal nodes, and five leaves:

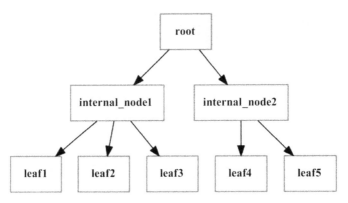

Figure 5.1: A tree with a root, internal nodes, and leaves

Trees have a property called **arity** that specifies the maximum number of children that occur for any node in the tree. An arity of 1 would give you a linked list. Perhaps the most common kinds of trees are binary trees (arity = 2). The kind of trees we need has as many children as there are symbols on the right-hand side of the rules in our grammar; these are so-called **n-ary trees**. While there is no arity bound for arbitrary context-free grammars, for any grammar we can just look and see which production rule has the most symbols on its right-hand side, and code our tree arity to that number if needed. In j0gram.y from the last chapter, the arity of Jzero is 9, although most non-leaf nodes will have two to four children. In the following subsections, we will dive deeper and learn how to define syntax trees and understand the difference between a parse tree and a syntax tree.

Defining a syntax tree type

Every node in a tree has several pieces of information that need to be represented in the class or data type used for tree nodes. This includes the following information:

- Labels or integer codes that uniquely identify the node and tell you what kind of node it is
- A data payload consisting of whatever information is associated with that node

- Information about that node's children, including how many children it has and references to those children (if any)

We use a class for this information in order to keep the mapping between Unicon and Java as simple as possible. Here is an outline of the `tree` class with its fields and constructor code in Unicon; the methods will be presented in the sections that follow within this chapter. The tree information can be represented in Unicon in a file named `tree.icn`, as follows:

```
class tree(id, sym, rule, nkids, tok, kids)
   … tree methods, to be presented later
initially(s,r,x[])
   id := serial.getid(); sym := s; rule := r
   if type(x[1]) == "token__state" then {
      nkids:=0; tok := x[1]
   } else {
      nkids := *x
     kids := x
   }
end
```

The `tree` class has the following fields:

- The `id` field is a unique integer identity or serial number that is used to distinguish tree nodes from each other. It is initialized by calling a `getid()` method in a singleton class named `serial`, which will be presented later in this section.
- The `sym` string is a human-readable description for debugging purposes.
- The member named `rule` holds the production rule (or, in the case of a leaf, the integer category) that the node represents. Yacc does not provide a numeric encoding for production rules, so you will have to make your own, whether you just count rules starting from 1 or get fancier. If you start at 1,000 or use negative numbers, you will never confuse a production rule number for a terminal symbol code.
- The member named `nkids` holds the number of child nodes underneath this node. Usually, it will be 0, indicating a leaf, or a number 2 or higher, indicating an internal node.
- The member named `tok` holds the lexical attributes of a leaf node, which comes to us via the `yylex()` function setting the parser's `yylval` variable, as discussed in *Chapter 2, Programming Language Design*.
- The member named `kids` is an array of tree objects holding references to the children of this tree node, if there are any.

The corresponding Java code looks like the following class tree in a file named tree.java. Its members match the fields in the Unicon tree class given previously:

```
package ch5;
class tree {
  int id;
  String sym;
  int rule;
  int nkids;
  token tok;
  tree kids[];
```

The tree.java file continues with class tree methods, which will be shown later, followed by two constructors for the tree class: one for leaves, which takes a token object as an argument, and one for internal nodes, which takes children. These can be seen in the following code snippet:

```
… tree methods, to be presented later
public tree(String s, int r, token t) {
  id = serial.getid();
  sym = s; rule = r; tok = t;
}
public tree(String s, int r, tree[] t) {
  id = serial.getid();
  sym = s; rule = r; nkids = t.length;
  kids = t;
  }
}
```

The previous pair of constructors initialize a tree's fields in an obvious way. You may be curious about the **identifiers** initialized from a serial class. These are used to give each node a unique identity required by the tool that draws the syntax trees for us graphically at the end of this chapter. Before we proceed with using these constructors, let's consider two different mindsets regarding the trees we are constructing.

Parse trees versus syntax trees

A **parse tree** is what you get when you allocate an internal node for each and every production rule used during the parsing of an input, even ones that introduce no branching. In contrast with syntax trees, which only introduce internal nodes when branching occurs, parse trees are a complete transcript of how the parser matched the input using the grammar.

They are unwieldy to use in practice. In real programming languages, there are lots and lots of non-terminal rules that build a non-terminal from a single non-terminal on their right-hand side. This results in a *weeping* tree appearance. *Figure 5.2* shows the height and shape of a parse tree for a trivial "Hello World" program. Solid boxes depict internal nodes, and boxes drawn with dashed lines depict leaf nodes. If you build a full parse tree, the unnecessary extra nodes will substantially slow down the rest of your compiler:

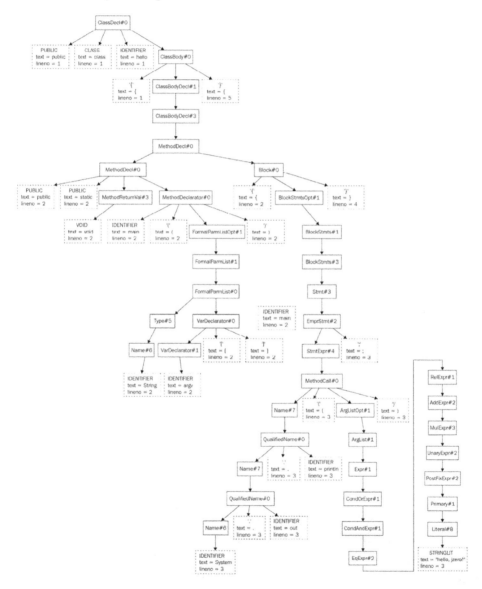

Figure 5.2: A parse tree for a "Hello World" program (67 nodes, height 27)

A **syntax tree** has an internal node whenever a production rule has two or more children on the right-hand side and the tree needs to branch out. *Figure 5.3* shows a syntax tree for the same hello. java program. Once again, solid boxes depict internal nodes, and boxes drawn with dashed lines depict leaf nodes. Note the differences in size and shape compared with the parse tree shown in *Figure 5.2*:

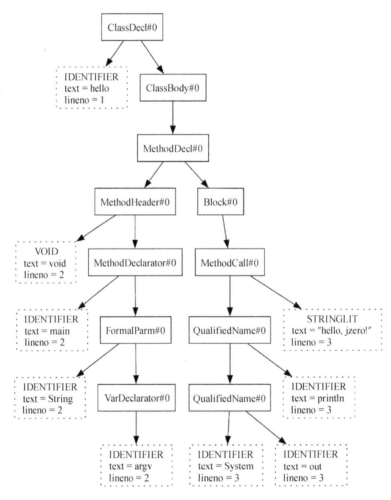

Figure 5.3: A syntax tree for a "Hello World" program (20 nodes, height 8)

While a parse tree may be useful for studying or debugging a parsing algorithm, a programming language implementation uses a much simpler tree. You will see this especially when we present the rules for building tree nodes for our example language, in the *Forming syntax trees for the Jzero language* section. Before we get there, let's consider what may be the trickiest part, creating leaves containing lexical analysis information and making them available to the parser.

Creating leaves from terminal symbols

Leaves make up a large percentage of the nodes in a syntax tree. The leaves in a syntax tree built by yacc come from the lexical analyzer. For this reason, this section discusses modifications to the code from *Chapter 2, Programming Language Design*. After you create leaves in the lexical analyzer, the parsing algorithm must pick them up somehow and plug them into the tree that it builds. This section describes that process in detail. First, you will learn how to embed token structures into tree leaves, and you will then learn how these leaves are picked up by the parser in its value stack. For Java, you will need to know about an extra type that is needed to work with the value stack. Lastly, the section provides some guidance as to which leaves are necessary and which can be safely omitted. Here is how to create leaves containing token information.

Wrapping tokens in leaves

The tree type presented earlier contains a field that is a reference to the token type introduced in *Chapter 2, Programming Language Design*. Every leaf will get a corresponding token and vice versa. Think of this as wrapping up the token inside a tree leaf. *Figure 5.4* is a **Unified Modeling Language** (**UML**) diagram that depicts each tree leaf containing a token:

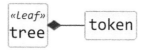

Figure 5.4: Diagram of a leaf containing a token

You could instead add the token type's member fields directly into the tree type. However, the strategy of allocating a token object, and then a separate tree node that contains a pointer to that token object, is reasonably clean and easy to understand. In Unicon, the code to create a leaf looks like this:

```
yylval := tree("token",cat, token(cat, yytext, yylineno))
```

In Java, the creation of a leaf node containing a token looks like the following code:

```
yylval = new tree("token",cat, new token(cat, yytext(), yylineno));
```

You could put this code within the j0.scan() method that is called for each token in the lexical analyzer. In Unicon, we are good at this point. In statically typed languages such as Java, what data type is yylval? In *Chapter 2, Programming Language Design*, yylval was of the type token; now, it looks like the type tree. But yylval is declared in the generated parser, and yacc doesn't know anything about your token or tree types. For a Java implementation, you must know the data type that the code generated by yacc uses for leaves, but first, you need to learn about the value stack.

Working with YACC's value stack

BYACC/J does not know about your tree class. For this reason, it generates its value stack as an array of objects whose type is named `parserVal`. If you rename BYACC/J's `parser` class to something else, such as `myparse`, using the `-Jclass=` command-line option, the value stack class will also automatically be renamed to `myparseVal`.

The `yylval` variable is part of the public interface of yacc. Every time yacc shifts the next terminal symbol onto its parse stack, it copies the contents of `yylval` onto a stack that it manages in parallel with the parse stack, called the **value stack**. BYACC/J declares the value stack elements as well as `yylval` in the parser class to be of the type `parserVal`.

Since a parse stack is managed in parallel with a value stack, whenever a new state is pushed on the parse stack, the value stack sees a corresponding push; the same goes for pop operations. Value stack entries whose parse state was produced by a shift operation hold tree leaves. Value stack entries whose parse state was produced by a reduce operation hold internal syntax tree nodes. *Figure 5.5* depicts a value stack in parallel with a parse stack:

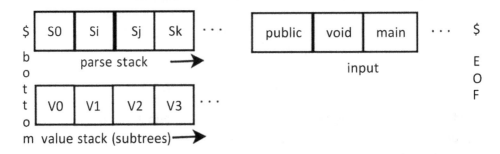

Figure 5.5: A parse stack and a value stack

In *Figure 5.5*, the dollar symbol $ on the left edge represents the bottom of the two stacks, which grow toward the right when values are pushed on the stack. The right side of the diagram depicts the sequence of terminal symbols whose tokens are produced by lexical analysis. Tokens are processed from left to right, with $ at the right edge of the screen representing the end of the file, also depicted as EOF. The ellipses (...) on the left side represent the room on the two stacks to process additional push operations during parsing, while those on the right side represent whatever additional input symbols remain after those that are depicted.

The parserVal type was briefly mentioned in *Chapter 4, Parsing*. To build syntax trees in BYACC/J, we must understand this type in detail. Here is the parserVal type, as defined by BYACC/J:

```
public class parserVal {
    public int ival;
    public double dval;
    public String sval;
    public Object obj;
    public parserVal() { }
    public parserVal(int val){ ival=val; }
    public parserVal(double val) { dval=val; }
    public parserVal(String val) { sval=val; }
    public parserVal(Object val) { obj=val; }
```

parserVal is a container that holds an int, a double, a String, and an Object, which can be a reference to any class instance at all. Having four fields here is a waste of memory for us since we will only use the obj field, but yacc is a generic tool. In any case, let's look at wrapping tree leaves within a parserVal object in order to place them in yylval.

Wrapping leaves for the parser's value stack

In terms of mechanics, parserVal is a third data type in the code that builds our syntax tree. BYACC/J requires that we use this type for the lexical analyzer to communicate tokens to the parser. For this reason, for the Java implementation, this chapter's class, j0, has a scan() method that looks like this:

```
public static int scan(int cat) {
    ch5.j0.par.yylval =
        new parserVal(
            new tree("token",0,
                new token(cat, yytext(), yylineno)));
    return cat;
}
```

In Unicon, scan() allocates two objects, as shown previously in *Figure 5.4*. In Java, each call to scan() allocates three objects, as shown in *Figure 5.6*:

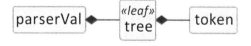

Figure 5.6: The three allocated objects: parserVal, tree *(leaf), and* token

OK—we wrapped tokens inside of tree nodes in order to represent leaf information, and then for Java, we wrapped leaf nodes inside `parserVal` in order to put them onto the value stack. Let's consider what putting a leaf on the value stack looks like in more detail. We will tell the story as it occurs in Java, recognizing that in Unicon it is a little bit simpler. Suppose you are at the beginning of your parse, and your first token is the reserved word `PUBLIC`. The scenario is shown in *Figure 5.7*. See the description of *Figure 5.5* if you need a refresher on how this diagram is organized:

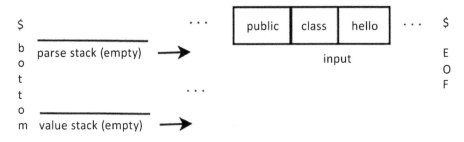

Figure 5.7: The parse stack state at the start of parsing

The first operation is a shift. An integer finite automaton state that encodes the fact that we saw `PUBLIC` is pushed onto the stack. `yylex()` calls `scan()`, which allocates a leaf wrapped in a `parserVal` instance and assigns `yylval` a reference to it, which `yylex()` pushes onto the value stack. The stacks are in lock-step, as shown in *Figure 5.8*:

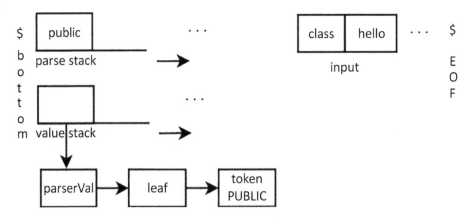

Figure 5.8: The parse and value stack state after a shift operation

Another of these wrapped leaves gets added to the value stack each time a shift occurs. Now, it's time to consider how all these leaves get placed into the internal nodes, and how internal nodes get assembled into higher-level nodes until you get back to the root. This all happens one node at a time when a production rule in the grammar is matched.

Determining which leaves you need

In most languages, punctuation marks such as semicolons and parentheses are only necessary for syntax analysis. They may help with human readability, force operator precedence, or make the grammar parse unambiguously. Once you successfully parse the input, you will never again need those leaves in your syntax tree for semantic analysis or code generation.

You can omit unnecessary leaves from the tree, or you can leave them in so that their source line number and filename information are in the tree in case these are needed for error message reporting. I usually omit them by default but add in specific punctuation leaves if I determine that the compiler needs a particular punctuation token's location in order to give a sufficiently precise error message.

The flip side of this equation is this: any leaf that contains a value, a name, or another semantic meaning of some kind in the language needs to be kept around in the syntax tree. This includes literal constants, IDs, and other reserved words or operators. Now, let's look at how and when to build internal nodes for your syntax tree.

Building internal nodes from production rules

In this section, we will learn how to construct the tree, one node at a time, during parsing. The internal nodes of your syntax tree, all the way back up to the root, are built from the bottom up, following the sequence of reduce operations with which production rules are recognized during the parse. The tree nodes used during the construction are accessed from the value stack.

Accessing tree nodes on the value stack

For every production rule in the grammar, there is a chance to execute some code called a **semantic action** when that production rule is used during a parse. As you saw in *Chapter 4*, *Parsing*, in the *Putting together the yacc context-free grammar* section, semantic action code comes at the end of a grammar rule, before the semicolon or vertical bar that ends a rule and starts the next one.

You can put any code you want in a semantic action. For us, the main purpose of a semantic action is to build a syntax tree node. Use the value stack entries corresponding to the right side of the production rule to construct the tree node for the symbol on the left side of the production rule. The left-side non-terminal that has been matched gets a new entry pushed into the value stack that can hold the newly constructed tree node.

For this purpose, yacc provides macros that refer to each position on the value stack during a reduce operation. $1, $2, ... $N refer to the current value-stack contents corresponding to the grammar rule's right-hand symbols 1 through N. By the time the semantic action code executes, these symbols have already been matched at some point in the recent past. They are the top N symbols on the value stack, and during the reduce operation they will be popped and a new value-stack entry will be pushed in their place. The new value-stack entry is whatever you assign to $$. By default, it will just be whatever is in $1; the default semantic action of yacc is $$=$1, and that semantic action is correct for production rules with one symbol (terminal or non-terminal) that is being reduced to the non-terminal on the left-hand side of the rule.

All of this is a lot to unpack. Here is a specific example. Suppose you are just finishing up parsing the hello.java input shown earlier, at the point where it is time to reduce the reserved words PUBLIC and CLASS, the class name, and the class body. The grammar rule that applies at this point is ClassDecl: PUBLIC CLASS IDENTIFIER ClassBody.

The preceding rule has four symbols on the right-hand side. The first three are terminal symbols, which means that on the value stack, their tree nodes will be leaves. The fourth symbol on the right side is a non-terminal, whose value stack entry will be an internal node, a subtree, which in this case happens to have three children. When it is time to reduce all that down to a ClassDecl production rule, we are going to allocate a new internal node. Since we are finishing parsing, in this case, it happens to be the root, but in any case, it will correspond to the class declaration that we have found, and it will have four children. *Figure 5.9* shows the contents of the parse stack and the value stack at the time of the reduce operation when the entire class is finally to be connected as one big tree:

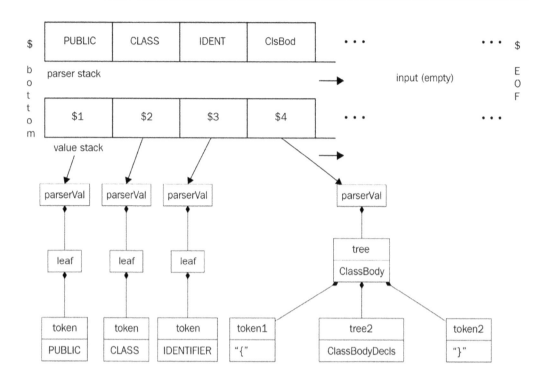

Figure 5.9: Parse and value stack right before a reduce operation

The mission of the semantic action for the `ClassDecl` production rule will be to create a new node, initialize its four children from $1, $2, $3, and $4, and assign it to $$. *Figure 5.10* shows how this looks after constructing the `ClassDecl` rule:

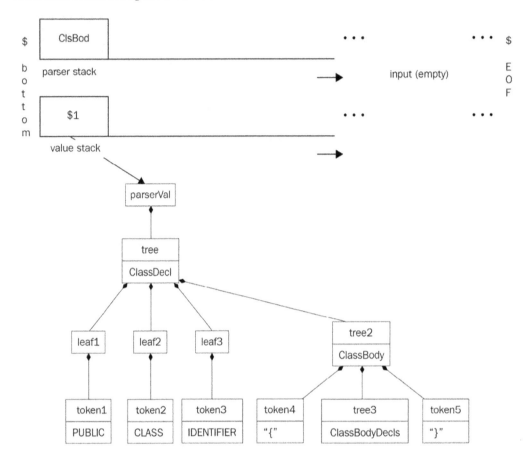

Figure 5.10: Subtrees are combined on the value stack during reduce operations

The entire tree is constructed very gradually, one node at a time, and the `parserVal` objects are removed at the point at which children get removed from the value stack and inserted into their parent node.

Using the tree node factory method

The `tree` class contains an important factory method named `node()`. A factory method is a method that allocates and returns an object. It is like a constructor, but it allocates an object of a different type from whatever class it is within. Factory methods are used heavily in certain design patterns.

In our case, the node( ) method takes a string describing the node, a production rule number, and any number of children, and returns an internal node to represent the production rule that was matched. The Unicon code for the node( ) method is shown in the following snippet:

```
method node(s,r,p[])
    return tree ! ([s,r] ||| p)
end
```

The Java code for the node( ) method is more complex due to the wrapping and unwrapping of the parserVal types. Wrapping a newly constructed internal node in a parserVal object is easy enough with a call to create a new parserVal object. The code is shown in the following snippet:

```
public static parserVal node(String s, int r, parserVal...p) {
    tree[] t = new tree[p.length];
    for(int i = 0; i < t.length; i++)
        t[i] = (tree)(p[i].obj);
    return new parserVal((Object)new tree(s,r,t));
}
```

The preceding Java code takes a variable number of arguments, unwraps them, and passes them into the constructor of the tree class. The unwrapping consists of selecting the obj field of the parserVal object and casting it to be of type tree.

Since the semantic actions for iyacc are Unicon code and for BYACC/J they are Java code, this requires some cheating. A semantic action is allowed in both Java and Unicon only if you limit it to common syntax such as method calls. If you start inserting other things in the semantic actions, such as if statements and other language-specific syntax, your yacc specification will become specific to one host language, such as Unicon or Java.

However, it was not quite possible for this book's examples to be crafted to use the same input file for both iyacc and BYACC/J as is. The reason for this is that semantic actions in yacc typically assign a value (a reference to a parse tree node) to a special variable named $$, and Unicon uses a := operator for assignment while Java uses =. This was addressed during the writing of this book by modifying iyacc so that semantic actions that start with $$= are accepted as a special operator that generates the Unicon equivalent assignment of $$:=.

The strategies that you need for building internal nodes in your syntax tree are simple: for every production rule, count how many children are either one of the following:

- A non-terminal
- A terminal that is not a punctuation mark

If the number of such children is more than 1, call the node() factory method to allocate a tree node, and assign it to be the value stack entry for the production rule. Now, it's time to demonstrate syntax tree construction in a non-trivial example: the Jzero language.

Forming syntax trees for the Jzero language

This section shows you how to build syntax trees for the Jzero language. The full j0gram.y file for this chapter is available on the book's GitHub site. The header is omitted here since the %token declarations are unchanged from how they appeared in the section titled *The Jzero Yacc specification* in the previous chapter. Although we are again presenting many of the grammar rules shown in the last chapter, the focus now is on the construction of new tree nodes associated with each production rule, if any.

As described earlier, the tree's internal nodes are constructed in semantic actions that are added at the ends of production rules. For each production rule that builds a new node, the new node is assigned to $$, the yacc value stack entry that corresponds to the new non-terminal symbol built by that production rule.

The starting non-terminal, which in the case of Jzero is a single class declaration, is the point at which the root of the entire tree is constructed. Its semantic action has extra work after assigning the constructed node to $$. At this top level, in this chapter, the code prints out the tree by calling the print() method in order to allow you to check whether it is correct. Subsequent chapters may assign the topmost tree node to a global variable named root for subsequent processing or call a different method here to translate the tree to machine code, or to execute the program directly by interpreting the statements in the tree.

The code is illustrated in the following snippet:

```
%%
ClassDecl: PUBLIC CLASS IDENTIFIER ClassBody {
    $$=j0.node("ClassDecl",1000,$3,$4);
    j0.print($$);
} ;
```

The non-terminal ClassBody consists of a pair of curly brackets that contains declarations (ClassBodyDecls) or is empty. In the empty case, it is an interesting question whether to assign an explicit leaf node indicating an empty ClassBody, as is done in the following code snippet, or whether the code should just say $$=null. Since the null value is different in Unicon and Java, it is simplest to create leaves to represent the empty parts of the tree.

```
ClassBody: '{' ClassBodyDecls '}' {
            $$=j0.node("ClassBody",1010,$2); }
          | '{' '}' { $$=j0.node("ClassBody",1011); };
```

The non-terminal ClassBodyDecls chains together as many fields, methods, and constructors as occur within the class. The first production rule terminates the recursions in the second production rule with a single ClassBodyDecl. Since there is no semantic action in the first production rule, it executes the default action, $$=$1; the subtree for ClassBodyDecl is promoted instead of creating a node for the parent. The code is illustrated in the following snippet:

```
ClassBodyDecls: ClassBodyDecl
              | ClassBodyDecls ClassBodyDecl {
            $$=j0.node("ClassBodyDecls",1020,$1,$2); };
```

There are three kinds of ClassBodyDecl to choose from. No extra tree node is allocated at this level as it can be inferred which kind of ClassBodyDecl each subtree is. The code is illustrated here:

```
ClassBodyDecl: FieldDecl | MethodDecl | ConstructorDecl ;
```

A field, or member variable, is declared with a base type followed by a list of variable declarations, as illustrated in the following code snippet:

```
FieldDecl: Type VarDecls ';' {
            $$=j0.node("FieldDecl",1030,$1,$2); };
```

The types in Jzero are very simple and include four built-in type names and a generic rule for names of classes, as illustrated in the following code snippet. No production rule has two children, so no new internal nodes are needed at this level. Arguably, String could be handled as a predefined Name and needn't be a special case:

```
Type: INT | DOUBLE | BOOL | STRING | Name ;
```

A name is either a single token called IDENTIFIER or a name with one or more periods in it, called QualifiedName, as illustrated in the following code snippet:

```
Name: IDENTIFIER | QualifiedName ;
QualifiedName: Name '.' IDENTIFIER {
            $$=j0.node("QualifiedName",1040,$1,$3);};
```

Variable declarations are a comma-separated list of one or more variable declarators. In Jzero, VarDeclarator is just IDENTIFIER unless it has square brackets following it that denote an array type. As the VarDeclarator internal node implies a set of square brackets, they are not represented explicitly in the tree. The code is illustrated in the following snippet:

```
VarDecls: VarDeclarator | VarDecls ',' VarDeclarator {
            $$=j0.node("VarDecls",1050,$1,$3); };
VarDeclarator: IDENTIFIER | VarDeclarator '[' ']' {
            $$=j0.node("VarDeclarator",1060,$1); };
```

In Jzero, a method can return a value of some return type, or it can return VOID, as illustrated in the following code snippet:

```
MethodReturnVal : Type | VOID ;
```

A method is declared by providing a method header followed by a block of code. All methods are public static methods. After the return value, the guts of a method header consisting of the method's name and parameters are MethodDeclarator, as illustrated in the following code snippet:

```
MethodDecl: MethodHeader Block {
            $$=j0.node("MethodDecl",1380,$1,$2); };
MethodHeader: PUBLIC STATIC MethodReturnVal
              MethodDeclarator {
            $$=j0.node("MethodHeader",1070,$3,$4); };
MethodDeclarator: IDENTIFIER '(' FormalParmListOpt ')' {
            $$=j0.node("MethodDeclarator",1080,$1,$3); };
```

An optional formal parameter list is either a non-empty FormalParmList or an empty production rule, the so-called **epsilon rule**, with no terminal or non-terminal symbols between the vertical bar and the semicolon. Epsilon rules are very handy but often require extra care to avoid problems, for correct parsing as well as correct syntax tree construction; this is discussed in the section titled *Avoiding common syntax tree bugs* later in this chapter. The production rule for optional formal parameter list looks like this.

```
FormalParmListOpt: FormalParmList | ;
```

A formal parameter list is a comma-separated list of formal parameters. This is a non-empty list, and the recursion is terminated by a lone formal parameter. Each formal parameter has a type followed by a variable name, possibly including square brackets for array types, as illustrated in the following code snippet:

```
FormalParmList: FormalParm | FormalParmList ',' FormalParm {
                $$=j0.node("FormalParmList",1090,$1,$3); };
FormalParm: Type VarDeclarator {
                $$=j0.node("FormalParm",1100,$1,$2); };
```

Constructors are declared similarly to methods, although they have no return type and do not use the public static prefix, as illustrated in the following code snippet:

```
ConstructorDecl: MethodDeclarator Block {
                $$=j0.node("ConstructorDecl",1110,$1,$2); };
```

A Block is a sequence of zero or more statements enclosed in curly braces. Although many of the tree nodes introduce the branching of two or more children, a few tree nodes have only one child because surrounding punctuation is unnecessary. Such nodes might themselves be unnecessary, but they may also make it easier to understand and process the tree. You can see an example in the following code snippet:

```
Block: '{' BlockStmtsOpt '}' {$$=j0.node("Block",1200,$2);};
BlockStmtsOpt: BlockStmts | ;
BlockStmts:  BlockStmt | BlockStmts BlockStmt {
                $$=j0.node("BlockStmts",1130,$1,$2); };
BlockStmt:   LocalVarDeclStmt | Stmt ;
```

Block statements can be either local variable declarations or statements. The syntax of LocalVarDeclStmt is indistinguishable from a FieldDecl rule. It may, in fact, be better to eliminate duplication by default. Whether you use another set of identical production rules or factor the common elements of the grammar, this may depend on whether it will be easier for you to write code that does the correct thing with various trees if they have recognizably different tree node labels and production rule numbers, or whether the differences will be recognized and handled properly due to the surrounding tree context. An example is given in the following code snippet:

```
LocalVarDeclStmt: LocalVarDecl ';' ;
LocalVarDecl: Type VarDecls {
                $$=j0.node("LocalVarDecl",1140,$1,$2); };
```

In the preceding case, a `LocalVarDecl` node is created, making it easy to distinguish local variables from class member variables in the syntax tree.

The many kinds of statements each result in their own unique tree nodes. Since they are one-child production rules, introducing another tree node here is unnecessary. The following code snippet illustrates this:

```
Stmt: Block | ';' | ExprStmt | BreakStmt | ReturnStmt |
    | IfThenStmt | IfThenElseStmt | IfThenElseIfStmt
    | WhileStmt | ForStmt ;
ExprStmt: StmtExpr ';' ;
StmtExpr: Assignment | MethodCall ;
```

Several non-terminals in Jzero exist to support common variations of if statements. Blocks are required for bodies of conditionals and loops in Jzero. This avoids a common ambiguity when conditionals are nested. The following code shows the tree nodes constructed for various forms of if statements in j0:

```
IfThenStmt: IF '(' Expr ')' Block {
    $$=j0.node("IfThenStmt",1150,$3,$5); };
IfThenElseStmt: IF '(' Expr ')' Block ELSE Block {
    $$=j0.node("IfThenElseStmt",1160,$3,$5,$7); };
IfThenElseIfStmt: IF '(' Expr ')' Block ElseIfSequence {
    $$=j0.node("IfThenElseIfStmt",1170,$3,$5,$6); }
| IF '(' Expr ')' Block ElseIfSequence ELSE Block {
    $$=j0.node("IfThenElseIfStmt",1171,$3,$5,$6,$8); };
ElseIfSequence: ElseIfStmt | ElseIfSequence ElseIfStmt {
    $$=j0.node("ElseIfSequence",1180,$1,$2); };
ElseIfStmt: ELSE IfThenStmt {
    $$=j0.node("ElseIfStmt",1190,$2); };
```

Tree nodes are generally created for these control structures, and they often introduce branching into the tree. Although while loops require only a child for the condition and a child for the loop body, the node for a for loop has four children. Did the language designers do that on purpose? You can see an example in the following code snippet:

```
WhileStmt: WHILE '(' Expr ')' Stmt {
    $$=j0.node("WhileStmt",1210,$3,$5); };
ForStmt: FOR '(' ForInit ';' ExprOpt ';' ForUpdate ')' Block {
    $$=j0.node("ForStmt",1220,$3,$5,$7,$9); };
```

```
ForInit: StmtExprList | LocalVarDecl | ;
ExprOpt: Expr |  ;
ForUpdate: StmtExprList | ;
StmtExprList: StmtExpr | StmtExprList ',' StmtExpr {
    $$=j0.node("StmtExprList",1230,$1,$3); };
```

A break statement is adequately represented by the leaf that says BREAK, as illustrated here:

```
BreakStmt: BREAK ';' ;
ReturnStmt: RETURN ExprOpt ';' {
    $$=j0.node("ReturnStmt",1250,$2); };
```

A return statement needs a tree node, since it may have an optional expression.

Primary expressions, including literals, do not introduce an additional layer of tree nodes above the content of their child. The only interesting action here is for parenthesized expressions, which discard the parentheses that were used for operator precedence and promote the second child without the need for an additional tree node at this level. Here is an example of this:

```
Primary:  Literal | FieldAccess | MethodCall |
          '(' Expr ')' { $$=$2; };
Literal: INTLIT | DOUBLELIT | BOOLLIT | STRINGLIT | NULLVAL ;
```

An argument list is one or more expressions, separated by commas. To allow zero expressions, a separate non-terminal is used, as illustrated in the following code snippet:

```
ArgList: Expr | ArgList ',' Expr {
                $$=j0.node("ArgList",1270,$1,$3); };
ArgListOpt:  ArgList | ;
```

Field accesses may be chained together since their left child, a Primary, can be another field access. When one non-terminal has a production rule that derives another non-terminal that has a production rule that derives the first non-terminal, the situation is called **mutual recursion** and it is normal and healthy. You can see an example of this in the following code snippet:

```
FieldAccess: Primary '.' IDENTIFIER {
                $$=j0.node("FieldAccess",1280,$1,$3); };
```

A method call has a defining syntax consisting of a method followed by a parenthesized list of zero or more arguments. Usually, this is a simple binary node in which the left child is pretty simple (a method name) and the right child may contain a large subtree of arguments...or it may be empty. Here is an example of this:

```
MethodCall: Name '(' ArgListOpt ')' {
                $$=j0.node("MethodCall",1290,$1,$3); }
   | Primary '.' IDENTIFIER '(' ArgListOpt ')' {
     $$=j0.node("MethodCall",1291,$1,$3,$5); } ;
```

As seen in the previous chapter, the expression grammar in Jzero has many recursive levels of non-terminals that are not all shown here. You should consult the book's website to see the full grammar with syntax tree construction. In the following code snippet, each operator introduces a tree node. After the tree is constructed, a simple walk of the tree will allow correct calculation (or correct code generation) of the expression:

```
PostFixExpr: Primary | Name ;
UnaryExpr: '-' UnaryExpr {$$=j0.node("UnaryExpr",1300,$1,$2);}
    | '!' UnaryExpr { $$=j0.node("UnaryExpr",1301,$1,$2); }
    | PostFixExpr ;
MulExpr: UnaryExpr
    | MulExpr '*' UnaryExpr {
      $$=j0.node("MulExpr",1310,$1,$3); }
    | MulExpr '/' UnaryExpr {
      $$=j0.node("MulExpr",1311,$1,$3); }
    | MulExpr '%' UnaryExpr {
      $$=j0.node("MulExpr",1312,$1,$3); };
AddExpr: MulExpr
    | AddExpr '+' MulExpr {$$=j0.node("AddExpr",1320,$1,$3); }
    | AddExpr '-' MulExpr {$$=j0.node("AddExpr",1321,$1,$3); };
```

In classic C language grammar, **comparison operators**, also called **relational operators**, are just another level of precedence for integer expressions. Java and Jzero are a bit more interesting in that the Boolean type is separate from integers and type-checked as such, but this distinction is covered in the chapters that follow, on semantic analysis and type checking. For the code shown in the following snippet, there are four relational operators. LESSTHANOREQUAL is the integer code that the lexical analyzer reports for <=, while GREATERTHANOREQUAL is returned for >=. For the < and > operators, the lexical analyzer returns their ASCII codes:

```
RelOp: LESSTHANOREQUAL | GREATERTHANOREQUAL | '<' | '>' ;
```

The relational operators are at a slightly higher level of precedence than the comparisons of whether values are equal or not equal to each other:

```
RelExpr: AddExpr | RelExpr RelOp AddExpr {
   $$=j0.node("RelExpr",1330,$1,$2,$3); };
EqExpr: RelExpr
       | EqExpr ISEQUALTO RelExpr {
         $$=j0.node("EqExpr",1340,$1,$3); }
       | EqExpr NOTEQUALTO RelExpr {
         $$=j0.node("EqExpr",1341,$1,$3); };
```

Below the relational and comparison operators, the && and || Boolean operators operate at different levels of precedence, as illustrated in the following code snippet:

```
CondAndExpr: EqExpr | CondAndExpr LOGICALAND EqExpr {
   $$=j0.node("CondAndExpr", 1350, $1, $3); };
CondOrExpr: CondAndExpr | CondOrExpr LOGICALOR CondAndExpr {
   $$=j0.node("CondOrExpr", 1360, $1, $3); };
```

The lowest level of precedence in many languages, as with Jzero, is the assignment operators. Jzero has += and -= but not ++ and --, which are deemed to be a can of worms for novice programmers and do not add a lot of value for teaching compiler construction. You can see these operators in use here:

```
Expr: CondOrExpr | Assignment ;
Assignment: LeftHandSide AssignOp Expr {
    $$=j0.node("Assignment",1370, $1, $2, $3); };
LeftHandSide: Name | FieldAccess ;
AssignOp: '=' | INCREMENT | DECREMENT ;
```

This section presented the highlights of Jzero syntax tree construction. Many production rules require the construction of a new internal node that serves as the parent of several children on the right-hand side of a production rule. However, the grammar has many cases where a non-terminal is constructed from only one symbol on the right-hand side, in which case the allocation of an extra internal node can usually be avoided. Now, let's look at how to check your tree afterward to make sure that it was assembled correctly.

Debugging and testing your syntax tree

The trees that you build must be rock solid. What this spectacular mixed metaphor means is if your syntax tree structure is not built correctly, you can't expect to be able to build the rest of your programming language. The most direct way of testing that the tree has been constructed correctly is to walk back through it and look at the tree that you have built. Actually, you need to do that for all possible trees!

This section contains two examples of traversing and printing the structure of syntax trees. You will print your tree first in a human-readable (more or less) ASCII text format, then you will learn how to print it out in a format that is easily rendered graphically using the popular open source Graphviz package, commonly accessed through PlantUML or the classic command-line tool called dot. First, consider some of the most common causes of problems in syntax trees.

Avoiding common syntax tree bugs

The most common problems with syntax trees result in program crashes when you print the tree out. Each tree node may hold references (pointers) to other objects, and when these references are not initialized correctly: boom! Debugging problems with references is difficult, even in higher-level languages.

The first major case is this: are your leaves being constructed and picked up by the parser? Suppose you have a lex rule like the one shown here:

```
";"                          { return 59; }
```

The ASCII code is correct. The parse will succeed but your syntax tree will be broken. You must create a leaf and assign it to yylval whenever you return an integer code in one of your Flex actions. If you do not, yacc will have garbage sitting around in yylval when yyparse() puts it on the value stack for later insertion into your tree. You should check that every semantic action that returns an integer code in your lex file also allocates a new leaf and assigns it to yylval. You can check each leaf to ensure it is valid on the receiving end by printing its contents when you first access it as a $1 or $2 rule, or whatever, in the semantic actions for the production rules of yacc.

The second major case is: are you constructing internal nodes correctly for all the production rules that have two or more children that are significant (and not just punctuation marks, for example)? If you are paranoid, you can print out each subtree to make sure it is valid before creating a new parent that stores pointers to the child subtrees. Then, you can print out the new parent that you've created, including its children, to make sure it was assembled correctly.

One weird special case that comes up in syntax tree construction has to do with epsilon rules: production rules where a non-terminal is constructed from an empty right-hand side. An example would be the following rule from the j0gram.y file:

```
FormalParmListOpt: FormalParmList | ;
```

For the second production rule in this example, there are no children. The default rule of yacc, $$=$1, does not look good since there is no $1 rule. You may construct a new leaf here, as in the following solution:

```
FormalParmListOpt: FormalParmList | { $$=
                      j0.node("FormalParamListOpt",1095); }
```

This leaf is different from normal since it has no associated token. Code that traverses the tree afterward had better not assume that all leaves have tokens. In practice, some people might just use a null pointer to represent an epsilon rule instead. If you use a null pointer, you may have to add checks for null pointers everywhere in your later tree traversal code, including the tree printers in the following subsections. If you allocate a leaf for every epsilon rule, your tree will be bigger without really adding any new information. Memory is cheap, so if it simplifies your code, it is probably OK to do this.

To sum up, and as a final warning: you may not discover fatal flaws in your tree construction code unless you write test cases that use every single production rule in your grammar! Such grammar coverage may be required for any serious language implementation project. Now, let's look at the actual methods to verify tree correctness by printing them.

Printing your tree in a text format

One way to test your syntax tree is to print out the tree structure as ASCII text. This is done via a tree traversal in which each node results in one or more lines of text output. The following print() method in the j0 class just asks the tree to print itself:

```
method print(root)
    root.print()
end
```

The equivalent code in Java must unpack the parserVal object and cast the object to a tree in order to ask it to print itself, as illustrated in the following code snippet:

```
public static void print(parserVal root) {
    ((tree)root.obj).print();
}
```

Trees generally print themselves recursively. A leaf just prints itself out, while an internal node prints itself and then asks its children to print themselves. For a text printout, indentation is used to indicate the nesting level or distance of a node from the root. The indentation level is passed as a parameter and incremented for each level deeper within the tree. The Unicon version of a tree class's `print()` method is shown in the following code snippet:

```
method print(level:0)
  writes(repl(" ",level))
  if \tok then
    write(id, "   ", tok.text, " (",tok.cat,          "): ",tok.lineno)
  else write(id, "  ", sym, " (", rule, "): ", nkids)
  every (!kids).print(level+1);
end
```

The preceding method indents some number of spaces given in a parameter and then writes a line of text describing the tree node. It then calls itself recursively, with one higher nesting level, on each of the node's children, if there are any. The equivalent Java code for the `tree` class text printout looks like this:

```
public void print(int level) {
  int i;
  for(i=0;i<level;i++) System.out.print(" ");
  if (tok != null)
    System.out.println(id + "    " + tok.text +
                        " (" + tok.cat + "): "+tok.lineno);
  else
    System.out.println(id + "   " + sym +
          " (" + rule + "): "+nkids);
  for(i=0; i<nkids; i++)
    kids[i].print(level+1);
}
public void print() {
  print(0);
}
```

The `build` and `run` commands were spelled out manually in a section in the previous chapter, but from this chapter on, we are relying on the make tool. To compile, use a terminal or Command Prompt window and type make, or on Windows:

```
make -f makefile.win
```

To run the j0 command, use either the command on the left (for Unicon builds) or the command on the right, for Java builds:

```
j0 filename.java                    java ch5.j0 filename.java
```

where filename.java is replaced with whatever j0 program you are providing as input. Note that on some Java versions and CLASSPATH settings, the correct command to run the Java version might be java j0 filename.java.

When you run the j0 command on the hello.java file with this tree print function in place, it produces the following output:

```
63    ClassDecl (1000): 2
6    hello (266): 1
62    ClassBody (1010): 1
   59   MethodDecl (1380): 2
    32   MethodHeader (1070): 2
    14   void (264): 2
    31   MethodDeclarator (1080): 2
    16   main (266): 2
    30   FormalParm (1100): 2
    20   String (266): 2
    27   VarDeclarator (1060): 1
    22   argv (266): 2
   58   Block (1200): 1
   53   MethodCall (1290): 2
    46   QualifiedName (1040): 2
    41   QualifiedName (1040): 2
    36   System (266): 3
    40   out (266): 3
    45   println (266): 3
    50   "hello, jzero!" (273): 3
no errors
```

Although the tree structure can be deciphered from studying this output, it is not exactly easy to understand. The next section shows a graphic way to depict the tree.

Printing your tree using dot

A more fun way to test your syntax tree is to print out the tree in a graphical form. As mentioned in the *Technical requirements* section, a tool called dot will draw syntax trees for us. Writing our tree in the input format of dot is done via another tree traversal in which each node results in one or more lines of text output. To draw a graphic version of the tree, change the j0.print() method to call the tree class's print_graph() method. In Unicon, this is trivial. The code is illustrated in the following snippet:

```
method print(root)
    root.print_graph(yyfilename || ".dot")
end
```

The equivalent code in Java must unpack the parserVal object and cast the object to a tree in order to ask it to print itself, as illustrated in the following snippet:

```
public static void print(parserVal root) {
    ((tree)root.obj).print_graph(yyfilename + ".dot");
}
```

As was true for a text-only printout, trees print themselves recursively. The Unicon version of a tree class's print_graph() method is shown in the following code snippet:

```
method print_graph(fw)
    if type(filename) == "string" then {
      fw := open(filename, "w") |
        stop("can't open ", image(filename), " for writing")
      write(fw, "digraph {")
      print_graph(fw)
      write(fw, "}")
      close(fw)
    }
    else if \tok then print_leaf(fw)
    else {
      print_branch(fw)
      every i := 1 to nkids do
        if \kids[i] then {
          write(fw, "N",id," -> N",kids[i].id,";")
          kids[i].print_graph(fw)
        } else {
```

```
            write(fw, "N",id," -> N",id,"_",j,";")
            write(fw, "N", id, "_", j,
                      " [label=\"Empty rule\"];")
            j +:= 1
         }
      }
   end
```

The Java implementation of print_graph() consists of two methods. The first is a public method that takes a filename, opens that file for writing, and writes the whole graph to that file, as illustrated in the following code snippet:

```
void print_graph(String filename){
   try {
      PrintWriter pw = new PrintWriter(
         new BufferedWriter(new FileWriter(filename)));
      pw.printf("digraph {\n");
      j = 0;
      print_graph(pw);
      pw.printf("}\n");
      pw.close();
      }
   catch (java.io.IOException ioException) {
      System.err.println("printgraph exception");
      System.exit(1);
      }
   }
```

In Java, function overloading allows public and private parts of print_graph() to have the same name. The two methods are distinguished by their different parameters. The public print_graph() above takes a String filename, opens the file, and passes the file as a parameter into the following method. This version of print_graph() prints a line or two about the current node, and calls itself recursively on each child:

```
void print_graph(PrintWriter pw) {
int i;
   if (tok != null) {
      print_leaf(pw);
      return;
```

```
        }
    print_branch(pw);
    for(i=0; i<nkids; i++) {
        if (kids[i] != null) {
            pw.printf("N%d -> N%d;\n", id, kids[i].id);
            kids[i].print_graph(pw);
        } else {
            pw.printf("N%d -> N%d%d;\n", id, kids[i].id, j);
            pw.printf("N%d%d [label=\"%s\"];\n", id, j, "Empty rule");
            j++;
        }
    }
}
```

The `print_graph()` method calls a couple of helper functions: `print_leaf()` for leaves and `print_branch()` for internal nodes. The `print_leaf()` method prints a dotted outline box containing the characteristics of a terminal symbol. The Unicon implementation of `print_leaf()` is shown here:

```
method print_leaf(pw)
    local s := parser.yyname[tok.cat]
    print_branch(pw)
    write(pw,"N",id,
          " [shape=box style=dotted label=\" ",s," \\n ")
    write(pw,"text = ",escape(tok.text),
            " \\l lineno = ", tok.lineno," \\l\"];\n")
end
```

The integer code for the token's terminal symbol is used as a subscript in an array of strings in the parser named yyname. This is generated by iyacc. The Java implementation of `print_leaf()` is similar to the Unicon version, as illustrated in the following code snippet:

```
void print_leaf(PrintWriter pw) {
    String s = parser.yyname[tok.cat];
    print_branch(pw);
    pw.printf("N%d [shape=box style=dotted label=\" %s \\n", id, s);
    pw.printf("text = %s \\l lineno = %d \\l\"];\n", escape(tok.text),
            tok.lineno);
}
```

The print_branch() method prints a solid box for internal nodes, including the name of the non-terminal represented by that node. The Unicon implementation of print_branch() is shown here:

```
method print_branch(pw)
   write(pw, "N",id," [shape=box label=\"",
         pretty_print_name(),"\"];\n");
end
```

The Java implementation of print_branch() is similar to its Unicon counterpart, as illustrated in the following code snippet:

```
void print_branch(PrintWriter pw) {
   pw.printf("N%d [shape=box label=\"%s\"];\n",
             id, pretty_print_name());
}
```

The escape() method adds escape characters when needed before double quotes so that dot will print the double quote marks. The Unicon implementation of escape() consists of the following code:

```
method escape(s)
   if s[1] == "\"" then
      return "\\" || s[1:-1] || "\\\""
   else return s
end
```

The Java implementation of escape() is shown here:

```
public String escape(String s) {
   if (s.charAt(0) == '\"')
      return "\\"+s.substring(0, s.length()-1)+"\\\"";
   else return s;
}
```

The pretty_print_name() method prints out the best human-readable name for a given node. For an internal node, that is its string label, along with a serial number to distinguish multiple occurrences of the same label. For a terminal symbol, it includes the lexeme that was matched. The code is illustrated in the following snippet:

```
method pretty_print_name() {
    if /tok then return sym || "#" || (rule%10)
    else return escape(tok.text) || ":" || tok.cat
end
```

The Java implementation of pretty_print_name() looks similar to the preceding code, as we can see here:

```
public String pretty_print_name() {
    if (tok == null) return sym +"#"+(rule%10);
    else return escape(tok.text)+":"+tok.cat;
}
```

After noting the code to generate the .dot file, you can invoke the make command as described in the previous section and then run this program again on the sample hello.java input file with the command in the left column for the Unicon implementation, and the command in the right column for the Java implementation:

```
j0 hello.java                          java ch5.j0 hello.java
```

The j0 program writes out a hello.java.dot file that is valid input for the dot program. Run the dot program with the following command to generate a PNG image:

```
dot -Tpng -Gdpi=300 hello.java.dot >hello.png
```

The following diagram shows a syntax tree for `hello.java`, as written to `hello.png`:

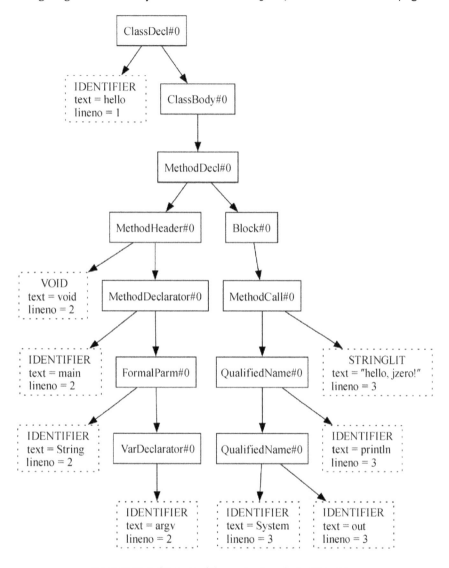

Figure 5.11: A diagram of the syntax tree for `hello.java`

If you do not write your tree construction code correctly, the program will crash when you run it, or the tree will be obviously bogus when you inspect the image. To test your lexical, parsing, and syntax tree construction code, you should run it on a wide variety of input programs and examine the resulting trees carefully.

In this section, you saw that only a few lines of code were needed to generate textual and graphical depictions of your syntax trees using tree traversals. The graphical rendering was provided by an external tool called dot. Tree traversals are a simple but powerful programming technique that will dominate the next several chapters of this book.

Summary

In this chapter, you learned about the crucial technical skills and tools needed to build a syntax tree while the input program is being parsed. A syntax tree is the main data structure used to represent source code internally to a compiler or interpreter.

You learned how to develop code that identifies which production rule was used to build each internal node so that we can tell what we are looking at later on. You learned how to add tree node constructors for each rule in the scanner. You learned how to connect tree leaves from the scanner to the tree built in the parser. You also learned how to check your trees and debug common tree construction problems.

You are done synthesizing the input source code to construct a data structure that you can use. Now, it is time to start analyzing the meaning of the program source code so that you can determine what computations it specifies. This is done by walking through the parse tree using tree traversals to perform semantic analysis.

The next chapter will start us off on that journey by walking the tree to build symbol tables that will enable you to track all the variables in the program and figure out where they were declared.

Questions

1. Where do the leaves of the syntax tree come from?

2. How are the internal nodes of a syntax tree created?

3. Where are leaves and internal nodes stored while a tree is being constructed?

4. Why are values wrapped and unwrapped when they are pushed and popped on the value stack?

Join our community on Discord

Join our community's Discord space for discussions with the authors and other readers:

`https://discord.com/invite/zGVbWaxqbw`

Section II

Syntax Tree Traversals

The heart of a compiler is the tree traversals. Upon completion of this section, you will have a compiler that performs semantic analysis and code generation.

This section comprises the following chapters:

- *Chapter 6, Symbol Tables*
- *Chapter 7, Checking Base Types*
- *Chapter 8, Checking Types on Arrays, Method Calls, and Structure Accesses*
- *Chapter 9, Intermediate Code Generation*
- *Chapter 10, Syntax Coloring in an IDE*

6

Symbol Tables

To understand the uses of names in program source code, your compiler must figure out what each use of a name refers to. If the program is reading from or writing to a variable, which variable is it? A local variable? A global variable? Or maybe a class member? You can look up symbols at each location they are used by using table data structures that are auxiliary to the syntax tree, called **symbol tables**. Performing operations to construct and then use symbol tables is the first step of **semantic analysis**. Semantic analysis is where the compiler studies the meaning of the input source code. *Chapter 7* and *Chapter 8* will build on this chapter and round out our discussion of semantic analysis.

Context-free grammars, explored in the preceding two chapters of this book, have terminal symbols and non-terminal symbols, and those are represented in tree nodes and token structures. When talking about a program's source code and its semantics, the word **symbol** is used for something very different from grammar symbols. In this and later chapters, a symbol refers to a name in the program source code. For example, a symbol can be the name of a variable, function, class, or package. In this book, the words symbol, name, and identifier are used interchangeably. The word variable refers to a subcategory of symbols whose values can be changed while the program executes.

This chapter will show you how to construct symbol tables, insert symbols into them, and use symbol tables to identify two kinds of semantic errors: undeclared and (illegally) redeclared variables. In later chapters, you will use symbol tables to check the types and store memory addresses assigned to symbols that will be used to generate code for the input program.

The examples in this chapter demonstrate how to use symbol tables by building them for the Jzero subset of Java that was described in *Chapters 1* and *2*. Symbol tables are important to be able to check types and generate code for your programming language. In this and the next few chapters, the main skill you will be learning is the art of **recursion** by writing many selective and specialized tree traversal functions.

This chapter covers the following main topics:

- Establishing the groundwork for symbol tables
- Creating and populating symbol tables for each scope
- Checking for undeclared variables
- Finding redeclared variables
- Handling package and class scopes in Unicon

It is time to learn about symbol tables and how to build them. First, however, you need to learn about some conceptual foundations required to do this work.

Technical requirements

The code for this chapter is available on GitHub: `https://github.com/PacktPublishing/Build-Your-Own-Programming-Language-Second-Edition/tree/master/ch6`

The Code in Action video for the chapter can be found here: `https://bit.ly/3ccYTZv`

Establishing the groundwork for symbol tables

In software engineering, you must go through requirements analysis and design before you start coding. Similarly, to build symbol tables, you first need to understand what they are for and how to go about writing the syntax tree traversals that do the work. For starters, you should review what kinds of information your compiler must store and recall different kinds of variables. The information will be stored in symbol tables from declarations in the program code, so let's look at those.

Declarations and scopes

The meaning of a computer program boils down to the meaning of the information being computed, and the actual computations to be performed. Symbol tables are all about the first part: defining what information the program is manipulating. We will begin by identifying what names are being used, what they are referring to, and how they are being used.

Consider a simple assignment statement such as the following:

```
x = y + 5;
```

In most languages, names such as x or y must be declared before they are used. A **declaration** specifies a name that will be used in the program and usually includes type information. An example declaration for x might look like this:

```
int x;
```

Each variable declaration has a **scope** that describes the region in the program where that variable is visible. In Jzero the user-defined scopes are the class scope and the local (method) scope. Jzero also must support scopes associated with a few predefined system packages, which is a small subset of the package scope functionality required of a full Java compiler. These predefined packages are discussed in the section titled *Creating symbol tables*. Other languages have additional, different kinds of scopes to deal with.

The example program shown below, which can be found in the xy5.java file at https://github. com/PacktPublishing/Build-Your-Own-Programming-Language-Second-Edition/tree/master/ch6, expands the preceding example to illustrate scopes. The light gray class scope surrounds the darker gray local scope:

```
public class xy5 {
    static double y = 5.0;
    public static void main(String argv[]) {
        int x;
        x = (int)y + 5;
        System.out.println("y + 5 = " + x);
    }
}
```

For any symbol, such as x or y, the same symbol may be declared in both scopes. A name that is declared within an inner scope overrides and hides the same name declared in an outer scope. Such nested scoping requires that a programming language creates multiple symbol tables. A common rookie error is to try and build your whole language with only a single symbol table because a symbol table sounds big and scary, and compiler books often talk about *the symbol table* instead of *symbol tables*. To avoid this mistake, plan on supporting multiple symbol tables from the start, and search for symbols by starting from the innermost applicable symbol table and working outward to enclosing tables. Now, let's think about the two basic ways that variables are used in programs to interact with a computer's memory: assignment and dereferencing.

Assigning and dereferencing variables

Variables are names for memory locations, and memory can be read or written. Writing a value to a memory location is called **assignment**. Reading a value from a memory location is called **dereferencing**. Most programmers have a rock-solid understanding of assignment. Assignment is one of the first things they learn about in programming; for example, the x=0 statement is an assignment to x. A lot of programmers are a bit fuzzy about dereferencing. Programmers write code that does dereferencing all the time, but they may not have heard of the term before. For example, the statement y=x+1 dereferences x to obtain its value before using that value when it performs the addition. Similarly, a call such as System.out.println(x) deferences x in the process of passing it into the println() method.

Both assignment and dereferencing are acts that use a memory address. They come into play in semantic analysis and code generation. But under what circumstances do assignment and dereferencing affect whether each particular use of a variable is legal? Assignments are not legal for things that were declared to be const, including names of methods. Are there any symbols that cannot be dereferenced? In most languages, undeclared variables cannot be dereferenced; they cannot be assigned, either. Anything else? Before we can generate code for an assignment or a dereference, we must be able to understand what memory location is used, and whether the requested operation is legal and defined in the language we are implementing.

So far, we have reviewed the concepts of assignment and dereferencing. Checking whether each assignment or dereference is legal requires storing and retrieving information about the names used in a program, and that is what symbol tables are for. There is one more bit of conceptual groundwork you need, and then you will be ready to build your symbol tables. You will be using a lot of syntax tree traversal functions in this and the next few chapters. Let's consider some of the varieties of tree traversal at your disposal.

Choosing the right tree traversal for the job

In the previous chapter, you printed out syntax trees using tree traversals where work at the current node was done, followed by recursively calling the traversal function on each child. This is called a **pre-order traversal**. The pseudocode template for this is as follows:

```
method preorder()
   do_work_at_this_node()
   every child := !kids do child.preorder()
end
```

Some examples in this chapter will visit the children and have them do their work first, and then use what they calculate to do the work at the current node. This is called **post-order traversal**. The pseudocode template for post-order traversal looks like this:

```
method postorder()
   every child := !kids do child.postorder()
   do_work_at_this_node()
end
```

Other traversals exist where the method does some work for the current node in between each child call – these are known as **in-order traversals**. Lastly, it is common to write a tree traversal consisting of several methods that work together and call each other as needed, possibly as many as one method for each kind of tree node. Although we will try to keep our tree traversals as simple as possible, the examples in this book will use the best tool for the job.

In a compiler, the best tool for the job is very often a post-order traversal. The reason for this is simple: the work it leaves is simple, and the work by parents often uses the information found in the children. Occasionally, however, traversals are more interesting. We will try to point out when we need to do something that is atypical.

In this section, you learned about several important concepts that will be used in the code examples in this and the following chapters. These included nested scopes, assignment and dereferencing, and different kinds of tree traversals. Now, it's time to use these concepts to create symbol tables. After that, you can consider how to populate your symbol tables by inserting symbols into them.

Creating and populating symbol tables for each scope

A symbol table contains a record of all the names that are declared for a scope. There is one symbol table for each scope. A symbol table provides a means of looking up symbols by their name to obtain information about them. If a variable was declared, the symbol table lookup returns a structure with all the information known about that variable: where it was declared, what its data type is, whether it is public or private, and so on. All this information can be found in the syntax tree. If we also place it in a table, the goal is to access the information directly, from anywhere else that information is needed.

The traditional implementation of a symbol table is a **hash table**, which provides a very fast information lookup. Your compiler could use any data structure that allows you to store or retrieve information associated with a symbol, even a linked list. But hash tables are the best for this, and they are standard in Unicon and Java, so we will use hash tables in this chapter.

Unicon provides hash tables with a built-in data type called a table. See *Appendix, Unicon Essentials*, for a description. Insertion and lookup in the table can be performed by subscripting, using a syntax similar to accessing elements in an array. For example, symtable[sym] looks up information associated with a symbol named sym, while symtable[sym] := x associates information contained in x with sym.

Java provides hash tables in standard library classes. We will use the Java library class known as HashMap for this. Information is retrieved from a HashMap with a method call such as symtable.get(sym) and stored in a HashMap via symtable.put(sym, x).

Both the Unicon table type and the Java HashMap type map elements from a domain to an associated range. In the case of a symbol table, the domain will contain the string names of the symbols in the program source code. For each symbol in the domain, the range will contain a corresponding instance of the symtab_entry class, a symbol table entry. In the Jzero implementations we will be presenting, the hash tables themselves will be wrapped in a class so that symbol tables can contain additional information about the entire scope, in addition to the symbols and symbol table entries.

Two major issues are: when are symbol tables created for each scope and how exactly is information inserted into them? The answer to both questions is: during a **syntax tree traversal**. But before we get to that, you need to learn about semantic attributes.

Adding semantic attributes to syntax trees

The tree type in the previous chapter was clean and simple. It contained a label for printing, a production rule, and some children. In real life, a programming language needs to compute and store a lot of additional information in various nodes of the tree. This information is stored in extra fields in tree nodes, commonly called **semantic attributes**. The values of these fields can sometimes be computed during parsing when we construct the tree nodes. More often, it is easier to compute the values of semantic attributes once the entire tree has been constructed. In that case, the attributes are constructed using a tree traversal.

There are two kinds of semantic attributes:

- **Synthesized attributes** are attributes whose values for each node can be constructed from the semantic attributes of their children.

- **Inherited attributes** are computed using information that does not come from the node's children.

The only possible path for information from elsewhere in the tree is through the parent, which is why the attribute is said to be inherited. In practice, inherited attributes may come from siblings or from far away in the syntax tree. This is generally accomplished incrementally, by pushing or pulling information up or down the tree one node at a time.

This chapter will add two attributes to the tree.icn and tree.java files from the previous chapter. The first attribute, the isConst Boolean, is a synthesized attribute that reports whether a given tree node contains only constant values known at compile time. The following diagram depicts a syntax tree for the expression x+1. The isConst (instance type) of a parent node (an addition) is computed from its children's isConst values:

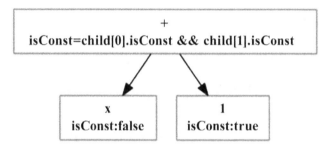

Figure 6.1: A synthesized attribute computes a node's value from its children

The preceding diagram shows a good example of a synthesized attribute being calculated from its children. In this example, the leaves for x and 1 already have isConst values, and those values must come from somewhere. It is easy to guess where the isConst value for the 1 token comes from: a language's literal constant values should be marked as isConst=true.

For a name like x, it is not so obvious where the isConst value comes from. As presented in the previous chapters, the Jzero language does not have Java's final keyword, which would designate a given symbol as being immutable. Your options are to either set isConst=false for every IDENTIFIER or extend Jzero to allow the final keyword, at least for variables. If you choose the latter, whether x is a constant or not should be found by looking up the symbol table information of x. The symbol table entry for x will only know whether x is a constant if we place that information there.

The second attribute, stab, is an inherited attribute containing a reference to the symbol table for the nearest enclosing scope that contains a given tree node. For most nodes, the stab value is simply copied from its parent; an exception is introduced when a node defines a new scope. The following diagram shows the stab attribute being copied from parents into children:

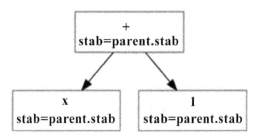

Figure 6.2: An inherited attribute computes a node's value from parent information

How will we get attributes pushed up to parents from children? Tree traversals. How will we get attributes pushed down to children from parents? Tree traversals. But first, we must make room in the tree nodes to store these attributes. This chapter's tree class header in Unicon has been revised to include these attributes, as follows:

```
class tree (id,sym,rule,nkids,tok,kids,isConst,stab)
```

This code doesn't do anything; the point is to add two fields for semantic attributes at the end. In Java, these class tree additions result in the following code:

```
class tree {
   int id, rule, nkids;
   String sym;
   token tok;
   tree kids[];
   Boolean isConst;
   symtab stab;
}
```

The tree class will have many methods added to it in this and coming chapters since most aspects of semantic analysis and code generation for your language will be presented as tree traversals. Now, let's look at the class types you need to render symbol tables and the symbol table entry class that contains the information that's held in symbol tables.

Defining classes for symbol tables and symbol table entries

Instances of the symtab class manage the symbols for one scope. For each symbol table, you will need to know what scope it is associated with, as well as what the enclosing scope is. An important method in this class, insert(), issues a semantic error if the symbol is already in the table. Otherwise, insert() allocates a symbol table entry and inserts it. The insert() method will be shown in the *Finding redeclared variables* section, later in this chapter. The Unicon code for the symtab class, which can be found in the symtab.icn file, starts as follows:

```
class symtab(scope, parent, t)
   method lookup(s)
      return \ (t[s])
   end
initially
   t := table()
end
```

The symtab class is almost just a wrapper around Unicon's built-in table data type. Within this class, the scope field is a string beginning with class, local, or package for user-declared scopes in Jzero. The corresponding Java class consists of the following code in symtab.java:

```
package ch6;
import java.util.HashMap;
public class symtab {
   String scope;
   symtab parent;
   HashMap<String,symtab_entry> t;
   symtab(String sc, symtab p) {
      scope = sc; parent = p;
      t = new HashMap<String,symtab_entry>();
   }
   symtab_entry lookup(String s) {
      return t.get(s);
   }
}
```

Each symbol table associates a name with an instance of the `symtab_entry` class. The `symtab_entry` class will hold all the information that we know about a given variable. The Unicon implementation of `symtab_entry` can be found in `symtab_entry.icn`:

```
class symtab_entry(sym,parent_st,st,isConst)
end
```

For now, the `symtab_entry` class contains no code; it just holds several data fields. The `sym` field is a string that holds the symbol that the entry denotes. The `parent_st` field is a reference to the enclosing symbol table. The `st` field is a reference to the new symbol table associated with this symbol's subscope, used only for symbols that have subscopes, such as classes and methods. In future chapters, the `symtab_entry` class will gain additional fields, both for semantic analysis and code generation purposes. The Java implementation of `symtab_entry` in `symtab_entry.java` looks as follows:

```
package ch6;
public class symtab_entry {
    String sym;
    symtab parent_st, st;
    boolean isConst;
    symtab_entry(String s, symtab p, boolean iC) {
      sym = s; parent_st = p, isConst = iC; }
    symtab_entry(String s, symtab p, boolean iC, symtab t) {
      sym = s; parent_st = p; isConst = iC; st = t; }
}
```

The preceding class contains no code other than two constructors. One is for regular variables, while the other is for classes and methods. Classes and method symbol table entries take a child symbol table as a parameter since they have a subscope. Having defined the class types for symbol tables and symbol table entries, it is time to look at how to create the symbol tables for the input program.

Creating symbol tables

You can create a symbol table for every class and every method by writing a tree traversal. Every node in the syntax tree needs to know what symbol table it belongs to. The brute-force approach presented here consists of populating the `stab` field of every tree node. Usually, the field is inherited from the parent, but nodes that introduce new scopes go ahead and allocate a new symbol table during the traversal.

The following Unicon mkSymTables() method constructs symbol tables. It is added to the tree class in the tree.icn file:

```
method mkSymTables(curr)
   stab := curr
   case sym of {
      "ClassDecl": { curr := symtab("class",curr) }k
      "MethodDecl": { curr := symtab("method",curr) }
   }
   every (!\kids).mkSymTables(curr)
end
```

The mkSymTables() method takes an enclosing symbol table named curr as a parameter. The corresponding Java method, mkSymTables(), in tree.java looks as follows:

```
void mkSymTables(symtab curr) {
   stab = curr;
   switch (sym) {
   case "ClassDecl": curr = new symtab("class", curr);
      break;
   case "MethodDecl": curr = new symtab("method", curr);
      break;
   }
   for (int i=0; i<nkids; i++) kids[i].mkSymTables(curr);
}
```

The root of the entire parse tree starts with a global symbol table with predefined symbols such as System and java. That begs the question: when and where is mkSymTables() called? The answer is after the root of the syntax tree has been constructed. Where the previous chapter was calling j0.print($$), it should now call j0.semantic($$) and all semantic analysis will be performed in that method of the j0 class. Therefore, the semantic action for the first production in j0gram.y becomes the following:

```
ClassDecl: PUBLIC CLASS IDENTIFIER ClassBody {
   $$=j0.node("ClassDecl",1000,$3,$4);
   j0.semantic($$);
} ;
```

The `semantic()` method in `j0.icn` looks as follows:

```
method semantic(root)
local out_st, System_st
    global_st := symtab("global")
    out_st := symtab("class")
    System_st := symtab("class")
    out_st.insert("println", false)
    System_st.insert("out", false, out_st)
    global_st.insert("System", false, System_st)
    root.mkSymTables(global_st)
    root.populateSymTables()
    root.checkSymTables()
    global_st.print()
end
```

This code creates a global symbol table and then predefines a symbol for the `System` class. `System` has a subscope in which a name, out, is declared to have a subscope in which `println` is defined. The corresponding Java code to initialize predefined symbols in the `j0.java` file looks like this. The `parserVal` from the BYACC/J value stack is unwrapped in order to obtain the `tree` node at the beginning:

```
public static void semantic(parserVal r) {
    Tree root = (tree)(r.obj);
    symtab out_st, System_st;
    global_st = new symtab("global");
    System_st = new symtab("class");
    out_st = new symtab("class");
    out_st.insert("println", false);
    System_st.insert("out", false, out_st);
    global_st.insert("System", false, System_st);
    root.mkSymTables(global_st);
    root.populateSymTables();
    root.checkSymTables();
    global_st.print();
}
```

Creating symbol tables is one thing; making use of them is another. Let's look at how symbols get put into the symbol tables. Then, we can start talking about how those symbol tables are used.

Populating symbol tables

Populating (inserting symbols into) symbol tables can be done during the same tree traversal in which those symbol tables are created. However, the code is simpler in a separate traversal. Every node knows what symbol table it lives within. The challenge is to identify which nodes introduce symbols.

For a class, the second child of `FieldDecl` has a list of symbols to be inserted. The first child of `MethodDeclarator` is a symbol to be inserted. For a method, the second child of `FormalParm` introduces a symbol. The second child of `LocalVarDecl` has a list of symbols to be inserted. These actions are shown in the following Unicon code:

```
method populateSymTables()
  case sym of {
    "ClassDecl": {
        stab.insert(kids[1].tok.text, , kids[1].stab)
        }
    "FieldDecl" | "LocalVarDecl" : {
        k := kids[2]
        while \k & k.label=="VarDecls" do {
          insert_vardeclarator(k.kids[2])
          k := k.kids[1]
          }
        insert_vardeclarator(k); return
        }
    "MethodDecl": {
        stab.insert(kids[1].kids[2].kids[1].tok.text, ,
                    kids[1].stab) }
    "FormalParm": { insert_vardeclarator(kids[2]); return }
    }
    every (!\kids).populateSymTables()
end
```

The corresponding Java code is as follows:

```
void populateSymTables() {
    switch(sym) {
    case "ClassDecl": {
        stab.insert(kids[0].tok.text, false, kids[0].stab);
```

```
        break;
    }
    case "FieldDecl": case "LocalVarDecl": {
        tree k = kids[1];
        while ((k != null) && k.sym.equals("VarDecls")) {
          insert_vardeclarator(k.kids[1]);
          k = k.kids[0];
          }
        insert_vardeclarator(k); return;
    }
    case "MethodDecl": {
        stab.insert(kids[0].kids[1].kids[0].tok.text, false,
                       kids[0].stab); }
    case "FormalParm": {
      insert_vardeclarator(kids[1]); return; }
    }
    for(int i = 0; i < nkids; i++) {
      tree k = kids[i];
      k.populateSymTables();
    }
  }
}
```

The `insert_vardeclarator(n)` method can be passed one of two possibilities: either an `IDENTIFIER` containing the symbol to be inserted or a `VarDeclarator` tree node that indicates an array is being declared. The Unicon implementation looks like this:

```
method insert_vardeclarator(vd)
    if \vd.tok then stab.insert(vd.tok.text)
    else insert_vardeclarator(vd.kids[1])
end
```

The Java implementation of the code looks as follows:

```
void insert_vardeclarator(tree vd) {
    if (vd.tok != null) stab.insert(vd.tok.text, false);
    else insert_vardeclarator(vd.kids[0]);
}
```

Populating symbol tables is necessary for later aspects of your programming language implementation, such as type checking and code generation. They will not be free to just skip down the subtree until they find the IDENTIFIER. Even in this first formulation, it is already good for checking certain common semantic errors such as undeclared variables. Now, let's look at how to compute a synthesized attribute, a skill you can use both when populating symbol tables with information and in later parts of semantic analysis and code generation.

Synthesizing the isConst attribute

isConst is a classic example of a synthesized attribute. Its calculation rules depend on whether a node is a leaf (following the base case) or an internal node (using the recursion step):

- **Base case**: For tokens, literals are isConst=true and everything else is isConst=false.
- **Recursion step**: For internal nodes, isConst is computed from children, but only through the expression grammar, where expressions have values.

If you are wondering which production rules are referred to by the expression grammar, it is pretty much those production rules derivable from the non-terminal named Expr. The Unicon implementation of this method is another traversal in tree.icn, as shown here:

```
method calc_isConst()
    every (!\kids).calc_isConst()
    case sym of {
        "INTLIT" | "DOUBLELIT" | "STRINGLIT" |
        "BOOLFALSE" | "BOOLTRUE": isConst := "true"
        "UnaryExpr": isConst := \kids[2].isConst
        "RelExpr": isConst := \kids[1].isConst &
          \kids[3].isConst
        "CondOrExpr" | "CondAndExpr" | "EqExpr" |
        "MULEXPR"|
        "ADDEXPR": isConst := \kids[1].isConst &
          \kid[2].isConst
        default: isConst := &null
    }
end
```

There are a couple of special cases in the preceding code. Whether binary relational operators such as the less than operator (<) are constant depends on the first and third children. Most other binary operators do not place the operator in the tree as a middle leaf; they are calculated from the isConst values of the first and second child. The Java implementation of the calc_isConst() method looks like this:

```java
void calc_isConst() {
    for(int i=0; i <nkids; i++)
        kids[i].calc_isConst();
    switch(sym) {
    case "INTLIT": case "DOUBLELIT": case "STRINGLIT":
    case "BOOLFALSE": case "BOOLTRUE": isConst = true;
        break;
    case "UnaryExpr": isConst = kids[1].isConst; break;
    case "RelExpr":
        isConst = kids[0].isConst && kids[2].isConst; break;
    case "CondOrExpr": case "CondAndExpr":
    case "EqExpr": case "MULEXPR": case "ADDEXPR":
        isConst = kids[0].isConst && kids[1].isConst; break;
    default: isConst = false;
    }
}
```

The whole method is a switch to handle the base case and set isConst, followed by a traversal of zero or more children. Java is arguably every bit as good as Unicon, or a bit better, at calculating the isConst synthesized attribute.

This concludes this section on creating and populating symbol tables. The main skill we practiced was the art of writing tree traversals, which are recursive functions. A regular tree traversal visits all the children and treats them identically. A programming language may traverse a tree selectively. It may ignore some children or do different things with different children. Now, let's look at an example of how symbol tables can be used to detect undeclared variables.

Checking for undeclared variables

To find undeclared variables, check the symbol table on each variable that's used for assignment or dereferencing. These reads and writes of memory occur in the executable statements and the expressions whose values are computed within those statements. Given a syntax tree, how do you find them?

The answer is to use tree traversals that look for IDENTIFIER tokens but only when they are in executable statements within blocks of code. To go about this, start from the top with a tree traversal that just finds the blocks of code. In Jzero, this is a traversal that finds the bodies of methods.

Identifying the bodies of methods

The check_codeblocks() method traverses the tree from the top to find all the method bodies, which is where the executable code is in Jzero. For every method declaration it finds, it calls another method called check_block() on that method's body. In tree.icn, the Unicon version is:

```
method check_codeblocks()
    if sym == "MethodDecl" then { kids[2].check_block() }
    else every k := !\kids do
            if k.nkids>0 then k.check_codeblocks()
end
```

The corresponding Java implementation of check_codeblocks() goes in the tree.java file:

```
void check_codeblocks() {
    tree k;
    if (sym.equals("MethodDecl")) { kids[1].check_block(); }
    else {
        for(int i = 0; i<=nkids; i++){
            k := kids[i];
            if (k.nkids>0) k.check_codeblocks();
        }
    }
}
```

The preceding method demonstrates the pattern of searching through the syntax tree while looking for one specific type of tree node. It does not call itself recursively on MethodDecl. Instead, it calls the more specialized check_block() method, which implements the work to be done when a method body has been found. This method knows it is in a method body, where the identifiers that it finds are uses of variables.

Spotting uses of variables within method bodies

Within a method body, any IDENTIFIER that is found is known to be inside a block of executable code statements. One exception is that new variables introduced by local variable declarations cannot possibly be undeclared variables:

```
method check_block()
    case sym of {
    "IDENTIFIER": {
      if not (stab.lookup(tok.text)) then
         j0.semerror("undeclared variable "||tok.text)
      }
    "FieldAccess" | "QualifiedName": kids[1].check_block()
    "MethodCall": {
       kids[1].check_block()
       if rule = 1290 then
          kids[2].check_block()
       else kids[3].check_block()
      }
    "LocalVarDecl": { } # skip
    default:  {
       every k := !kids do {
           k.check_block()
          }
        }
      }
    }
  end
```

The preceding check_block() method is handling several special-case tree shapes. Refer to the j0gram.y grammar file to examine the uses of IDENTIFIER that are not looked up in the local symbol table due to their syntactic context. In the case of FieldAccess or QualifiedName, the second child is an IDENTIFIER that is a field name, not a variable name. It can be checked once type information is added over the next few chapters. Similarly, the number 1290 denotes a production rule for an ordinary method call where both children are checked, but *rule 1291*, the second production rule of MethodCall, is a qualified method call that skips checking its second child because it is an IDENTIFIER that follows a dot and must be handled differently. The corresponding Java method is as follows:

```
void check_block() {
```

```
   switch (sym) {
   case "IDENTIFIER": {
     if (stab.lookup(tok.text) == null)
        j0.semerror("undeclared variable " + tok.text);
     break;
     }
   case "FieldAccess": case "QualifiedName":
     kids[0].check_block();
     break;
 case "MethodCall": {
     kids[0].check_block()
     if (rule == 1290)
        kids[1].check_block();
     else kids[2].check_block();
     break;
     }
   case "LocalVarDecl": break;
   default:
     for(i=0;i<nkids;i++)
          kids[i].check_block();
   }
 }
```

Despite the break statements, the Java implementation is equivalent to the Unicon version described earlier. The main idea you learned in this section was how to split up an overall tree traversal task into a general traversal that looks for a node of interest, and then a specialized traversal that performs its work at nodes found by the general traversal. Now, let's look at detecting a variable redeclaration semantic error, which occurs when symbols are being inserted into the symbol tables.

Finding redeclared variables

When a variable has been declared, most languages report an error if the same variable is declared again in the same scope. The reason for this is that within a given scope, the name must have a single, well-defined meaning. Trying to declare a new variable would entail allocating some new memory and from then on, mentioning that name would be ambiguous. If the x variable is defined twice, it is unclear to which x any given use refers. You can identify such redeclared variable errors when you insert symbols into the symbol table.

Inserting symbols into the symbol table

The insert() method in the symbol table class calls the language's underlying hash table API. The method takes a symbol, a Boolean isConst flag, and an optional nested symbol table, for symbols that introduce a new (sub)scope. The Unicon implementation of the symbol table's insert() method is shown here. If you go to https://github.com/PacktPublishing/Build-Your-Own-Programming-Language-Second-Edition/tree/master/ch6, this can be found in symtab.icn, along with the other class symtab methods:

```
method insert(s, isConst, sub)
    if \ (t[s]) then j0.semerror("redeclaration of "||s)
    else t[s] := symtab_entry(s, self, sub, isConst)
end
```

A symbol table lookup is performed before insertion. If the symbol is already present, a redeclaration error is reported. The corresponding Java implementation of the symbol table's insert() methods looks as follows:

```
void insert(String s, Boolean iC, symtab sub) {
    if (t.containsKey(s)) {
        j0.semerror("redeclaration of " + s);
    } else {
        sub.parent = this;
        t.put(s, new symtab_entry(s, this, iC, sub));
    }
}
void insert(String s, Boolean iC) {
    if (t.containsKey(s)) {
        j0.semerror("redeclaration of " + s);
    } else {
        t.put(s, new symtab_entry(s, this, iC));
    }
}
```

This code is crude but effective. The use of the underlying hash table Java API is long-winded but readable. Now, let's look at the semerror() method.

Reporting semantic errors

The semerror() method in the j0 class must report the error to the user, as well as making a note that an error has occurred so that the compiler will not attempt code generation. The code for reporting semantic errors is similar to the code for reporting lexical or syntax errors, although sometimes, it is harder to pinpoint what line in what file is to blame. For now, it is OK to treat these errors as fatal and stop compilation when one occurs. In later chapters, you will make this error non-fatal and report additional semantic errors after one is found. The Unicon code for the j0 class's semerror() method is as follows:

```
method semerror(s)
   stop("semantic error: ", s)
end
```

The Java code for the j0 class's semerror() method is shown here:

```
void semerror(String s) {
   System.out.println("semantic error: " + s);
   System.exit(1);
}
```

Identifying redeclaration errors occurs most naturally while the symbol table is being populated, that is, when an attempt is being made to insert a declaration. Unlike an undeclared symbol error, where all nested symbol tables must be checked before an error can be reported, a redeclaration error is reported immediately, but only if the symbol has already been declared in the current inner-most scope. Now, let's look at how a real programming language deals with other symbol table issues that did not come up in this discussion.

Handling package and class scopes in Unicon

Creating symbol tables for Jzero considers two scopes: class and local. Since Jzero does not do instances, Jzero's class scope is static and lexical. A larger, real-world language must do more work to handle scopes. Java, for example, must distinguish when a symbol declared in the class scope is a reference to a variable shared across all instances of the class, and when the symbol is a normal member variable that's been allocated separately for each instance of the class. In the case of Jzero, an isMember Boolean can be added to the symbol table entries to distinguish member variables from class variables, similar to the isConst flag.

Unicon's implementation is a lot different than Jzero's. A summary of its symbol tables and class scopes allows for a fruitful comparison. Whatever it does similarly to Jzero might also be how other languages handle things. What Unicon does differently than Jzero, each language might do in its own unique way. How Unicon handles these topics is being presented here for its quirky real-world insights, not because it is somehow exemplary or ideal.

One basic difference between Unicon and the Jzero example in this chapter is that Unicon's syntax tree is a heterogeneous mixture of different types of tree node objects. In addition to a generic tree node type, there are separate tree node types to represent classes, methods, and a few other semantically significant language constructs. The generic tree node type lives in a file named tree.icn, while the other classes live in a file named idol.icn that is descended from Unicon's predecessor, a language called **Idol**. Now, let's look at another difference between Unicon and Jzero that comes up in Unicon's implementation of packages. This is known as **name mangling**.

Mangling names

Scope checks may state that a symbol has been found in a package. A lot of programming languages – historic C++ is a prime example – use **name mangling** in generated code. Name mangling refers to the practice of altering a name in order to distinguish different uses of it, or to add information. In Unicon, some scoping rules are resolved via name mangling. A name such as foo, if it is found to be in package scope for a package bar, is written out in the generated code as bar__foo.

The mangle_sym(sym) method from the Unicon implementation has been presented in its partial form here and has been abstracted a bit for readability. This method takes a symbol (a string) and mangles it according to which imported package it belongs to, including the declared package of the current file, which takes precedence over any imports:

```
procedure mangle_sym(sym)
  ...
    if member(package_level_syms, sym) then
      return package_mangled_symbol(sym)
    if member(imported, sym) then {
      L := imported[sym]
      if *L > 1 then
         yyerror(sym || " is imported from multiple packages")
      else return L[1] || "__" || sym
    }
    return sym
  end
```

In the `mangle_sym()` method, a Unicon table named `package_level_syms` stores entries for symbols declared in the package associated with the current file. Another table, called `imported`, tracks all the symbols defined in other packages. This table returns a list of the other packages in which a symbol is found. The size of that list is given by `*L`. If a symbol is defined in two or more imported packages, using that symbol in this file is ambiguous and generates an error. The use of packages is a relatively simple compile-time mechanism for making separate namespaces for different scopes. More difficult scoping rules must be handled at runtime. For example, accessing class members in Unicon requires the compiler to generate code that uses a reference to a current object named `self`.

Inserting self for member variable references

Scoping rules can come back with the answer that a symbol is a class member variable. In Unicon, all methods are non-static and method calls always have an implicit first parameter named `self`, which is a reference to the object upon which the method has been invoked. A class scope is implemented by prefixing the name with a dot operator to reference the variable within the `self` object. This code, extracted from a method named `scopeck_expr()` in Unicon's `idol.icn` semantic analysis file, illustrates how `self.` can be prefixed onto member variable references:

```
"token": {
    if node.tok = IDENT then {
        if not member(\local_vars, node.s) then {
            if classfield_member(\self_vars, node.s)then
                node.s := "self." || node.s
            else
                node.s := mangle_sym(node.s)
        }
    }
}
```

This code modifies the contents of the existing syntax tree field in place. The use of the `self.` string prefix is possible because the code is written out in a source code-like form and further compiled to C or virtual machine bytecode by a subsequent code generator. The use of `self` as a reference to the current object is needed not only to access the member variables within the object but also to access calls to the object's methods. On the flip side of that, let's look at how Unicon provides the `self` variable when methods are called.

Inserting self as the first parameter in method calls

When an identifier appears in front of parentheses, the syntax indicates that it is the name of a function or method being called. In this case, additional special handling is required. The code for calling a method must look up the method name in an auxiliary structure called the methods vector. The methods vector is referenced via self.__m. For example, for a method named meth, instead of becoming self.meth, the reference to the method becomes self.__m.meth.

In addition to using the methods vector, __m, a method call requires self to be inserted as a first parameter into the call. In Unicon's predecessor, this was explicit in the generated code. A call such as meth(x) would become self.__m.meth(self, x). In the Unicon implementation, this insertion of the object into the parameter list of the call is built into the implementation of the dot operator in the runtime system. When the dot operator is asked to perform self.meth, it looks up meth to see whether it is a regular member variable. If it finds that it is not, it checks whether self.__m.meth exists, and if it does, the dot operator both looks up that function and pushes self onto the stack as its first parameter.

To summarize: the Unicon virtual machine was modified to make code generation for method calls simpler. Consider the call to o.m() in the following example. The semantics of the o.m(3,4) call are equivalent to o.__m.m(o,3,4) but the compiler just generates the instructions for o.m(3,4) and the Unicon dot operator does all the work:

```
class C(…)
    method m(c,d); … end
end
procedure main()
    o := C(1,2)
    o.m(3,4)
end
```

One of the nice parts about building a programming language is that you can make the runtime system that runs your generated code do anything you want. Now, let's consider how to test and debug your symbol tables to tell whether they are correct and working.

Testing and debugging symbol tables

You can test your symbol tables by writing many test cases and verifying whether they obtain the expected undeclared or redeclared variable error messages. But nothing says confidence like an actual visual depiction of your symbol tables.

If you have built your symbol tables correctly by following the guidance in this chapter, then there should be a tree of symbol tables. You can print out your symbol tables using the same tree printing techniques that were used to verify your syntax trees in the previous chapter, using either a textual representation or a graphical one.

Symbol tables are slightly more work to traverse than syntax trees. To output the symbol table, you need to output information for the table and then visit all the children, not just look one up by name. Also, there are two classes involved: symtab and symtab_entry. Suppose you start at the root symbol table. In Unicon, to iterate through all the symbol tables, use the following method in symtab.icn:

```
method print(level:0)
  writes(repl(" ",level))
  write(scope, " - ", *t, " symbols")
  every (!t).print(level+1);
end
```

Notice that although the children are being invoked with a method of the same name, the print() method in symtab_entry is a different method than the one on symtab. The Java code for the symbol table's print() method looks like this:

```
void print() { print(0); }
void print(int level) {
    for(int i=0;i<level;i++) System.out.print(" ");
    System.out.print(scope + " - " + t.size()+" symbols");
    for (symtab_entry se: t.values()) se.print(level+1);
}
```

For the print() method of symtab_entry, an actual symbol is printed out. If that symbol table entry has a subscope, it is then printed and indented more deeply to show the nesting of the scopes. In symtab_entry.icn, the Unicon code is:

```
method print(level:0)
  writes(repl(" ",level), sym)
  if \isConst then writes(" (const)")
  write()
  (\st).print(level+1);
end
```

The mutually recursive call to print the nested symbol table is skipped if it is null. In Java, the code is longer but more explicit:

```
void print(level:0) {
    for(int i=0;i<level;i++) System.out.print(" ");
    System.out.print(sym);
    if (isConst) System.out.print(" (const)");
    System.out.println("");
    if (st != null) st.print(level+1);
}
```

Printing out symbol tables doesn't take many lines of code. You may find that it's worth adding additional lexical information, such as filenames and line numbers where variables were declared. In future chapters, it will be logical to also extend these methods with type information.

To run the Jzero compiler with the symbol table output shown in this chapter, download the code from this book's GitHub repository, go into the ch6/ subdirectory, and build it with the Make program. By default, Make will build both the Unicon and Java versions. When you run the j0 command with the symbol table output in place, it produces the following output. In this case, the Java implementation is being shown:

```
C:\Users\clint\books\byopl\github\Build-Your-Own-Programming-Language\ch6>set CLASSPATH=".;C:\Users
\clint\books\byopl\github\Build-Your-Own-Programming-Language"

C:\Users\clint\books\byopl\github\Build-Your-Own-Programming-Language\ch6>java ch6.j0 hello.java
yyfilename hello.java
global - 2 symbols
 hello
  class - 2 symbols
   main
    method - 0 symbols
   System
 System
  class - 1 symbols
   out
    class - 1 symbols
     println
no errors

C:\Users\clint\books\byopl\github\Build-Your-Own-Programming-Language\ch6>
```

Figure 6.3: Symbol table output from the Jzero compiler

You must read the hello.java input file pretty carefully to ascertain whether this symbol table output is correct and complete. The more complicated your language's scoping and visibility rules, the more complicated your symbol table's output will be.

For example, this output does not print anything for a variable's public and private status, but for a full Java compiler, we would want that. When you are satisfied that the symbols are all present and accounted for in the correct scopes, you can move on to the next phase of semantic analysis.

Summary

In this chapter, you learned about the crucial technical skills and tools used to build symbol tables that track all the variables in all the scopes in the input program. You create a symbol table for every scope in the program and insert entries into the correct symbol table for each variable. All of this is accomplished via traversals of the syntax tree.

You learned how to write tree traversals that create symbol tables for each scope, as well as how to create inherited and synthesized attributes for the symbol table associated with the current scope for each node in your syntax tree. You then learned how to insert symbol information into the symbol tables associated with your syntax tree and detect when the same symbol is redeclared illegally. You learned how to write tree traversals that look up information in symbol tables and identify any undeclared variable errors. These skills will enable you to take your first steps in enforcing the semantic rules associated with your programming language. In the rest of your compiler, both semantic analysis and code generation will rely upon and add to the symbol tables that you established in this chapter.

Now that you have built symbol tables by walking through the parse tree using tree traversals, it is time to start considering how to check the program's use of data types. The next chapter will start us off on that journey by showing you how to check basic types such as integers and real numbers.

Questions

1. What is the relationship between the various symbol tables that are created within the compiler and the syntax tree that was created in the previous chapter?

2. What is the difference between synthesized semantic attributes and those that are inherited? How are they computed and where are they stored?

3. How many symbol tables do we need in the Jzero language? How are symbol tables organized?

4. Suppose our Jzero language allowed multiple classes, compiled separately in separate source files. How would that impact our implementation of symbol tables in this chapter?

Join our community on Discord

Join our community's Discord space for discussions with the authors and other readers:

https://discord.com/invite/zGVbWaxqbw

7

Checking Base Types

This is the first of two chapters about type checking. In most mainstream programming languages, **type checking** is a key aspect of semantic analysis that must be performed before you can generate code. Type checking uses the syntax trees from *Chapter 5* and the symbol tables from *Chapter 6*.

This chapter will show you how to do simple type checks for the base types included in the Jzero subset of Java. A by-product of checking the types is adding type information to the syntax tree. Knowing the types of operands in the syntax tree enables you to generate correct instructions for various operations. For example, if your compiler sees the code x + y, should it generate code for integer addition? Floating-point addition? Something else?

This chapter covers the following main topics:

- Type representation in the compiler
- Assigning type information to declared variables
- Determining the type at each syntax tree node
- Runtime type checks and type inference – a Unicon example

It is time to learn about type checking, starting with base types. Some of you may be wondering, why do type checking at all? If your compiler does not do any type checking, it must generate code that works, no matter what types of operands are used. Python, Lisp, BASIC, and Unicon are examples of languages with this design approach. Often, this makes a language user-friendly, but it runs slower because the types of operands must be checked at runtime. There is also a whole broad category of fatal type errors that can be avoided if types are checked at compile time. For these two reasons, speed and safety, we will cover type checking. We will begin by looking at how to represent the type information that you extract from the source code.

Technical requirements

The code for this chapter is available on GitHub: `https://github.com/PacktPublishing/Build-Your-Own-Programming-Language-Second-Edition/tree/master/ch7`

The Code in Action video for the chapter can be found here: `https://bit.ly/3cgvkWT`

Type representation in the compiler

Frequently, our compiler will need to do things such as compare the types of two variables to see whether they are compatible. Program source code represents types with string data, which is incorporated in our syntax tree. In some languages, it might be possible to use little syntax subtrees to represent the types that are used in type checking, but in general, type information does not exactly correspond to a subtree within our syntax tree. This is because part of the type information is pulled in from elsewhere, such as another type. For this reason, we need a new data type specifically designed to represent the type information associated with any given value that is declared or computed in the program.

It would be nice if we could just represent types with an atomic value such as an integer code or a string type name. For example, we could use 1 for an integer, 2 for a real number, or 3 for a string. If a language had only a small, fixed set of built-in types, an atomic value would suffice. However, most real programs use types that are more complex. The representation of a compound type such as an array, a class, or a method is more involved. We will start with a base class capable of representing atomic types.

Defining a base class for representing types

The type information associated with any name or value in your language can be represented within a new class named `typeinfo`. The `typeinfo` class is not called `type` because some programming languages use that as a reserved word or built-in name. In Unicon, it is the name of a built-in function, so declaring a class with that name would cause an error.

The `typeinfo` class has a `basetype` member for storing what kind of data type is represented. Complex types have additional information as needed. For example, a type whose `basetype` indicates that it is an array has an additional `element_type`. With this extra information, we will be able to distinguish an array of integers from an array of strings or an array of some class type. In some languages, array types also have an explicit size or starting and ending indices.

There are many ways that you could handle this variation in the information needed for different types. A classic object-oriented representation of these differences is to use subclasses. For Jzero, we will add arraytype, methodtype, and classtype as subclasses of typeinfo. First, there is the superclass itself, which can be found in the Unicon typeinfo.icn file, as shown in the following code:

```
class typeinfo(basetype)
   method str()
      return string(basetype)|"unknown"
   end
end
```

In addition to the basetype member, the typeinfo class has methods to facilitate debugging. Types need to be able to print themselves in a human-readable format. The Java version, in the typeinfo.java file, looks like this:

```
public class typeinfo {
   String basetype;
   public typeinfo() { basetype = "unknown"; }
   public typeinfo(String s) { basetype = s; }
   public String str() { return basetype; }
}
```

An extra constructor taking no arguments is required for the subclasses to compile properly in Java. Having a class, and not just an integer, to encode the type information allows us to represent more complex types by subclassing the base class.

Subclassing the base class for complex types

The Unicon code for the subclasses of typeinfo is also stored in typeinfo.icn since the subclasses are short and closely related. In Jzero, the arraytype class only has an element_type; in other languages, an array type might require additional fields to hold the array size or the type and range of valid indices. The Unicon representation of the array type in Jzero is as follows:

```
class arraytype : typeinfo(element_type)
initially
   basetype := "array"
end
```

The arraytype.java file contains the corresponding Java implementation of the arraytype class:

```java
public class arraytype extends typeinfo {
    typeinfo element_type;
    public arraytype(typeinfo t) {
        basetype = "array"; element_type = t; }
}
```

The representation for methods, also called **class member functions,** is a class named methodtype that includes a signature consisting of the method's parameters and return type. For now, all the methodtype class does is allow methods to be identified as such. The Unicon implementation of the methodtype class is as follows:

```
class methodtype : typeinfo(parameters,return_type)
initially
    basetype := "method"
end
```

The Unicon constructor for the methodtype class takes a list of zero or more parameters and a return type; in Java, the parameters are passed via an array. In both Unicon and Java, the parameters are passed by reference, so take care that the caller does not subsequently modify the list or array that was passed. The parameters and return type will be used in the next chapter to check the types when methods (functions) are called. The Java representation of methods looks as follows and can be found in the methodtype.java file:

```java
public class methodtype extends typeinfo {
    parameter [] parameters;
    typeinfo return_type;
    methodtype(parameter [] p, typeinfo rt){
        parameters = p; return_type = rt;
    }
}
```

The parameters member variable could be an array of typeinfo. Instead, a separate class is defined for parameters here to allow languages to include parameter names along with their types to represent methods. The Unicon implementation of the parameter class is found in typeinfo. icn and reads as follows:

```
class parameter(name, param_type)
end
```

Some of these classes are pretty empty so far. They are placeholders that will include more code in subsequent chapters or require more substantial treatments in other languages. The corresponding Java implementation of the parameter class in the parameter.java file is shown here:

```
public class parameter {
    String name;
    typeinfo param_type;
    parameter(String s, typeinfo t) { name=s; param_type=t; }
}
```

The class for representing classes includes a class name, its associated symbol table, and lists of zero or more fields, methods, and constructors. In some languages, this might be more complex than Jzero, including superclasses, for example. The Unicon implementation of the classtype class in typeinfo.icn is shown here:

```
class classtype : typeinfo(name, st, fields, methods, constrs)
    method str()
        return name
    end
initially
    basetype := "class"
end
```

You might be wondering about the st field, which holds a symbol table. In *Chapter 6, Symbol Tables,* symbol tables were constructed and stored in syntax tree nodes, where they formed a logical tree corresponding to the program's declared scopes. References to those same symbol tables need to be placed in the types so that we can compute the type resulting from the use of the dot operator, which references a scope that is not associated with the syntax tree. The classtype.java file contains the Java implementation of the classtype class, as shown in the following code:

```
public class classtype extends typeinfo {
    String name;
    symtab st;
    parameter [] methods;
    parameter [] fields;
    typeinfo [] constrs;
    public classtype(String s) { name = s; }
    public classtype(String s, symtab stab) { name = s; st = stab; }
}
```

Given a typeinfo class, it is appropriate to add a member field of this type to both the tree class and the symtab_entry class so that type information can be represented for expressions and variables. We will call it typ in both classes:

```
class tree  (id,sym,rule,nkids,tok,kids,isConst,stab,typ)
class symtab_entry(sym,parent_st,st,isConst,typ)
```

We are not repeating the classes here in their entirety; the code for this can be found in the ch7/ subdirectory at https://github.com/PacktPublishing/Build-Your-Own-Programming-Language-Second-Edition. In Java, the respective classes are amended as follows:

```
class tree { . . .
    typeinfo typ; . . . }
class symtab_entry { . . .
    typeinfo typ; . . . }
```

Given a typ field, it is possible to write the mini tree traversals needed to place type information in the symbol tables with the variables as they are declared. Let's look at assigning this type information to declared variables.

Assigning type information to declared variables

Type information is constructed during a tree traversal and then stored with the associated variables in the symbol table. This would usually be part of the traversal that populates the symbol table, as presented in the previous chapter. In this section, we will traverse the syntax tree looking for variable declarations, as we did previously, but this time, we need to propagate type information by using synthesized and/or inherited attributes.

For type information to be available at the time that we are inserting variables into the symbol table, the type information must be computed at some prior point in time. This type information is computed either by a preceding tree traversal or during parsing when the syntax tree is constructed. Consider the following grammar rule and semantic action from *Chapter 5, Syntax Trees*:

```
FieldDecl: Type VarDecls ';' {
    $$=j0.node("FieldDecl",1030,$1,$2); };
```

The semantic action builds a tree node connecting Type with VarDecls under a new node called FieldDecl. Your compiler must synthesize type information from Type and inherit it into VarDecls. The information flowing up from the left subtree and going down into the right subtree can be seen in the following diagram:

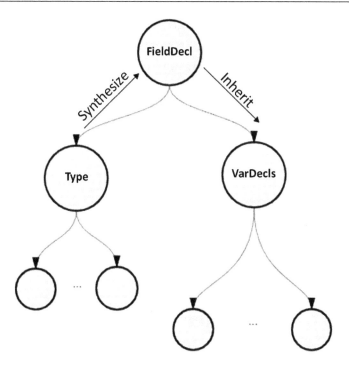

Figure 7.1: The flow of type information in variable declarations

We can embed this into the syntax tree construction process via *mini traversals* of the subtrees. The following code adds a call to a method named `calctype()`, which is where this semantic analysis will be conducted within `j0gram.y`, as shown in the previous example:

```
FieldDecl: Type VarDecls ';' {
   $$=j0.node("FieldDecl",1030,$1,$2);
   j0.calctype($$);
};
```

From examining the grammar, you may note that a similar call to `calctype()` is needed for non-terminal `FormalParm`, and that there are a few additional places in the grammar where a type is associated with an identifier or list of identifiers. The `j0` class's `calctype()` method turns around and calls two tree traversals on the two children of `FieldDecl`. The Unicon version of this method in `j0.icn` looks as follows:

```
method calctype(t)
   t.kids[1].calctype()
   t.kids[2].assigntype(t.kids[1].typ)
end
```

The j0 class's `calctype()` method calls the class tree's `calctype()` method, which calculates the synthesized typ attribute in the left child. The type is then passed down as an inherited attribute into the right child. The Java version of this method in j0.java looks like this:

```
void calctype(parserVal pv){
   tree t = (tree)pv.obj;
   t.kids[0].calctype();
   t.kids[1].assigntype(t.kids[0].typ);
}
```

Methods within the `tree` class, such as `calctype()` and `assigntype()`, are special case tree traversals whose tree shape and the kinds of possible tree nodes that they might be invoked on are a small subset of the possibilities that must be handled by more general tree traversals. The traversal code can be specialized to take advantage of this. By way of example, we first consider the `calctype()` method.

Synthesizing types from reserved words

The `calctype()` method calculates the synthesized typ attribute. The recursive work of calculating the value for the children is done first, followed by the calculation for the current node. This form of traversal is called a **post-order traversal** and it is common in compilers. In Unicon, the `calctype()` method in the tree class in tree.icn looks like this:

```
method calctype()
  every (!\kids).calctype()
  case sym of {
    "FieldDecl": typ := kids[1].typ
    "token": {
      case tok.cat of {
        parser.IDENTIFIER:{return typ := classtype(tok.text) }
        parser.INT:{ return typ := typeinfo(tok.text) }
        default:
          stop("don't know the type of ", image(tok.text))
        }
      }
    default:
      stop("don't know the type of ", image(sym))  }
end
```

This code constructs the current tree node's typ value using information from its children; in this case, by directly accessing a child's public typ field. Alternatively, information obtained from a child could be obtained by calling a method that returns the child type as its return value, such as the return value of calctype(). In this code, the number of case branches is small because the Jzero grammar for non-terminal Type is minimal. In other languages, it would be richer. The corresponding Java code is shown in the following example's calctype() method in tree.java:

```
typeinfo calctype() {
  for (tree k : kids) k.calctype();
  switch (sym) {
    case "FieldDecl": return typ = kids[0].typ;
    case "token": {
      switch (tok.cat) {
        case parser.IDENTIFIER:{
          return typ=new classtype(tok.text); }
        case parser.INT: { return typ=new typeinfo(tok.text); }
        default:
          j0.semerror("don't know the type of " + tok.text);
      }
    }
    default:
      j0.semerror("don't know the type of " + sym);}
}
```

Having synthesized the type from the left child of FieldDecl, let's look at how to inherit that type into the variable nodes in the right child subtree of FieldDecl.

Inheriting types into a list of variables

Passing type information into a subtree is performed in the assigntype(t) method. Inherited attributes are generally coded via a **pre-order traversal**, in which the current node does its work and then calls the children with information they are inheriting. The Unicon implementation of the assigntype(t) method is as follows:

```
method assigntype(t)
  typ := t
  case sym of {
  "VarDeclarator": {
    kids[1].assigntype(arraytype(t))
```

```
      return
    }
  "token": {
    case tok.cat of {
      parser.IDENTIFIER: return
      default: stop("eh? ", image(tok.cat))
      }
    }
  default:
    stop("don't know how to assign the type of ", image(sym))
    }
  every (!\kids).assigntype(t)
end
```

Since the information is coming down from a parent into children, it is natural to pass this information as a parameter to the child, who then assigns it as their type via typ := t. It would also be possible to copy it down via an explicit assignment into a child's public field. The corresponding Java implementation of the assigntype(t) method is shown here:

```
void assigntype(typeinfo t) {
    typ = t;
    switch (sym) {
    case "VarDeclarator": {
      kids[0].assigntype(new arraytype(t));
      return;
    }
    case "token": {
      switch (tok.cat) {
        case parser.IDENTIFIER:{ return; }
        default: j0.semerror("eh? " + tok.cat);
      }
    }
    default:
        j0.semerror("don't know how to assigntype " + sym);
    }
    for(tree k : kids) k.assigntype(t);
  }
```

Attaching type information to variable names where they are declared is important, and it is not too difficult, especially for a simple language such as Jzero. Now, it is time to look at the main task of this chapter: how to calculate and check type information in the expressions that comprise the executable code in the bodies of functions and methods.

Determining the type at each syntax tree node

Within the syntax tree, the nodes associated with actual code expressions in the method bodies have a type associated with the value that the expression computes. For example, if a tree node corresponds to the sum of adding two numbers, the tree node's type is determined by the types of the operands and the rules of the language for the addition operator. Our goal for this section is to spell out how this type information can be calculated.

As you saw in the *Type representation in the compiler* section, the class for syntax tree nodes has an attribute to store that node's type, if there is one. The type attribute is calculated bottom-up, during a post-order tree traversal. There is a similarity here to checking for undeclared variables, which we did in the previous chapter, in that the expressions that need types checked only occur in the bodies of functions. The call to invoke this type checking tree traversal, starting at the root of the syntax tree, is added at the end of the semantic() method, within the j0 class. In Unicon, the invocation consists of the following:

```
root.checktype()
```

There isn't a parameter here, but that is the same thing as passing in a null value or a false. In Java, the following statement is added:

```
root.checktype(false);
```

In both cases, the parameter indicates whether a given node is within the body of an executable statement. At the root, the answer is false. It will turn true when the tree traversal reaches the bodies of methods that contain code. To perform the tree traversal, you must consider what to do regarding the leaves of the tree.

Determining the type at the leaves

At the leaves, the types of literal constant values are self-evident from their lexical category. To begin, we must add a typ field to the class token. For literals, we must initialize typ in the constructor. In Unicon, the first line and initial section of token.icn becomes the following:

```
class token(cat, text, lineno, colno, ival, dval, sval,typ)
    . . .
```

```
initially
   case cat of {
   parser.INTLIT:{ ival := integer(text); typ := typeinfo("int")}
   parser.DOUBLELIT:{dval:=real(text); typ := typeinfo("double")}
   parser.STRINGLIT:{
      sval := deEscape(text); typ := typeinfo("String") }
   parser.BOOLLIT: { typ := typeinfo("boolean") }
   parser.NULLVAL: { typ :=  typeinfo("null") }
   ord("="|"+"|"-"): { typ := typeinfo("n/a") }
   }
end
```

The code here assigns the "n/a" type to operators (the code for calculating the types of expressions using those operators from their operand types will be shown later, in the *Calculating and checking the types at internal nodes* section). Note that the three operators shown here are for illustration purposes and a full language implementation should represent the type information for all supported operators. In Java, the corresponding change to the class token for literal types looks as follows:

```
package ch7;
public class token {

    . . .
   public typeinfo typ;
   public token(int c, String s, int l) {
      cat = c; text = s; lineno = l;
      id = serial.getid();
      switch (cat) {
      case parser.INTLIT: typ = new typeinfo("int"); break;
      case parser.DOUBLELIT:typ = new typeinfo("double");
          break;
      case parser.STRINGLIT: typ= new typeinfo("String");
          break;
      case parser.BOOLLIT: typ = new typeinfo("boolean");
          break;
      case parser.NULLVAL: typ = new typeinfo("null"); break;
      case '=': case '+': case '-':
          typ = new typeinfo("n/a"); break;
      }
   }
```

The types of variables are looked up in the symbol table. This implies that symbol table population must occur before type checking. The symbol table lookup for the Unicon version is performed by a type() method and added to a class token in token.icn. It takes the symbol table that the token is scoped within as a parameter:

```
method type(stab)
  if \typ then return typ
  if cat === parser.IDENTIFIER then
    if rv := stab.lookup(text) then return typ := rv.typ
  stop("cannot check the type of ",image(text))
end
```

This first line in this method returns the type for this token immediately if it has been determined previously. If not, the rest of this method just checks whether we have an identifier, and if so, looks it up in the symbol table. The corresponding addition to token.java looks as follows:

```
public typeinfo type(symtab stab) {
  symtab_entry rv;
  if (typ != null) return typ;
  if (cat == parser.IDENTIFIER)
      if ((rv = stab.lookup(text)) != null) return typ=rv.typ;
  j0.semerror("cannot check the type of " + text);
}
```

Having shown the code to calculate the type of syntax tree leaves, it is now time to examine how to check the types at the internal nodes. This is the core function of type checking.

Calculating and checking the types at internal nodes

The internal nodes are only checked within the executable statements and expressions in the code bodies of the program. This is a post-order traversal where work at children is done first and then work is done at the parent node. The process of visiting the children is delegated by the checktype() method in the tree class to the checkkids() helper function. The selection of which nodes to visit varies depending on the tree node, and the work that's done at the parent depends on whether it is in a block of code. The Unicon implementation of these methods in tree.icn is as follows:

```
method checktype(in_codeblock)
  if checkkids(in_codeblock) then return
  if /in_codeblock then return
```

```
    case sym of {
      "Assignment": typ := check_types(kids[1].typ,
                                         kids[3].typ)
      "AddExpr": typ := check_types(kids[1].typ, kids[2].typ)
      "Block" | "BlockStmts": { typ := &null }
      "MethodCall": { }
      "QualifiedName": {
         if type(kids[1].typ) == "classtype__state" then {
            typ := (kids[1].typ.st.lookup(
                     kids[2].tok.text)).typ
            } else stop("illegal . on ",kids[1].typ.str())
      }
      "token": typ := tok.type(stab)
      default: stop("cannot check type of ", image(sym))
    }
end
```

In addition to the `checkkids()` helper method, this code relies on a helper function called `check_types()`, which determines the result type, given operands. The corresponding Java implementation of `checktype()` is shown here:

```
void checktype(boolean in_codeblock) {
  if (checkkids(in_codeblock)) return;
  if (! in_codeblock) return;
  switch (sym) {
  case "Assignment":
    typ = check_types(kids[0].typ, kids[2].typ); break;
  case "AddExpr":
    typ = check_types(kids[0].typ, kids[1].typ); break;
  case "Block": case "BlockStmts": typ = null; break;
  case "MethodCall": break;
  case "QualifiedName": {
    if (kids[0].typ instanceof classtype) {
      classtype ct = (classtype)(kids[0].typ);
      typ = (ct.st.lookup(kids[1].tok.text)).typ;
    } else j0.semerr("illegal . on  " + kids[0].typ.str());
    break;
    }
```

```
    case "token": typ = tok.type(stab); break;
    default: j0.semerror("cannot check type of " + sym);
    }
  }
```

By default, the checkkids() helper function calls checktype() on every child, but in some cases, it does not. On method declaration, for example, the method header has no executable code expressions and is skipped; only the block of code is visited, and in that block, the in_codeblock Boolean parameter is set to true. Similarly, within a block of code where a local variable declaration is encountered, only the list of variables is visited, and within that list, in_codeblock is turned off (only to be turned back on again in initializers). As another example, identifiers on the right-hand side of a period operator are not looked up in the regular symbol table; instead, they are looked up relative to the type of expression on the left-hand side of the period and thus require special handling. The Unicon implementation of checkkids() is shown here:

```
  method checkkids(in_codeblock)
    case sym of {
      "MethodDecl": { kids[2].checktype(1); return }
      "LocalVarDecl": { kids[2].checktype(); return }
      "FieldAccess": { kids[1].checktype(in_codeblock);
          return }
      "QualifiedName": {
        kids[1].checktype(in_codeblock);
        }
      default: { every (!\kids).checktype(in_codeblock) }
      }
  end
```

The corresponding Java implementation of this helper function is shown here:

```
  public boolean checkkids(boolean in_codeblock) {
    switch (sym) {
    case "MethodDecl": kids[1].checktype(true); return true;
    case "LocalVarDecl": kids[1].checktype(false); return true;
    case "FieldAccess": kids[0].checktype(in_codeblock);
                        return true;
    case "QualifiedName":
                        kids[0].checktype(in_codeblock);
                        break;
```

```
    default: if (kids != null)
                for (tree k : kids) k.checktype(in_codeblock);
    }
    return false;
}
```

The check_types() helper method calculates the type of the current node from the types of up to two operands. Its calculation varies, depending on what operator is being performed, as well as the rules of the language. Its answer might be that the type is the same as one or both operands, or it may be some new type or an error. The Unicon implementation of check_types() in tree.icn is as follows. Note that the code here is illustrative but incomplete, showing only integer type checks on a few operators due to space considerations. A full implementation, even for a toy language like Jzero, has additional base types and operators.

```
method check_types(op1, op2)
   operator := get_op()
   case operator of {
      "="|"+"|"-" : {
         if tok := findatoken() then
            writes("line ", tok.tok.lineno, ": ")
         if op1.basetype === op2.basetype === "int" then {
            write("typecheck ",operator," on a ",
                  op2.str(), " and a ", op1.str(), " -> OK")
            return op1
         }
         else stop("typecheck ",operator," on a ",
                   op2.str(), " and a ", op1.str(),
                   " -> FAIL")
      }
      default: stop("cannot check ", image(operator))
   }
end
```

This method relies on two helper methods. The get_op() method reports which operator is being performed. The findatoken() method seeks out the first token in the source code represented by a given syntax tree node; it is used to report the line number. The corresponding Java implementation of check_types() is shown here:

```
public typeinfo check_types(typeinfo op1, typeinfo op2) {
  String operator = get_op();
  switch (operator) {
  case "=": case "+": case"-": {
    tree tk;
    if ((tk = findatoken())!=null)
      System.out.print("line " + tk.tok.lineno + ": ");
    if ((op1.basetype.equals(op2.basetype)) &&
        (op1.basetype.equals("int"))) {
      System.out.println("typecheck "+operator+" on a "+
              op2.str() + " and a "+ op1.str()+
                " -> OK");
      return op1;
  }
    else j0.semerror("typecheck "+operator+" on a "+
            op2.str()+ " and a "+ op1.str()+
              " -> FAIL");
    }
  default: j0.semerror("cannot check " + operator);
  }
return null;
}
```

The operator that the current syntax node represents can usually be ascertained from the node's corresponding non-terminal symbol. In some cases, the actual production rule must also be used. The Unicon implementation of get_op() found in tree.icn is shown here:

```
method get_op()
  return case sym of {
      "Assignment" : "="
      "AddExpr": if rule=1320 then "+" else "-"
      default: fail
  }
end
```

Unicon allows us to return the result that's produced by a case expression. Additive expressions designated by "AddExpr" include both addition and subtraction. The production rule is used to disambiguate. The corresponding Java implementation of get_op() is similar, as given here:

```java
public String get_op() {
  switch (sym) {
  case "Assignment" : return "=";
  case "AddExpr": if (rule==1320) return "+";
                  else return "-";
  }
  return sym;
}
```

The findatoken() method is used from an internal node in the syntax tree to chase down one of its leaves. It recursively dives into the children until it finds a token. The Unicon implementation of findatoken() is as follows:

```
method findatoken()
   if sym==="token" then return self
   return (!kids).findatoken()
end
```

The corresponding Java implementation of findatoken() is shown here:

```java
public tree findatoken() {
  tree rv;
  if (sym.equals("token")) return this;
  for (tree t : kids)
     if ((rv=t.findatoken()) != null) return rv;
  return null;
}
```

Even the basics of type checking, all of which have been shown in this section, have required you to learn a lot of new ways to traverse trees. The fact is, building a programming language or writing a compiler is a big, complex job, and if we showed a complete one for a mainstream language, this book would be much thicker, and longer than our page limit allows.

This chapter presented how to add roughly half of a type checker to Jzero. Running j0 with these additions is not very glamorous; it just lets you see simple type errors get detected and reported.

If you want to see that, download the code from this book's GitHub site, go into the ch7/ subdirectory, and build code with the Make program. By default, Make will build both the Unicon and Java versions. When you run the j0 command with preliminary type checking in place, it produces an output similar to the following. In this case, the Unicon implementation is shown:

```
D:\Users\Clinton Jeffery\books\byopl\github\Build-Your-Own-Programming-Language\ch7>type hello.java

public class hello {
    public static void main(String argv[]) {
        int x;
        x = 0;
        x = x + "hello";
        System.out.println("hello, jzero!");
    }
}

D:\Users\Clinton Jeffery\books\byopl\github\Build-Your-Own-Programming-Language\ch7>j0 hello.java
line 4: typecheck = on a int and a int -> OK
line 5: typecheck + on a String and a int -> FAIL

D:\Users\Clinton Jeffery\books\byopl\github\Build-Your-Own-Programming-Language\ch7>
```

Figure 7.2: The output from the type checker produces OK or FAIL on various operators

Of course, if the program has no type errors, you will see nothing but lines ending with OK. Now, let's consider an aspect of type checking that is encountered when implementing some programming languages, including Unicon: runtime type checks.

Runtime type checks and type inference in Unicon

The Unicon language handles types a lot differently than the Jzero type system described in this chapter. In Unicon, types are not associated with declarations but with values. The Unicon virtual machine code generator does not place type information in symbol tables or do compile-time type checking. Instead, types are represented explicitly at runtime and checked everywhere before a value is used. Explicitly representing type information at runtime is common in interpreted and object-oriented languages, and optional in some semi-object-oriented languages such as C++.

Consider the write() Unicon function. Every argument to write() that isn't a file specifying where to write to must be a string, or be able to be converted into a string. In the Unicon virtual machine, the type information is created and checked at runtime as needed. The pseudocode for the Unicon write() function looks like this:

```
for (n = 0; n < nargs; n++) {
    if (is:file(x[n])) {
        // set the current output file
    } else if (cnv:string(x[n])) {
```

```
        // output the string to the output file
    } else runtime_error("string or file expected")
}
```

For every argument to write(), the preceding code says to either set the current file, convert the argument into a string and write it, or stop with a runtime error. Checking types of things at runtime provides extra flexibility but slows down execution. Keeping type information around at runtime also consumes memory – potentially a lot of memory. To perform a runtime type check, every value in the Unicon language is stored in a descriptor. A descriptor is a struct that contains a value plus an extra word of memory that encodes its type, called the **d-word**. A Boolean expression such as is:file(x) on some Unicon value, x, boils down to performing a check to see whether the d-word says the value is of the file type.

Unicon also has an optimizing compiler that generates C code. The optimizing compiler performs **type inference**, which determines a unique type more than 90% of the time, eliminating the need for most runtime type checks. Consider the following trivial Unicon program:

```
procedure main()
    s := "hello" || read()
    write(s)
end
```

The optimizing compiler knows that "hello" is a string, and read() only returns strings. It can infer that the s variable holds only string values, so this particular call to write() is passed a value that is already a string and does not need to be checked or converted. Type inference is beyond the scope of this book, but it is valuable to know that it exists and that for some languages, it is an important bridge that allows flexible higher-level languages to run at speeds comparable to those of lower-level compiled languages.

Summary

In this chapter, you learned how to represent base types and check the type safety of common operations, such as preventing adding an integer to a function. All of this can be accomplished by traversing the syntax tree.

You learned how to represent types in a data structure and add an attribute to the syntax tree nodes to store that information. You also learned how to write tree traversals that extract type information about variables and store that information in their symbol table entries. You then learned how to calculate the correct type at each tree node, checking whether the types are used correctly in the process. Finally, you learned how to report type errors that you found.

The process of type checking may seem like a thankless job that just results in a lot of error messages, but really, the type information that you compute at each of the operators and function calls in the syntax tree will be instrumental in determining what machine instructions to generate for those tree nodes. Now that you have built a type representation and implemented simple type checks, it is time to consider some more complex operations necessary to check compound types, such as function calls and classes. You will do this in the next chapter.

Questions

1. What purpose does type checking serve, besides just frustrating tired programmers?

2. Why is a structure type (in our case, a class) needed to represent type information? Why can't we just use an integer to represent each type?

3. The code in this chapter outputs lines that report every successful type check with OK. This is very reassuring. Why don't other compilers report successful type checks like this?

4. Java is pickier about types than its ancestor, the C programming language. What are the advantages of being pickier about types, instead of automatically converting them on demand?

Join our community on Discord

Join our community's Discord space for discussions with the authors and other readers:

`https://discord.com/invite/zGVbWaxqbw`

8

Checking Types on Arrays, Method Calls, and Structure Accesses

This is the second of two chapters regarding type checking. The previous chapter introduced type checking for built-in atomic types. In comparison, this chapter will cover more complex type-checking operations.

This chapter will show you how to perform type checks for the arrays, parameters, and return types of method calls in the Jzero subset of Java. Additionally, it covers the type checking of structured types such as classes.

In this chapter, we will cover the following main topics:

- Type-checking arrays
- Checking method calls
- Checking structured type accesses

By the end of the chapter, you will be able to write more sophisticated tree traversals to check types that themselves contain one or more other types. Being able to support such composite types in your programming language is necessary for you to go beyond toy programming languages and into the realm of languages that are useful in the real world. It is time to learn more about type checking. We will begin with the simplest composite type: arrays.

Technical requirements

The code for this chapter is available on GitHub: `https://github.com/PacktPublishing/Build-Your-Own-Programming-Language-Second-Edition/tree/master/ch8`

The Code in Action video for the chapter can be found here: `https://bit.ly/30w1V8I`

Checking operations on array types

An **array** is a sequence of elements that are all the same type. Up to this point, the Jzero language hasn't really supported array types, other than to allow enough syntax for `main()` to declare its array of the `String` parameter. Now, it is time to add support for the remainder of the Jzero array operations, which are a small subset of what Java arrays can do. Jzero arrays are limited to single-dimension arrays created without initializers. To check array operations properly, we will modify the code from the previous chapters so that we can recognize array variables when they are declared, and then check all uses of these arrays to only allow legal operations. Let's begin with array variable declarations.

Handling array variable declarations

The idea that a variable will hold a reference to an array is attached to the variable's type in the recursive grammar rule, in `j0gram.y`, for the non-terminal `VarDeclarator`. The rule in question is the second production rule, which appears after the vertical bar, as follows:

```
VarDeclarator: IDENTIFIER | VarDeclarator '[' ']' {
  $$=j0.node("VarDeclarator",1060,$1); };
```

For this rule, the corresponding code in the class tree's `assigntype()` method adds an `arraytype()` on top of the type that is being inherited, as `assigntype()` recurses into the `VarDeclarator` child node. The Unicon code for this, in the `tree.icn` file, appears as follows:

```
method assigntype(t)
  . . .
    "VarDeclarator": {
      kids[1].assigntype(arraytype(t))
      return
    }
```

The t type being inherited is not discarded. It becomes the element type of the array type that is constructed here. The corresponding Java code in `tree.java` is almost identical:

```
    void assigntype(typeinfo t) {
      . . .
      case "VarDeclarator": {
        kids[0].assigntype(new arraytype(t));
        return;
      }
```

Because it is recursive, this code works for multiple-dimension arrays represented by a chain of VarDeclarator nodes in the syntax tree, although for the sake of brevity, the rest of Jzero will not. Even for single-dimension arrays, things get interesting when you consider how type information is checked when arrays are used in executable code. The first point within the code where you will need to check array types is when an array is created.

Checking types during array creation

Arrays in Java are created with the new expression; this is something that, up to this point, was omitted from Jzero. This entails a new token added to javalex.l for the reserved new word, as shown in the following code:

```
"new"    { return j0.scan(parser.NEW); }
```

Additionally, it entails a new kind of primary expression, called an ArrayCreation expression. This is added in the grammar within j0gram.y, as shown in the following code:

```
Primary:  Literal | FieldAccess | MethodCall |
          '(' Expr ')' { $$=$2;} | ArrayCreation ;
ArrayCreation: NEW Type '[' Expr ']' {
    $$=j0.node("ArrayCreation", 1260, $2, $4); };
```

Having added the new reserved word and defined a tree node for it, it is time to consider how a type is assigned for that expression. Let's consider the creation of an array in the new int [3] Java expression. The int token is being used in an executable expression for the first time, and initially, the code that creates the int token inside token.icn should allocate its type as follows:

```
class token(cat, text, lineno, colno, ival, dval, sval, typ)
  . . .
  initially
    case cat of {
        parser.INT:     typ := typeinfo("int")
        parser.DOUBLE:  typ := typeinfo("double")
        parser.BOOLEAN: typ := typeinfo("boolean")
        parser.VOID:    typ := typeinfo("void")
```

As you can see, the same additions are needed for the other atomic scalar types. The corresponding Java code in the constructor in token.java is shown here:

```
case parser.INT: typ = new typeinfo("int"); break;
case parser.DOUBLE: typ = new typeinfo("double"); break;
case parser.BOOLEAN: typ = new typeinfo("boolean"); break;
case parser.VOID: typ = new typeinfo("void"); break;
```

These additions to the class token take care of the leaves that are providing our base types. The ArrayCreation node's type is calculated with an addition to the checktype() method. In tree. icn, the addition to checktype(), which primarily consists of a call to arraytype(), is shown here:

```
class tree (id,sym,rule,nkids,tok,kids,isConst,stab,typ)

    . . .

    method checktype(in_codeblock)

    . . .

    "ArrayCreation": typ := arraytype(kids[1].typ)
```

The Java code that corresponds to this in the tree.java file is as follows:

```
case "ArrayCreation":
    typ = new arraytype(kids[0].typ); break;
```

So, when a newly created array is used, usually, in an assignment, its array type must match the type that is allowed by the surrounding expression. For example, in the following two lines, the assignment operator on the second line must allow arrays when its type is being checked:

```
int x[];
x = new int[3];
```

The code to allow the assignment of an array variable from an array value is added to the check_types() method in the tree.icn file, as shown here:

```
method check_types(op1, op2)

    . . .

    else if (op1.basetype===op2.basetype==="array") &
            operator==="=" &
            check_types(op1.element_type,
                        op2.element_type) then {
        return op1
        }
```

The code checks that both op1 and op2 are arrays, that we are doing an assignment, and that the element types are OK. Here, a `write()` statement in the then part might be useful for the purposes of testing this chapter's code. However, in a compiler, only type errors will be shown. The corresponding Java addition to the check_types() method in the tree.java file is as follows:

```
else if (op1.basetype.equals("array") &&
        op2.basetype.equals("array") &&
        operator.equals("=") &&
        (check_types(((arraytype)op1).element_type,
            ((arraytype)op2).element_type) != null)) {
    return op1;
}
```

From the examples in this section, it might appear as though type checking is just a bunch of nit-picky attention to detail, but with all of that detail, we prevent many coding errors and ensure faster execution. The recursive call to check_types() on the arrays' element types prevents a program from accidentally assigning an array of string to a variable of type array of int, for example. Now, it is time to consider type checking for array element accesses.

Checking types during array accesses

Array accesses consist of read and write operations on an array's elements using the subscript operator. Here, we need to add syntax support for these operations and build syntax tree nodes before we can perform any type checking on them. Adding array accesses to the grammar consists of adding a non-terminal ArrayAccess and then adding two production rules that use this non-terminal symbol:

- One for assignments that store a value in an array element
- One for expressions that fetch the value from an array element

The changes to the j0gram.y file will appear as follows. They have been reordered in the grammar for clarity:

```
ArrayAccess: Name '[' Expr ']' {
  $$=j0.node("ArrayAccess",1390,$1,$3); };
LeftHandSide: Name | FieldAccess | ArrayAccess ;
Primary:  Literal | FieldAccess | MethodCall | ArrayAccess
    |'(' Expr ')' { $$=$2;} | ArrayCreation ;
```

The square bracket operator that is used to access array elements must check the types of its operands and use them to calculate the result type. The result of an array subscript removes one level of array from the type of the left operand, thereby producing its element type. The addition to the checktype() method in the tree.icn file looks like this:

```
method checktype(in_codeblock)
. . .

    "ArrayAccess": {
        if match("array ", kids[1].typ.str()) then {
            if kids[2].typ.str()=="int" then
                typ := kids[1].typ.element_type
            else stop("subscripting array with ",
                        kids[2].typ.str())
            }
        else stop("illegal subscript on type ",
                    kids[1].typ.str())
        }
```

The preceding code checks that the type of kids[1] is an array type and the type of kids[2] is an integer type. If those are good, the value assigned to this node's typ is the array's element_type. The corresponding Java addition to the checktype() method in the tree.java file is shown here:

```
case "ArrayAccess":
if (kids[0].typ.str().startsWith("array ")) {
    if (kids[1].typ.str().equals("int"))
    typ = ((arraytype)(kids[0].typ)).element_type;
    else j0.semerror("subscripting array with " +
                        kids[1].typ.str());
    }
else j0.semerror("illegal subscript on type " +
                    kids[0].typ.str());
break;
```

In this section, we have demonstrated how to type-check arrays. Fortunately, non-terminal symbols in the grammar, and hence in the syntax tree, make it easy to find the spots where this form of type checking is needed. Now, it is time to look at perhaps the most challenging part of type checking. We will learn how to check the parameters and return types of method calls next.

Checking method calls

The **function call** is the fundamental building block of both imperative and functional programming paradigms. In object-oriented languages, functions are called methods, but they can play
all the same roles that functions can. In addition to this, a set of methods provides an object's
public interface. To type check a method call, both the number and the type of the parameters
must be verified along with the return type.

Calculating the parameters and return type information

The type representation introduced in the previous chapter, *Chapter 7, Checking Base Types*, included a methodtype class that had fields for the parameters and the return type; however, we
haven't yet presented the code to extract that information from the syntax tree and place it into
the type. The parameters and return type of a method are called its **signature**. The grammar rule
where a method signature is declared is the one that builds a MethodHeader node. To calculate
the return type, we need to synthesize it from the MethodReturnVal node. To calculate the parameters, we need to walk to the FormalParmList subtree within MethodDeclarator. You can do
this by adding a call to j0.calctype() to the grammar rule for MethodHeader in j0gram.y; this
is similar to the ones we added earlier for variable declarations:

```
MethodHeader: PUBLIC STATIC MethodReturnVal MethodDeclarator {
    $$=j0.node("MethodHeader",1070,$3,$4);
    j0.calctype($$);
    };
```

The calctype() method in the j0 class has not been modified, but the methods it calls over
in tree.icn have been extended to add more type information, as needed, to handle method
signatures. The calctype() method in the tree class gets a small upgrade to synthesize a leaf's
type from its contained token type, if present. In Unicon, it is the following line added to tree.
icn that assigns typ from tok.typ:

```
method calctype()
    . . .
    "token": {
        if typ := \(tok.typ) then return
```

The corresponding Java addition to calctype() in tree.java is shown here:

```
if ((typ = tok.typ) != null) return;
```

The modifications to the `assigntype()` method for constructing method signatures are more substantial. For variable declarations, you are simply passing the type as an inherited attribute down a list to the individual variables' leaf identifiers. For a method, the type to be associated with the identifier is constructed from the inherited attribute, which is the return type, plus the remainder of the method's signature obtained from the subtree associated with the parameter list. The code in `tree.icn` is:

```
method assigntype(t)
   case sym of {

   . . .

   "MethodDeclarator": {
      parmList := (\(kids[2]).mksig()) | []
      kids[1].typ := typ := methodtype(parmList , t)
      return
   }
```

In this code, the `parmList` parameter list is constructed as a list of types. If the parameter list is not empty, it is constructed by calling the `mksig()` method on that non-empty tree node. If the parameter list is empty, `parmList` is initialized to the empty list, `[]`. The parameter list and the return type of t are passed in to construct the method type that is assigned to the `MethodDeclarator` node and its first child, that is, the identifier method name that will be inserted into the class symbol table. The corresponding Java addition to the `assigntype()` method in `tree.java` is shown here:

```
case "MethodDeclarator":
    typeinfo parmList[];
    if (kids[1] != null) parmList = kids[1].mksig();
    else parmList = new typeinfo [0];
    kids[0].typ = typ = new methodtype(parmList , t);
    return;
```

The `mksig()` method constructs a list of the types of parameters of a method. The `mksig()` method is an example of a very specialized tree method. It is a subtree traversal that only traverses a very narrow subset of all tree nodes. It is only ever called on a formal parameter list and only needs to consider the `FormalParmList` and `FormalParm` nodes as it walks down the parameter list, picking up the types of each parameter. The Unicon code for `mksig()` in `tree.icn` is as follows:

```
method mksig()
    case sym of {
       "FormalParm": return [ kids[1].typ ]
```

```
           "FormalParmList":
              return kids[1].mksig() ||| kids[2].mksig()
           }
   end
```

The `FormalParm` case returns a list of size 1. The `FormalParmList` case returns the concatenation of two recursive calls on its children. The corresponding Java code in `tree.java` is shown here:

```
typeinfo [] mksig() {
  switch (sym) {
  case "FormalParm": return new typeinfo[]{kids[0].typ};
  case "FormalParmList":
    typeinfo ta1[] = kids[0].mksig();
    typeinfo ta2[] = kids[1].mksig();
    typeinfo ta[] = new typeinfo[ta1.length + ta2.length];

    System.arraycopy(ta1, 0, ta, 0, ta1.length);
    System.arraycopy(ta2, 0, ta, ta1.length, ta2.length);
    return ta;
  }
  return null;
}
```

The Java implementation uses arrays. The majority of the preceding code concatenates the two arrays returned from the calls to `mksig()` on the children. This concatenation could be performed by importing `java.util.Arrays` and using utility methods there, but the `Arrays` code is not much shorter or clearer. There is one last tweak to the code that is required to connect all this method-type information and make it usable. When the method is inserted into the symbol table in the `populateSymTables()` method, its type information needs to be stored there. In Unicon, the change in `tree.icn` is shown here:

```
method populateSymTables()
  case sym of {
    . . .
  "MethodDecl": {
      stab.insert(kids[1].kids[2].kids[1].tok.text, ,
                  kids[1].stab, kids[1].kids[2].typ)
```

Compared to previous chapters, the addition of type information is just one extra parameter being passed into the symbol table's `insert()` method. The corresponding Java code in `tree.java` is shown here:

```
stab.insert(s, false, kids[0].stab, kids[0].kids[1].typ);
```

We have constructed the type information for methods when they are declared and made that type information available in the symbol table. Now, let's take a look at how to use type information from various methods to check the types of the actual parameters when they are called.

Checking the types at each method call site

The method call sites can be found in the syntax tree by looking for the two production rules in `j0gram.y` that build a non-terminal `MethodCall`. The rule where a `MethodCall` is a `Name` followed by a parenthesized list of zero or more parameters is shown here. It includes the classic function syntax, which is primarily used to call methods within the same class, as well as qualified names with the `object.function` syntax to invoke a method within another class. This section focuses on type checking for the classic function syntax. The `object.function` syntax is covered in the *Checking structured type accesses* section. The code given here has been amended in that section.

The code to check the types of method calls is added to the `checktype()` method. The Unicon additions to `tree.icn` appear as follows:

```
method checktype(in_codeblock)
   . . .
  "MethodCall": {
     if rule = 1290 then {
       if kids[1].sym ~== "token" then
         stop("can't check type of Name ", kids[1].sym)
       if kids[1].tok.cat == parser.IDENTIFIER then {
         if (\(rv:=stab.lookup(kids[1].tok.text))) then {
           rv := rv.typ
           if not match("method ", rv.str()) then
             stop("method expected, got ", rv.str())
             cksig(rv)
         }
       }
     else stop("can't typecheck token ", kids[1].tok.cat)
     }
     else stop("Jzero does not handle complex calls")
```

```
    }

    . . .
```

In the preceding code, the method is looked up in the symbol table and its type is retrieved. If there are no parameters, the type is checked to ensure that its parameter list is empty. If there are actual parameters in the call, they are checked against the formal parameters via a call to the cksig() method. If that check succeeds, the typ field for this node is assigned from the return_type, which was specified for the method that was called. The corresponding Java code in tree.java is shown here:

```java
case "MethodCall":
  if (rule == 1290) {
    symtab_entry rve;
    methodtype rv;
    if (!kids[0].sym.equals("token"))
      j0.semerror("can't check type of " + kids[0].sym);
    if (kids[0].tok.cat == parser.IDENTIFIER) {
      if ((rve = stab.lookup(kids[0].tok.text)) != null) {
        if (! (rve.typ instanceof methodtype))
          j0.semerror("method expected, got " + rve.typ.str());
        rv = (methodtype)rve.typ;
        cksig(rv);
      }
    }
    else j0.semerror("can't typecheck " + kids[0].tok.cat);
  }
  else j0.semerror("Jzero does not handle complex calls");
  break;
```

The method that is used to check a function's signature and apply its return type is the cksig() method. The Unicon implementation of cksig() in tree.icn is shown here:

```
method cksig(sig)
local i:=*sig.parameters, nactual := 1, t := kids[2]
  if /t then {
    if i ~= 0 then stop("0 parameters, expected ", i)
  }
  else {
    while t.sym == "ArgList" do {
      nactual +:= 1; t:=t.kids[1] }
```

```
      if nactual ~= i then
        stop(nactual " parameters, expected ", i)
      t := kids[2]
      while t.sym == "ArgList" do {
        check_types(t.kids[-1].typ, sig.parameters[i])
        t := t.kids[1]; i-:=1
        }
      check_types(t.typ, sig.parameters[1])
    }
  typ := sig.return_type
end
```

This method first handles zero parameters as a special case; however, aside from that, it checks one parameter at a time in a `while` loop. For each parameter, it calls `check_types()` to check the formal and actual types. Because of the way the syntax tree is constructed, parameters are encountered in reverse order during the tree traversal here. The first parameter is found when you hit a tree node that is not an `ArgList`. After processing the arguments, `cksig()` sets the `MethodCall` node's type to the type returned by the method. The corresponding Java code in `tree.java` appears as follows:

```
void cksig(methodtype sig) {
  int i = sig.parameters.length, nactual = 1;
  tree t = kids[1];
  if (t == null) {
    if (i != 0) j0.semerror("0 params, expected ",i);
  }
  else {
    while (t.sym.equals("ArgList")){
      nactual++;
      t=t.kids[0];
    }
    if (nactual != i)
      j0.semerror(nactual + " parameters, expected "+ i);
    t = kids[1];
    i--;
    while (t.sym.equals("ArgList")) {
      check_types(t.kids[1].typ, sig.parameters[i--]);
      t = t.kids[0];
```

```
    }
  check_types(t.typ, sig.parameters[0]);
}
  typ = sig.return_type;
}
```

The check_types() method and its get_op() helper method need to be tweaked in order to handle parameter type checking. The Unicon implementation of these changes appears in tree.icn as follows:

```
method get_op()
  return case sym of { …
    "MethodCall" : "param"
. . .
method check_types(op1, op2)
    operator := get_op()
    case operator of {
        "param"|"return"|"="|"+"|"-" : {
          . . .
```

The corresponding Java changes to get_op() and check_types() in tree.java are as follows:

```
public String get_op() {
  switch (sym) {
  case "MethodCall" : return "param";
    . . .
public typeinfo check_types(typeinfo op1, typeinfo op2) {
  String operator = get_op();
  switch (operator) {
  case "param": case "return": case "=": case "+":    case"-":
```

So, you have learned how to check the types of parameters passed into method calls, which is one of the most challenging aspects of type checking. Now it is time to check the return types that come out of the function call via its return statements.

Checking the type at return statements

The type of the expressions in the method's return statements must match the method's declared return type. These two locations are quite some distance apart in the syntax tree. There are lots of different ways in which you might connect them.

For example, you could add a `return_type` attribute to all the tree nodes and inherit the type from the `MethodHeader` into the `Block` and down through the code into the return statements. However, that approach is a waste of time for a relatively sparsely used piece of information. The symbol table is the most convenient way to connect remote locations. We can insert a dummy symbol into the symbol table that can hold a function's return type. This dummy symbol can be looked up and checked against the type at every return statement. The dummy symbol named `return` is ideal. It is easy to remember and is a reserved word that will never conflict with a real symbol in user code. The code to insert the return type into the method's symbol table is an addition to the `populateSymTables()` method. The Unicon implementation in `tree.icn` is as follows:

```
method populateSymTables()
case sym of {

  . . .

  "MethodDecl": {
    stab.insert(kids[1].kids[2].kids[1].tok.text, ,
              kids[1].stab, kids[1].kids[2].typ)
    kids[1].stab.insert("return", , ,
         kids[1].kids[1].typ)
    }
```

In this code, `kids[1]` is the `MethodHeader` node. Its `stab` field is the local symbol table being inserted as a subscope inside the enclosing class scope. The `kids[1].kids[1]` expression is the `MethodReturnVal` node, which is usually just the token denoting the return type. The pair of blank spaces separated by commas between `"return"` and the type are `null` values. They are being passed into the second and third parameters of the `insert()` symbol table. The corresponding Java code that is added to the `populateSymTables()` method in `tree.java` is as follows:

```
kids[0].stab.insert("return", false, null,
              kids[0].kids[0].typ);
```

The type-checking code that makes use of this return type information within the return statements is added to the `checktype()` method, which is also in the tree class. The Unicon implementation in `tree.icn` appears as follows:

```
method checktype(in_codeblock)

  ...

  case sym of {

  ...

  "ReturnStmt": {
```

```
          if not (rt := ( \ (stab).lookup("return")).typ) then
              stop("stab did not find a returntype")
          if \ (kids[1].typ) then
              typ := check_types(rt, kids[1].typ)
          else {
            if rt.str() ~== "void" then
              stop("void return from non-void method")
            typ = rt;
          }
      }
```

The corresponding Java code is presented here:

```
  void checktype(Boolean in_codeblock) {
     ...
     switch (sym) {
     ...
     case "ReturnStmt":
        symtab_entry ste;
        if ((ste=stab.lookup("return")) == null)
           j0.semerror("stab did not find a returntype");
        typeinfo rt = ste.typ;
        if (kids[0].typ != null)
          typ = check_types(rt, kids[0].typ);
        else {
          if (!rt.str().equals("void"))
             j0.semerror("void return from non-void method");
        typ = rt;
          }
        break;
```

So, you have learned how to check return statements. Now, it is time to learn how to check the accesses to the fields and methods of a class instance.

Checking structured type accesses

In this book, the phrase **structured type** will denote composite objects that can hold a mixture of types whose elements are accessed by name. This contrasts with arrays, whose elements are accessed by their position and whose elements are of the same type.

In some languages, there are struct or record types for this kind of data. In Jzero and most object-oriented languages, classes are used as the principal structured type.

This section discusses aspects of how to check the types for operations on classes and, more specifically, class instances. This organization mirrors the presentation of array types at the beginning of this chapter, starting with what is needed to process declarations of class variables.

The original intent of Jzero was to support a tiny Java subset that was somewhat comparable to Wirth's PL/0 language. Such a language does not require class instances or object orientation, and space limitations prevent us from covering many of the bells and whistles needed for a feature-rich object-oriented language such as Java or C++. However, we will present some of the highlights. The first thing to consider is how to declare instance variables for class types.

Handling instance variable declarations

Variables of class types are declared by giving a class name and then a comma-separated list of one or more identifiers. For example, our compiler needs to handle declarations that are similar to the following declaration of three strings:

```
String a, b, c;
```

Jzero had to handle such declarations from the beginning since the main() procedure takes an array of strings. Although our Jzero compiler already supports class variable declarations, a few additional considerations are in order.

In many object-oriented languages, variable declarations will have accompanying visibility rules such as public and private. In Jzero, all methods are public and all variables are private, but you could go ahead and implement an isPublic attribute anyway. A similar consideration applies to static variables. Jzero has no static variables, but you could implement an isStatic attribute if you decide you want them. Extending our example to include these two considerations will look like the following:

```
private static String a, b, c;
```

To support these Java attributes, you can add them to tokens, tree nodes, and the symbol table entry type. You can propagate them from the reserved word over to where the variables are declared, just as we did for type information.

Checking types at instance creation

Objects, also called **class instances**, are created using the new reserved word, as was the case for arrays, which we discussed in the *Checking operations on array types* section. The additions to the grammar in j0gram.y are shown here:

```
Primary:  Literal | FieldAccess | MethodCall | ArrayAccess
    |'(' Expr ')' { $$=$2;} | ArrayCreation | InstanceCreation;
InstanceCreation: NEW Name '(' ArgListOpt ')' {
  $$=j0.node("InstanceCreation", 1261, $2, $4); };
```

This added syntax enables instance creation. To calculate the type of the expression so that it can be checked, we need to look up the type of the class in the symbol table. For that to work, at an earlier point in time, we must construct the corresponding classtype object and associate it with the class name in the enclosing symbol table.

Instead of embedding code to construct the class type with subtree traversals during parsing, as we did in the preceding sections to construct the signature for a method, for a class, it is easier to wait until after parsing and populating the symbol table, that is, just before type checking. That way, all the information for constructing the class type is ready for us in the class symbol table. A call to a new mkcls() method is added to the semantic() method in j0.icn, after the symbol table processing and before type checking, as follows:

```
method semantic(root)
   .  .  .
  root.checkSymTables()
  root.mkcls()
  root.checktype()
```

The corresponding Java addition to j0.java is shown here:

```
  root.mkcls();
```

The mkcls() method stands for *make class*. When it sees a class declaration, it looks up the class name and goes through the class symbol table, putting entries into the correct category. There is one list for fields, one for methods, and one for constructors. The Unicon implementation of mkcls() from tree.icn is shown here:

```
method mkcls()
    if sym == "ClassDecl" then {
```

```
          rv := stab.lookup(kids[1].tok.text)
          flds := []; methds := []; constrs := []
          every k := key(rv.st.t) do
             if match("method ", rv.st.t[k].typ.str()) then
                put(methds, [k, rv.st.t[k].typ])
             else put(flds, [k, rv.st.t[k].typ])
        /(rv.typ) := classtype(kids[1].tok.text, rv.st,
                                 flds, methds, constrs)
     }
   else every k := !kids do
      if k.nkids>0 then k.mkcls()
 end
```

When this traversal hits a class declaration, it looks up the class name and fetches the symbol table for that class. Every symbol is checked, and if it is a method, it goes on the list of methods named methds; otherwise, it goes on the list of fields, named flds. The class type in the class's symbol table entry is assigned an instance of a class type that holds all of this information. You might notice that constructors are not identified and placed on the constructor list. It is OK for Jzero to not support constructors, but a larger subset of Java would support at least one constructor for each class. In any case, the corresponding Java version is shown as follows:

```
void mkcls() {
  symtab_entry rv;
  if (sym.equals("ClassDecl")) {
    int ms=0, fs=0;
    rv = stab.lookup(kids[0].tok.text);
    for(String k : rv.st.t.keySet()) {
      symtab_entry ste = rv.st.t.get(k);
      if ((ste.typ.str()).startsWith("method ")) ms++;
      else fs++;
    }
    parameter flds[] = new parameter[fs];
    parameter methds[] = new parameter[ms];
    fs=0; ms=0;
    for(String k : rv.st.t.keySet()) {
    symtab_entry ste = rv.st.t.get(k);
    if ((ste.typ.str()).startsWith("method "))
       methds[ms++] = new parameter(k, ste.typ);
```

```
        else flds[fs++] = new parameter(k, ste.typ);
      }
      rv.typ = new classtype(kids[0].tok.text,
                    rv.st, flds, methds, new typeinfo[0]);
    }
    else for(int i = 0; i<nkids; i++)
      if (kids[i].nkids>0) kids[i].mkcls();
  }
```

There is one more piece of code that is needed to complete the handling of instance creation. The type field has to be set for the InstanceCreation nodes in the checktype() method. After all the work of placing the type of information for the class in the symbol table, this is a simple lookup. The Unicon implementation in tree.icn looks like this:

```
method checktype(in_codeblock)
  . . .
  "InstanceCreation": {
    if not (rv := stab.lookup(kids[1].tok.text)) then
      stop("unknown type ",kids[1].tok.text)
    if not (typ := \ (rv.typ)) then
      stop(kids[1].tok.text, " has unknown type")
  }
```

The preceding code is just a *symbol table lookup* that includes the fetching of the type from the symbol table entry, plus lots of error checking. The corresponding Java additions in tree.java appear as follows:

```
void checktype(boolean in_codeblock) {
  ...
  case "InstanceCreation":
    symtab_entry rv;
    if ((rv = stab.lookup(kids[0].tok.text))==null)
      j0.semerror("unknown type " + kids[0].tok.text);
    if ((typ = rv.typ) == null)
      j0.semerror(kids[0].tok.text +
              " has unknown type");
    break;
```

So, you have learned how to construct type information for classes and use it to produce the correct type at instance creation. Now, let's explore what it will take to support access to names defined within the instance.

Checking types of instance accesses

Instance accesses refer to references to the fields and methods of an object. There are implicit accesses, where a field or method of the current object is referenced directly by name, and explicit accesses, where the dot operator is used to access an object through its public interface. Implicit accesses are handled by regular symbol table lookups in the current scope, which will automatically try to enclose scopes, including the class scope where the current object's class methods and variables can be found. This section is about explicit access using the dot operator. In the j0gram.y grammar, these are called QualifiedName nodes. Adding support for qualified names begins by modifying the MethodCall code in the class tree's checktype() method. The code presented earlier in this chapter for method signature checking on simple names is put into an else clause. The Unicon implementation in tree.icn adds the following lines:

```
method checktype(in_codeblock)

  . . .

  "MethodCall": {
    if rule = 1290 then {
      if kids[1].sym == "QualifiedName" then {
        rv := kids[1].dequalify()
        cksig(rv)
      }
      else {
        if kids[1].sym ~== "token" then
          ...
        else stop("can't check type of ",
                kids[1].tok.cat)
      }
```

The code in checktype() recognizes qualified names when used as the name of the method being called, and it calls a dequalify() method to obtain the type of the dotted name. It then uses the signature-checking method, cksig(), as presented earlier, to check the types at the call. The corresponding Java code in tree.java is as follows:

```
void checktype(boolean in_codeblock) {
  ...
```

```
          if (kids[0].sym.equals("QualifiedName")) {
            rv = (methodtype)(kids[0].dequalify());
            cksig(rv);
            }
          else {
              …
          }
```

`kids[0]` is a tree node with two children. The type of the left child contains the symbol table within which we look up the right child to find its method type. The `dequalify()` method does this dirty work. The Unicon implementation in `tree.icn` looks like this:

```
    method dequalify()
    local rv, ste
      if kids[1].sym == "QualifiedName" then
        rv := kids[1].dequalify()
      else if kids[1].sym=="token" &
              kids[1].tok.cat=parser.IDENTIFIER then {
        if not \ (rv := stab.lookup(kids[1].tok.text)) then
            stop("unknown symbol ", kids[1].tok.text)
        rv := rv.typ
      }
      else stop("can't dequalify ", sym)
      if rv.basetype ~== "class" then
        stop("can't dequality ", rv.basetype)
      if \ (ste := rv.st.lookup(kids[2].tok.text)) then
        return ste.typ
      else stop(kids[2].tok.text, " is not in ", rv.str())
    end
```

This method first calculates the type for the left-hand side operand. This requires a recursion if the left operand is another qualified name. Otherwise, the left operand must be an identifier that can be looked up in the symbol table. Either way, the left operand's type is checked to make sure it is a class, and if so, the identifier on the right-hand side of the dot is looked up in that class and its type is returned. The corresponding Java implementation in `tree.java` is shown here:

```
    public typeinfo dequalify() {
        typeinfo rv = null;
```

```
    symtab_entry ste;
    if (kids[0].sym.equals("QualifiedName"))
      rv = kids[0].dequalify();
    else if (kids[0].sym.equals("token") &
        (kids[0].tok.cat==parser.IDENTIFIER)) {
    if ((ste = stab.lookup(kids[0].tok.text)) != null)
      j0.semerror("unknown symbol " + kids[0].tok.text);
    rv = ste.typ;
    }
    else j0.semerror("can't dequalify " + sym);
    if (!rv.basetype.equals("class"))
      j0.semerror("can't dequalify " + rv.basetype);
    ste = ((classtype)rv).st.lookup(kids[1].tok.text);
    if (ste != null) return ste.typ;
    j0.semerror("couldn't lookup " + kids[1].tok.text +
        " in " + rv.str());
    return null;
  }
}
```

In this section, you learned how to handle structure accesses. We included a type-checking consideration where variables of a class type were declared and instantiated. Then, you learned how to calculate the types of qualified names within objects.

After all of this type checking, the output is, once again, a bit anticlimactic. You can download the code from the book's Github site, navigate to the ch8/ subdirectory, and build it with the make program. This will build both the Unicon and the Java versions. As a reminder, you will have to configure installed software and/or set your CLASSPATH to the directory where you unpacked the book examples, as discussed from *Chapter 2*, *Programming Language Design*, to *Chapter 5*, *Syntax Trees*. When you run the j0 command with type checking in place, it produces an output that is similar to the following:

```
>type funtest.java
public class funtest {
    public static int foo(int x, int y, String z) {
        return 0;
    }
    public static void main(String argv[]) {
        int x;
        x = foo(0,1,"howdy");
        x = x + 1;
        System.out.println("hello, jzero!");
    }
}


>java ch8.j0 funtest.java
line 3: typecheck return on a int and a int -> OK
line 7: typecheck param on a String and a String -> OK
line 7: typecheck param on a int and a int -> OK
line 7: typecheck param on a int and a int -> OK
line 7: typecheck = on a int and a int -> OK
line 8: typecheck + on a int and a int -> OK
line 8: typecheck = on a int and a int -> OK
line 9: typecheck param on a String and a String -> OK
no errors
```

Figure 8.1: Type checking on parameters and return types

If the program has no type errors, all the lines will end with OK. In later chapters, Jzero will not bother to output when successful type checks occur, so this will be the last you see of these OK lines.

Summary

This chapter was the second of two chapters covering various aspects of type checking. You learned how to represent compound types. For example, you learned how to build method signatures and use them to check method calls. All of this is accomplished via traversals of the syntax tree, and much of it involves adding minor extensions to the functions presented in the previous chapter.

This chapter also showed you how to recognize array declarations and build the appropriate type representations for them. You learned how to check whether correct types are being used for array creation and access and to build type signatures for method declarations. You also learned how to check that the correct types are being used for method calls and returns.

Congratulations on making it this far! At this point, we have completed our discussion of the front half of a programming language implementation, in which input source code is analyzed. Notations and tools helped greatly with lexical and syntax analysis, after which almost everything has been tree construction and traversals.

While writing fancier tree traversal functions is a valuable skill in its own right, representing type information and propagating it around the syntax tree to where it is needed also makes excellent practice of the skills you will need for the next steps in your compiler. Now that you have implemented type checking, you are ready to move on to code generation. This denotes the midpoint in your programming language implementation. So far, you have been gathering information about the program. The next chapter begins working toward the translated output of the input program, starting with intermediate code generation.

Questions

1. What are the main differences between checking the types of array accesses and checking the types of struct or class member accesses?

2. How do a function's return statements know what type they are returning? They are often quite far away in the tree from the location where the function's return type is declared.

3. How are types checked during a function call? How does this compare with type checking operators such as plus and minus?

4. Besides accesses via the [ ] and . operators, what other forms of type checking are necessary for arrays, structures, or class types?

Join our community on Discord

Join our community's Discord space for discussions with the authors and other readers:

https://discord.com/invite/zGVbWaxqbw

9

Intermediate Code Generation

After the semantic analysis is complete, you can contemplate how to execute the program. For compilers, the next step is to produce a sequence of machine-independent instructions called intermediate code. This chapter starts by defining intermediate code and then shows you how to generate intermediate code, by looking at examples for the Jzero language. After the preceding chapters, where you learned how to write tree traversals that analyze and add information to the syntax tree constructed from the input, in this chapter we finally begin the process of constructing the compiler's output. Intermediate code generation is usually followed by an optimization phase and final code generation for a target machine.

This chapter covers the following main topics:

- What is intermediate code?
- An intermediate code instruction set
- Generating code for expressions
- Generating code for control flow

We will start by gaining some perspective on what intermediate code is, and why it is so useful. You can think of intermediate code generation as the process of preparing for final code generation.

Technical requirements

The code for this chapter is available on GitHub: https://github.com/PacktPublishing/Build-Your-Own-Programming-Language-Second-Edition/tree/master/ch9

The Code in Action video for the chapter can be found here: https://bit.ly/30t3gNQ

What is intermediate code?

At this point, you may be asking, what is intermediate code, and what will I need to generate it? We previously defined it as a sequence of machine-independent instructions, but what does that mean? A sequence can be represented by either an array or a linked list, or possibly something fancier; in Unicon, it will just be a list, while in Java, it will be an ArrayList. But what are the elements – these machine-independent instructions? Like a machine-dependent instruction, a machine-independent instruction has an opcode and zero or more data values used as operands. The difference is that instructions for a real machine have a very specific and precise binary layout in one or more bytes, using machine-specific registers and memory addressing modes. In contrast, machine-independent instructions refer to data values in a more abstract way, as values to be found at some location in memory.

The act of generating intermediate code produces enough information to enable the task of generating the final code that can be run. Like many things in life, a daunting task becomes possible when you prepare well. Eager developers might want to skip this phase and jump straight to final code generation, so let's consider why intermediate code generation is so advantageous. Generating final machine code is a complex task, and most compilers use intermediate code to break the work up into stages to complete it successfully. This section will show you the details of what and why, starting with some specific technical motivations to generate intermediate code.

Why generate intermediate code?

The goal of this phase of your compiler is to produce a list of machine-independent instructions for each method in the program. Generating preliminary machine-neutral code as an intermediate representation of a program's instructions has the following benefits:

- It allows you to identify the memory regions and byte offsets in which variables will be stored before worrying about machine-specific details, such as registers and addressing modes.

- It allows you to work out most of the details of the control flow, such as identifying where labels and go-to instructions will be needed.

- Including intermediate code in a compiler reduces the size and scope of the CPU-specific code, improving the portability of your compiler to new architectures.

- It allows you to check your work up to this point and provides output in a human-readable format, before you get bogged down in low-level machine code.

- Generating intermediate code allows for a wide range of optimizations to be applied before machine-specific final code generation. Optimizations that are made to intermediate code benefit all final code generators that you target after this point.

Now, let's look at the memory regions that are introduced and manipulated in the intermediate code.

Learning about the memory regions in the generated program

In an interpreter, an address in the user's program refers to memory within the interpreter's address space and can be manipulated directly. A compiler has the more difficult challenge of reasoning about addresses that are abstractions of memory locations in future executions of the generated program. At compile time, the user program's address space does not exist yet, but when it does, it will be organized in a similar way to the following:

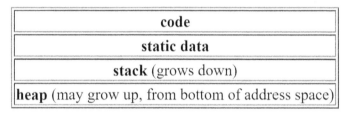

Figure 9.1: Runtime memory regions

Some addresses will be in static memory, some on the stack, some in the heap, and some in the code. In general terms, the uses of these regions are summarized as follows:

- code: Code regions are usually read-only static regions that contain nothing but code.
- static: Static data regions contain things like global variables and constants.
- stack: Stack regions contain parameters, saved registers, and local variables associated with function calls and returns. They allocate and deallocate memory by growing and shrinking at one end.
- heap: Heap regions contain data objects that are allocated on demand and returned and reused when they are no longer needed.

In practice, these fundamental regions are further divided into subregions for specific purposes. In the final code, the means by which these regions are accessed often differs, but for intermediate code addresses, we just need a way to tell what region each address lives in.

We could use integer codes to represent these different memory regions, but in Unicon and Java, a string name is a direct human-readable way to designate them. The (sub)regions we will use for intermediate code and their interpretations are shown here:

- "loc": In the local region, the offset is relative to the top of the stack. For example, it will probably be accessed relative to a stack frame pointer register.

- "global": The global region holds statically allocated variables. The offset is relative to the start of a data region that is fixed at load time. Depending on your final code, it may be resolved to an absolute address.

- "strings": The strings region holds statically allocated read-only values. Aside from being read-only, its properties are like those of the global region. It typically holds strings and other constant structured data; small constants belong in the "imm" pseudo region, defined below.

- "lab": A unique integer label is used to abstract an offset relative to the start of the code region, which is usually a read-only static region. Labels are resolved to an absolute address in the final code, but we let the assembler do the work of calculating the byte offsets. In intermediate code, as in assembler code, labels are just names for machine instructions.

- "class": The offset is relative to the start of some object allocated from the heap, meaning it will be accessed relative to another address. For example, an object-oriented language might address instance variables as offsets relative to a self or this pointer. The self or this pointer might in turn be stored in a dedicated register, a fixed offset from the frame pointer register, or a global variable.

- "imm": The pseudo-region for immediate values denotes that the offset is the actual value, not an address. Most CPUs have limits on the size of values they can store in the instruction itself. The mapping of the immediate pseudo-region into the final code will typically attempt to map values to the instruction if the instruction set supports it; otherwise, they will be mapped into memory as additions to the strings region.

Regions are not very difficult once you are used to them. Now, let's look at how they are used in the data structure that the compiler uses to represent addresses in the generated code.

Introducing data types for intermediate code

The most common form of intermediate code used in compilers is **three-address code**. Each instruction will contain an opcode and from zero to three operands, which are the values that are used by that instruction, usually an address.

You might wonder, why is the number three, and not four, or two? Skipping the obligatory Monty Python joke here, human arithmetic is largely based on binary operators in which two source values are used to produce a third value as a result. Actual hardware CPU instruction sets may use three addresses per instruction or some other number, but for machine-independent intermediate code, three is a reasonable number of operands; hence, we will stick with three-address code, which is mainstream.

For the Jzero compiler, we will define a class called address that represents an address as a region and an offset. The Unicon implementation of the address class begins in the address.icn file, as shown here:

```
class address(region, offset)
end
```

The corresponding Java version requires us to decide what types to use for the region and offset. We are using strings to represent regions, while an offset is typically a distance in bytes from the start of a region, so it can be represented by an integer. The Java implementation of the address class in address.java is as follows:

```
public class address {
    public String region;
    public int offset;
    address(String s, int x) { region = s; offset = x; }
}
```

We will add methods to this class for use later. Given this representation of addresses, we can define our three-address code in a class called tac, which consists of an opcode and up to three addresses. Not all opcodes will use all three addresses. The Unicon implementation of the tac class in tac.icn is shown here:

```
class tac(op, op1, op2, op3)
end
```

The corresponding Java implementation in tac.java is as follows. Several overloaded constructors are provided for different numbers of operands:

```
public class tac {
  String op;
  address op1, op2, op3;
    tac(String s) { op = s; }
```

```
    tac(String s, address o) { op = s; op1 = o; }
    tac(String s, address o1, address o2) {
        op = s; op1 = o1; op2 = o2; }
    tac(String s, address o1, address o2, address o3) {
        op = s; op1 = o1; op2 = o2; op3 = o3; }
}
```

To make it convenient to assemble lists of three-address instructions, we will add a factory method named gen() to the class tree, which creates a single three-address instruction and returns a new list of size one that contains it. One interesting question that comes to mind at this point is whether to use Unicon's built-in list type and Java's ArrayList class, or implement an explicitly linked list representation. An explicitly linked list representation would keep the Unicon and Java code closer in sync and facilitate some sharing of sublists. Plus, to be honest, I am a little bit ashamed at the thought of using Java's ArrayList get() and set(), length versus length() versus size(), and so forth.

On the other hand, if we roll our own linked lists, we will be wasting space and time on relatively low-level code for basic operations that the implementation language should provide. So, we will use the built-in list type in Unicon and ArrayList in Java and see how well they perform.

The Unicon implementation in tree.icn is shown here:

```
method gen(o, o1, o2, o3)
    return [ tac(o, o1, o2, o3) ]
end
```

The Unicon version does not have to do anything to allow arguments to be omitted and initializes omitted o1...o3 parameters to the null value. The corresponding Java implementation in tree.java uses variable argument method syntax. It looks like this:

```
ArrayList<tac> gen(String o, address ... a) {
    ArrayList<tac> L = new ArrayList<tac>();
    tac t = null;
    switch(a.length) {
        case 3: t = new tac(o, a[0], a[1], a[2]); break;
        case 2: t = new tac(o, a[0], a[1]); break;
        case 1: t = new tac(o, a[0]); break;
        case 0: t = new tac(o); break;
        default: j0.semerr("gen(): wrong # of arguments");
    }
```

```
    L.add(t);
    return L;
}
```

The preceding examples demonstrate two ways that Java awkwardly supports methods with a variable number of arguments. First, there is method overloading: the tac class has four different constructors to accommodate a different number of arguments. On the other hand, the gen() method uses Java's variable argument syntax, which provides a weird array-like thing that is not an array to hold the arguments to the method.

Three-address code instructions are easily mapped down into short sequences of 1–2 native instructions, and computers with complex instruction sets often have instructions that have three operands and more direct correspondence to three-address code. Now, let's look at how to augment tree nodes to include information needed for intermediate code, including these three-address instructions.

Adding the intermediate code attributes to the tree

In the previous two chapters, we added symbol table scope and type information to the syntax tree nodes. Now, it is time to add representations for several pieces of information needed for code generation.

For every tree node that contains intermediate code, a field named icode will denote the list of code instructions that correspond to executing the code for that subtree.

For expressions, a second attribute named addr will denote the address where the computed value of the expression can be found after that expression executes.

For every tree node that contains intermediate code, the first field denotes the label to use as a target when the control flow should execute at the beginning of that code, and the follow field denotes the label to use as a target when the control flow executes whatever instruction logically comes immediately after that code.

Lastly, for every tree node that represents a Boolean expression, the onTrue and onFalse fields will hold labels to use as targets when that Boolean expression is found to be true or false, respectively. These names were chosen to avoid the reserved words true and false in Java.

In Unicon, adding these attributes to the class tree in tree.icn leaves us with the following:

```
class tree (id,sym,rule,nkids,tok,kids,isConst,stab,
            typ,icode,addr,first,follow,onTrue,onFalse)
```

Our tree nodes are getting fatter and fatter. While we may have to allocate thousands of them to compile a program, on a machine with gigabytes of main memory, the memory cost of these added attributes generally will not be noticeable. The corresponding Java additions to tree.java look like this:

```
class tree {
    . . .
    typeinfo typ;
    ArrayList<tac> icode;
    address addr, first, follow, onTrue, onFalse;
```

At this point, you might be wondering how we plan to calculate these new attributes. The answer is mostly that they are synthesized via a post-order tree traversal, which we will look at in the following sections. However, there will be a few wrinkles.

Generating labels and temporary variables

A couple of helper methods will prove instrumental during intermediate code generation. You can think of them as factory methods if you want; a factory method is a method that constructs an object and returns it. In any case, we need one for labels, to facilitate control flow, and one for temporary variables. Let's call them genlabel() and genlocal().

The label generator, genlabel(), generates a unique label. A unique integer can be obtained from serial.getid(), so genlabel() can, for example, concatenate the string "L" with the result from a call to that method. It is an interesting question whether genlabel() should return the label as an integer or string, an address, a three-address instruction, or a list containing a three-address instruction. The right answer is pr obably an address. The Unicon code for genlabel() in tree.icn might look like this:

```
method genlabel()
    return address("lab", serial.getid())
end
```

The corresponding Java method in tree.java is as follows:

```
address genlabel() {
    return new address("lab", serial.getid());
}
```

The temporary variable generator, `genlocal()`, needs to reserve a chunk of memory in the local region. Logically, this entails memory allocation on the top of the stack in some future address space when the generated program executes at a later date. This is heady stuff. In practice, a stack allocation is made in a big chunk whenever a method is called. The compiler calculates how big that chunk needs to be for each method, including all the local variables within the program, as well as the temporary variables that are used to calculate the partial results during the various operators, when the expressions in the method are executed.

Each local variable requires some number of bytes, but for this book, the units allocated are full 64-bit words whose address is an even multiple of eight; some CPUs require this property. Offsets are reported in bytes, but if you need a byte, you round up and allocate a word. The symbol table is where Jzero allocates local variables. In the tree class code, methods can invoke `genlocal()` from the symbol table with the `stab.genlocal()` expression. To implement `genlocal()`, symbol table entries are extended to keep track of the address that each new variable occupies, and the symbol table itself tracks how many bytes have been allocated in total. Whenever a request for a new variable comes in, we allocate the number of words it requires, and we increment a counter by that amount.

As given, `genlocal()` allocates a single word and produces an address for it. For a language that allocates multi-word entities on the stack, `genlocal()` can be extended to take a parameter that specifies the number of words to allocate, but since Jzero allocates arrays and class instances from the heap, Jzero's `genlocal()` can get away with allocating one eight-byte word each call.

Symbol table entries are extended with an address field named `addr`. The Unicon addition to `symtab_entry.icn` is shown here:

```
class symtab_entry(sym,parent_st,st,isConst,typ,addr)
```
The Java addition to **symtab_entry.java** looks like this:
```
public class symtab_entry {
    . . .
    address addr;
    . . .
    symtab_entry(String s, symtab p, boolean iC,
        symtab t, typeinfo ti, address a) {
      sym = s; parent_st = p; isConst = iC;
      st = t; typ = ti; addr = a;
```

The symbol table class gets a byte counter for how many bytes have been allocated within the region corresponding to the symbol table. Symbol table insertion places an address in the symbol table entry and increments the counter. A call to `genlocal()` inserts a new variable. As written, this method allows only for the creation of temporary variables to hold the results of operations that produce integers. A language larger than Jzero might need methods that create additional types of local variables. The Unicon implementation in `symtab.icn` is shown here:

```
class symtab(scope, parent, t, count)
   . . .
  method insert(s, isConst, sub, typ)
      . . .
      t[s] := symtab_entry(s, self, sub, isConst, typ,
                         address(scope,count))
            count +:= 8
      . . .
  end
  method genlocal()
  local s := "__local$" || count
     insert(s, false, , typeinfo("int"))
     return t[s].addr
  end
initially
  t := table()
  count := 0
end
```

The preceding change to the `insert()` method passes in the address at the top of the region to the `symtab_entry` constructor whenever a variable is allocated, and then increments the counter to allocate space for it. The addition of the `genlocal()` method consists of inserting a new variable and returning its address. The temporary variable has a dollar symbol in it, $, so that the name cannot appear as a regular variable name in the source code. The Java implementation of this addition to `symtab.java` consists of the following changes:

```
public class symtab {
   . . .
  int count;
```

```
    . . .
  void insert(String s, Boolean iC, symtab sub,
             typeinfo typ){
      . . .
      t.put(s, new symtab_entry(s, this, iC, sub, typ,
                            new address(scope,count))));
      count += 8;
    }
  }
  address genlocal() {
          String s = "__local$" + count;
          insert(s, false, null, new typeinfo("int"));
          return t.get(s).addr;
  }
```

With the helper methods for generating labels and temporary variables in place, let's look at an intermediate code instruction set.

An intermediate code instruction set

Intermediate code is like machine-independent assembler code for an abstract CPU. The instruction set defines a set of opcodes. Each opcode specifies its semantics, including how many operands it uses and what state changes occur from executing it. Because this is intermediate code, we do not have to worry about registers or addressing modes – we can just define state changes in terms of what modifications must occur in the main memory. The intermediate code instruction set includes both regular instructions and pseudo instructions, as is the case for other assembler languages. Let's look at a set of opcodes for the Jzero language. There are two categories of opcodes: instructions and declarations.

Instructions

Except for immediate mode, the operands of instructions are addresses. Based on their operand position, most instructions implicitly dereference (read) and assign (write) values in memory located at those addresses. On typical modern machines, units of words are 64 bits. Offsets are given in bytes:

Opcode	C equivalent	Description
ADD,SUB,MUL,DIV	x=y op z	Store result of binary operation on y and z to x
NEG	x = -y	Store result of unary operation on y to x
ASN	x = y	Store y to x
ADDR	x = &y	Store address of y to x
LCON	x = *y	Store contents pointed to by y to x
SCON	*x = y	Store y to location pointed to by x
GOTO	goto L	Unconditional jump to L
BLT,BLE,BGT,BGE	if(x rop y)goto L	Test relation and conditionally jump to L
BIF	if (x) goto L	Conditionally jump to L if x != 0
BNIF	if (!x) goto L	Conditionally jump to L if x == 0
PARM		Store x as a parameter (push onto call stack)
CALL	x=p(...)	Call procedure p with n words of parameters
RET	return x	Return from function with result x

Figure 9.2: Different opcodes, the C equivalents, and their descriptions

Next, we will have a look at some of the declarations.

Declarations

Declarations and other pseudo-instructions typically associate a name with some amount of memory in one of the memory regions of the program. The following are some declarations and their descriptions:

Declaration	Description
glob x,n	Declare a global variable named x that refers to offset n in the global region
proc x,n1,n2	Declare a procedure x with n1 words of parameters and n2 words of locals
loc x,n	Declare a local variable named x that refers to offset n in the local region
lab Ln	Declare a label Ln that will be a name for an instruction in the code region
end	Declare the end of the current procedure

Figure 9.3: Declarations and their descriptions

These instructions and declarations are general and able to express a variety of computations. Input/output could be modeled by adding instructions or by making runtime system calls. We will make use of this instruction set substantially later in this chapter, starting in the *Generating code for expressions* section. But first, we must compute some more attributes in our syntax tree that are needed for control flow.

Annotating syntax trees with labels for control flow

The code for some tree nodes will be sources or targets of control flow. To generate code, we need a way to generate the labels at the targets and propagate that information to the instructions that will go to those targets. It makes sense to start with the attribute named first. The first attribute holds a label to which branch instructions can jump to execute a given statement or expression. It can be synthesized by brute force if need be; if you had to, you could just allocate a unique label to each tree node. The result would be replete with redundant and unused labels, but it would work. For most nodes, the first label can be synthesized from one of their children, instead of allocating a new one.

Consider the additive expression e1 + e2, which builds a non-terminal named AddExpr. If there was any code in e1, it would have a first field, and that would be the label to use for the first field of the entire AddExpr. If e1 had no code, for example, because it was a simple variable or constant, e2 might have some code and supply the first field for the parent. If neither subexpression has any code, then we need to generate a new label for whatever code we generate in the AddExpr node that performs the addition. Similar logic applies to other operators. The Unicon implementation of the genfirst() method in tree.icn looks like this:

```
method genfirst()
   every (!\kids).genfirst()
   case sym of {
   "UnaryExpr": first := \kids[2].first | genlabel()
   "AddExpr"|"MulExpr": first := \kids[1|2].first |
                                  genlabel()
   . . .
   default: first := (!\kids).first
   }
end
```

The case branches in the preceding code rely on Unicon's goal-directed evaluation. A non-null test is applied to children's first fields for those children that may have code. If those non-null tests fail, genlabel() is called to assign first if this node generates an instruction.

The default, which is good for a lot of non-terminals higher up in the grammar, is to assign `first` if a child has one but not to call `genlabel()`. The corresponding Java code in `tree.java` looks like this:

```java
void genfirst() {
   if (kids != null) for(tree k:kids) k.genfirst();
   switch (sym) {
     ...
     case "AddExpr": case "MulExpr": case "RelExpr": {
        if (kids[0].first != null) first = kids[0].first;
        else if (kids[1].first != null)
                  first = kids[1].first;
        else first = genlabel();
     }
     . . .
   }
}
```

In addition to the `first` attribute, we need an attribute named `follow` that denotes the label to jump to for whatever code immediately comes after a given block. This will help implement statements such as if-then, as well as break statements. The `follow` attribute propagates information from ancestors and siblings rather than children. The implementation must use an inherited attribute, instead of a synthesized one. Instead of a simple bottom-up post-order traversal, information is copied down in a pre-order traversal, as was seen previously for copying type information into variable declaration lists. The `follow` attribute uses `first` attribute values and must be computed after `genfirst()` has been run.

Consider the most straightforward grammar rule where you might define a `follow` attribute. In the Jzero grammar, the basic rule for statements executing in sequence consists of the following:

```
BlockStmts : BlockStmts BlockStmt ;
```

For an inherited attribute, the parent (`BlockStmts`, which is to the left of the colon) is responsible for providing the `follow` attribute for the two children. The left child's `follow` will be the first instruction in the right child, so the attribute is moved from one sibling to the other. The right child's `follow` will be whatever follows the parent, so it is copied down. Once these values have been set, the parent must have the children do the same for their children, if any. The Unicon implementation in `tree.icn` is shown here:

```
method genfollow()
```

```
    case sym of {
    . . .
    "BlockStmts": {
       kids[1].follow := kids[2].first
       kids[2].follow := follow
       }
    . . .
    }
    every (!\kids).genfollow()
  end
```

The corresponding Java code in tree.java looks like this:

```
void genfollow() {
  switch (sym) {
  . . .
  case "BlockStmts": {
     kids[0].follow = kids[1].first;
     kids[1].follow = follow;
     break;
     }
  . . .
  }
  if (kids != null) for(tree k:kids) k.genfollow();
}
```

Computing these attributes enables the generation of instructions for the control flow that goes to these various labels. You may have noticed that a lot of these first-and-follow labels might never be used. We can either generate them all anyway, or we can devise a mechanism to only emit them when they are an actual target of a branch instruction. Before we move on to code generation for the challenging control flow instructions that use these labels, let's consider the simpler problem of generating code for ordinary arithmetic and similar expressions.

Generating code for expressions

The easiest code to generate is straight-line code consisting of statements and expressions that execute in sequence with no control flow. As described earlier in this chapter, there are two attributes to compute for each node: the attribute for where to find an expression's value is called addr, while the intermediate code necessary to compute its value is called icode.

The values to be computed for these attributes for a subset of the Jzero expression grammar are shown in the following table. The ||| operator refers to list concatenation:

Production	Semantic Rules

Assignment : IDENT '=' AddExpr	`Assignment.addr = IDENT.addr`
	`Assignment.icode = AddExpr.icode |||`
	`                    gen(ASN, IDENT.addr, AddExpr.addr)`

AddExpr : AddExpr$_1$ '+' MulExpr	`AddExpr.addr = genlocal()`
	`AddExpr.icode = AddExpr`$_1$`.icode ||| MulExpr.icode |||`
	`                    gen(ADD,AddExpr.addr,AddExpr`$_1$`.addr,MulExpr.addr)`

AddExpr : AddExpr$_1$ '-' MulExpr	`AddExpr.addr = genlocal()`
	`AddExpr.icode = AddExpr`$_1$`.icode ||| MulExpr.icode |||`
	`                    gen(SUB,AddExpr.addr,AddExpr`$_1$`.addr,MulExpr.addr)`

MulExpr : MulExpr$_1$ '*' UnaryExpr	`MulExpr.addr = genlocal()`
	`MulExpr.icode = MulExpr`$_1$`.icode ||| UnaryExpr.icode |||`
	`                    gen(MUL,MulExpr.addr,MulExpr`$_1$`.addr,UnaryExpr.addr)`

MulExpr : MulExpr$_1$ '/' UnaryExpr	`MulExpr.addr = genlocal()`
	`MulExpr.icode = MulExpr`$_1$`.icode ||| UnaryExpr.icode |||`
	`                    gen(DIV,MulExpr.addr,MulExpr`$_1$`.addr,UnaryExpr.addr)`

UnaryExpr : '-' UnaryExpr$_1$	`UnaryExpr.addr = genlocal()`
	`UnaryExpr.icode = UnaryExpr`$_1$`.icode |||`
	`                    gen(NEG,UnaryExpr.addr,UnaryExpr`$_1$`.addr)`

UnaryExpr : '(' AddExpr ')'	`UnaryExpr.addr = AddExpr.addr` `UnaryExpr.icode = AddExpr.icode`
UnaryExpr : IDENT	`UnaryExpr.addr = IDENT.addr` `UnaryExpr.icode = emptylist()`

Figure 9.4: Semantic rules for expressions

The main intermediate code generation algorithm is a post-order traversal of the syntax tree. To present it in small chunks, the traversal is broken into the main method, gencode( ), and helper methods for each non-terminal. In Unicon, the gencode( ) method in tree.icn looks as follows:

```
method gencode()
  every (!\kids).gencode()
  case sym of {
  . . .
    "AddExpr": { genAddExpr() }
    "MulExpr": { genMulExpr() }
  . . .
    "token":   { gentoken() }
    default: {
      icode := []
      every icode |||:= (!\kids).icode
```

```
      }
    }
  end
```

The default case for tree nodes that do not know how to generate code consists of just concatenating the code of the children. The corresponding Java code in tree.java looks like this:

```java
void gencode() {
  if (kids != null) for(tree k:kids) k.gencode();
  switch (sym) {
  . . .
  case "AddExpr": { genAddExpr(); break; }
  case "MulExpr": { genMulExpr(); break; }
  . . .
  case "token": { gentoken(); break; }
  default: {
    icode = new ArrayList<tac>();
    if (kids != null) for(tree k:kids)
        icode.addAll(k.icode);
    }
  }
}
```

The methods that are used to generate code for specific non-terminals must occasionally generate different instructions, depending on the production rule. The Unicon code for genAddExpr() is shown here:

```
method genAddExpr()
      addr := genlocal()
      icode := kids[1].icode ||| kids[2].icode |||
            gen(if rule=1320 then "ADD" else "SUB",
                addr, kids[1].addr, kids[2].addr)
  end
```

After generating a temporary variable to hold the result, the code is constructed by adding the appropriate arithmetic instruction to the end of the children's code. In this method, rule 1320 refers to an addition; otherwise, the operation is a subtraction. The corresponding Java code looks like this:

```java
void genAddExpr() {
```

```
      addr = genlocal();
      icode = new ArrayList<tac>();
      icode.addAll(kids[0].icode);
      icode.addAll(kids[1].icode);
      icode.addAll(gen(((rule==1320)?"ADD":"SUB"), addr,
                   kids[0].addr, kids[1].addr));
   }
```

The gentoken() method generates code for terminal symbols. The icode attribute for terminal symbols is empty. In the case of a variable, the addr attribute is a symbol table lookup, while in the case of a literal constant, the addr attribute is a reference to a value in the constant region, or an immediate value. In Unicon, the gentoken() method looks like this:

```
method gentoken()
   icode := []
   case tok.cat of {
      parser.IDENTIFIER: { addr := stab.lookup(tok.text).addr }
      parser.INTLIT: { addr := address("imm", tok.ival) }

      . . .

      }
end
```

The icode attribute is an empty list, while the addr attribute is obtained via a symbol table lookup. In Java, gentoken() looks like this:

```
void gentoken() {
   icode = new ArrayList<tac>();
   switch (tok.cat) {
      case parser.IDENTIFIER: {
         addr = stab.lookup(tok.text).addr; break; }
      case parser.INTLIT: {
         addr = new address("imm", tok.ival); break; }

      . . .

      }
}
```

You may observe from all this that generating intermediate code for expressions in straight-line code is mainly a matter of concatenating the operands' code, followed by adding one or more new instructions per operator. This work is made easier by allocating space in the form of the addresses of temporary variables ahead of time. The code for control flow is a bigger challenge.

Generating code for control flow

Generating code for control structures such as conditionals and loops is more challenging than code for arithmetic expressions, as shown in the preceding section. Instead of using synthesized attributes in a single bottom-up pass, code for control flow uses label information that must be moved to where it is needed using inherited attributes. This may involve multiple passes through the syntax tree. We will start with the conditional expression logic needed for even the most basic control flow, such as if statements, and then show you how to apply that to loops, followed by the considerations needed for method calls.

Generating label targets for condition expressions

We have already set up for control flow by assigning the first and follow attributes, as described in the *Annotating syntax trees with labels for control flow* section. Consider what role the first and follow attributes play, starting with the simplest control flow statement, the if statement. Consider a code fragment such as the following:

```
if (x < 0) x = 1;
y = x;
```

The syntax tree for these two statements is shown here:

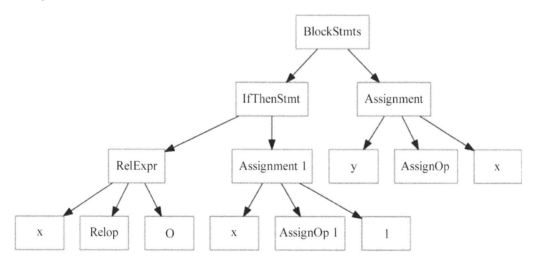

Figure 9.5: Syntax tree illustrating control flow

The BlockStmts node assigns the follow attribute of the IfThenStmt node to the first attribute of the y=x assignment. The code that is generated for RelExpr should go to the first label of the then part, pictured here as , if RelExpr is true.

It should follow the whole IfThenStmt if RelExpr is false. To implement this, label values computed from IfThenStmt can be inherited down into two new attributes of RelExpr. As discussed earlier, we do not call them true and false because those are Java-reserved words. The attribute for where to go when an expression is true is called onTrue, and the attribute for where to go when an expression is false is called onFalse. The semantic rules we want to implement are shown in the following table:

Production	Semantic Rules

IfThenStmt :
  if '(' Expr ')' Stmt
	Expr.onTrue = Stmt.first
	Expr.onFalse = IfThenStmt.follow
	Stmt.follow = IfThenStmt.follow
	IfThenStmt.icode = (Expr.icode != null) ? Expr.icode
	                             : gen(BIF, Expr.onFalse, Expr.addr, con:0)
	IfThenStmt.icode |||:= gen(LABEL, Expr.onTrue) ||| Stmt.icode

IfThenElseStmt :
  if '(' Expr ')' $Stmt_1$ else $Stmt_2$
	Expr.onTrue = $Stmt_1$.first
	Expr.onFalse = $Stmt_2$.first
	$Stmt_1$.follow = IfThenElseStmt.follow;
	$Stmt_2$.follow = IfThenElseStmt.follow;
	IfThenElseStmt.icode = (Expr.icode != null) ? Expr.icode
	                                  : gen(BIF, Expr.onFalse, Expr.addr, con:0)
	IfThenElseStmt.icode |||:= gen(LABEL, Expr.onTrue) ||| $Stmt_1$.icode |||
	    gen(GOTO, IfThenElseStmt.follow) ||| gen(LABEL, Expr.onFalse) ||| $Stmt_2$.icode

Figure 9.6: Semantic rules for the if-then and if-then-else statements

As we can see, the condition in IfThenStmt is an Expr that inherits onTrue from Stmt, which is its then part, and inherits onFalse from the parent's follow attribute – whatever code follows the whole IfThenStmt. These attributes must be inherited down into Boolean subexpressions through operators such as logical AND and OR. The semantic rules for the Boolean operators are shown in the following table:

Production	Semantic Rules
AndExpr : $AndExpr_1$ && EqExpr	EqExpr.first = genlabel(); $AndExpr_1$.onTrue = EqExpr.first; $AndExpr_1$.onFalse = AndExpr.onFalse; EqExpr.onTrue = AndExpr.onTrue; EqExpr.onFalse = AndExpr.onFalse; AndExpr.icode = $AndExpr_1$.icode ||| gen(LABEL, EqExpr.first) ||| EqExpr.icode;
OrExpr : $OrExpr_1$ || AndExpr	AndExpr.first = genlabel(); $OrExpr_1$.onTrue = OrExpr.onTrue; $OrExpr_1$.onFalse = AndExpr.first; AndExpr.onTrue = OrExpr.onTrue; AndExpr.onFalse = OrExpr.onFalse; OrExpr.icode = $OrExpr_1$.icode ||| gen(LABEL, AndExpr.first) ||| AndExpr.icode;
UnaryExpr : ! $UnaryExpr_1$	$UnaryExpr_1$.onTrue = UnaryExpr.onFalse $UnaryExpr_1$.onFalse = UnaryExpr.onTrue UnaryExpr.icode = UnaryExpr1.icode

Figure 9.7: Semantic rules for Boolean expressions

The code to compute the onTrue and onFalse attributes is placed in a method called gentargets().
The Unicon implementation in tree.icn looks like this:

```
method gentargets()
   case sym of {
   "IfThenStmt": {
      kids[1].onTrue := kids[2].first
      kids[1].onFalse := follow
      }
   "CondAndExpr": {
      kids[1].onTrue := kids[2].first
      kids[1].onFalse := onFalse
      kids[2].onTrue := onTrue
      kids[2].onFalse := onFalse
      }
   . . .
   }
   every (!\kids).gentargets()
end
```

The corresponding Java method looks like this:

```
void gentargets() {
   switch (sym) {
   case "IfThenStmt": {
      kids[0].onTrue = kids[1].first;
      kids[0].onFalse = follow;
      break;
      }
   case "CondAndExpr": {
      kids[0].onTrue = kids[1].first;
      kids[0].onFalse = onFalse;
      kids[1].onTrue = onTrue;
      kids[1].onFalse = onFalse;
      break;
      }
   . . .
   }
```

```
        if (kids!=null) for(tree k:kids) k.gentargets();
    }
```

Having seen how the onTrue and onFalse attributes get assigned, perhaps the last piece of the puzzle is the code that's generated for relational operators, such as the x < y test. On these operators, it would be possible to generate code that computes a true (1) or false (0) result and store it in a temporary variable in the same way that results are computed for arithmetic operators. However, the point of computing the onTrue and onFalse labels was to generate code that would jump directly to the correct label, depending on whether a test was true or false. This is helpful for implementing the short-circuit semantics for the Boolean operators that Jzero inherits from Java and C. Here is the Unicon implementation of the genRelExpr() method, which is called from gencode() to generate intermediate code for relational expressions:

```
method genRelExpr()
  op :=  case kids[2].tok.cat of {
    ord("<"): "BLT"; ord(">"): "BGT";
    parser.LESSTHANOREQUAL: "BLE"
    parser.GREATERTHANOREQUAL: "BGE"
    }
  icode := kids[1].icode ||| kids[3].icode |||
            gen(op, onTrue, kids[1].addr, kids[3].addr) |||
            gen("GOTO", onFalse)
end
```

This code starts by setting the op variable to the three-address opcode that corresponds to the integer category of the operator, extracted from kids[2].tok.cat. Then, it constructs code by concatenating the left and right operands, followed by a conditional branch if the operator evaluates to true, followed by an unconditional branch if the operator was false. The corresponding Java implementation looks like this. Although we know the shape of the tree node on which we

```
void genRelExpr() {
  String op;
  switch (kids[1].tok.cat) {
    case '<': op="BLT"; break; case ';': op="BGT"; break;
    case parser.LESSTHANOREQUAL: op="BLE"; break;
    default: op="BGE";
    }
```

```
    icode = new ArrayList<tac>();
    icode.addAll(kids[0].icode); icode.addAll(kids[2].icode);
    icode.addAll(gen(op, onTrue, kids[0].addr, kids[2].addr));
    icode.addAll(gen("GOTO", onFalse));
}
```

Compared to the code that is generated for ordinary arithmetic, the code for control structures such as `if` statements passes a lot of label information around. Now, let's look at what must be added to the code to support loop control structures.

Generating code for loops

This section presents ideas for generating intermediate code for `while` loops and `for` loops. The `while` loop code should be almost identical to an `if-then` statement, with the sole additions of a label at the top, and a goto at the bottom to jump to that label. A `for` loop is just a `while` loop with a couple of additional expressions thrown in. The following table shows the semantic rules for these two control structures:

Production	Semantic Rules

WhileStmt : while '(' Expr ')' Stmt	Expr.onTrue = genlabel();
	Expr.first = genlabel();
	Expr.false = WhileStmt.follow;
	Stmt.follow = Expr.first;
	WhileStmt.icode = gen(LABEL, Expr.first) |||
	  Expr.icode ||| gen(LABEL, Expr.true) |||
	  Stmt.icode ||| gen(GOTO, Expr.first)

ForStmt : for(ForInit; Expr; ForUpdate)
  Stmt
 	
a.k.a.
 	
ForInit;
while (Expr) {
  Stmt
  ForUpdate
}
	Expr.true = genlabel();
	Expr.first = genlabel();
	Expr.false = S.follow;
	Stmt.follow = ForUpdate.first;
	S.icode = ForInit.icode |||
	  gen(LABEL, Expr.first) |||
	  Expr.icode ||| gen(LABEL, Expr.true) |||
	  Stmt.icode |||
	  ForUpdate.icode |||
	  gen(GOTO, Expr.first)

Figure 9.8: Semantic rules for the intermediate code generation of loops

The genWhileStmt() method is representative of similar control flow code generation methods such as genIfStmt() and genForStmt(). Most of the work is done while computing the first, follow, onTrue, and onFalse attributes. The Unicon implementation of genWhileStmt() is as follows:

```
method genWhileStmt()
   icode := gen("LAB", kids[1].first) ||| kids[1].icode |||
            gen("LAB", kids[1].onTrue) |||
            kids[2].icode ||| gen("GOTO", kids[1].first)
end
```

The Java implementation of genWhileStmt() is shown here:

```
void genWhileStmt() {
   icode = new ArrayList<tac>();
   icode.addAll(gen("LAB", kids[0].first));
   icode.addAll(kids[0].icode);
   icode.addAll(gen("LAB", kids[0].onTrue));
   icode.addAll(kids[1].icode);
   icode.addAll(gen("GOTO", kids[0].first));
}
```

There is one remaining aspect of control flow to present. Method (or function) calls are fundamental building blocks in all forms of imperative code and object-oriented code.

Generating intermediate code for method calls

The intermediate code instruction set provides three opcodes related to method calls: PARM, CALL, and RET. To invoke a method, the generated code executes several PARM instructions, one for each parameter, followed by a CALL instruction. The called method then executes until it reaches a RET instruction, at which time it returns to the caller. This intermediate code is an abstraction of several different ways that hardware supports method (or function) abstractions.

On some CPUs, parameters are mostly passed in registers, while on others, they are all passed on the stack. At the intermediate code level, we must worry about whether PARM instructions occur in the order actual parameters appear in the source code or in reverse order. In object-oriented languages such as Jzero, we also worry about how a reference to an object is accessible inside a called method. Programming languages have answered these questions in different ways on different CPUs, but for our purposes, we'll use the following calling conventions: parameters are given in reverse order, followed by the object instance (a self or this pointer) as an implicit extra parameter, followed by the CALL instruction.

When gencode() gets to a MethodCall, which is a type of primary expression in our grammar, it will call genMethodCall(). Its Unicon implementation is shown here:

```
method genMethodCall()
  local nparms := 0
  if k := \ kids[2] then {
    icode := k.icode
    while k.sym === "ArgList" do {
      icode |||:= gen("PARM", k.kids[2].addr)
      k := k.kids[1]; nparms +:= 1
    }
    icode |||:= gen("PARM", k.addr); nparms +:= 1
  }
  else icode := [ ]
  if kids[1].sym === "QualifiedName" then {
    icode |||:= kids[1].icode
    icode |||:= gen("PARM", kids[1].kids[1].addr)
  }
  else icode |||:= gen("PARM", "self")
  icode |||:= gen("CALL", kids[1].addr, nparms)
end
```

The generated code starts with the code to compute the values of the parameters. Then, it issues PARM instructions in reverse order, which comes for free from the way the context-free grammar constructed the syntax tree for argument lists. The trickiest parts of this method have to do with how the intermediate code knows the address to use for the current object. The Java implementation of genMethodCall() is shown here:

```
void genMethodCall() {
  int nparms = 0;
  icode = new ArrayList<tac>();
  if (kids[1] != null) {
    icode.addAll(kids[1].icode);
    tree k = kids[1];
    while (k.sym.equals("ArgList")) {
      icode.addAll(gen("PARM", k.kids[1].addr));
      k = k.kids[0]; nparms++;
    }
    icode.addAll(gen("PARM", k.addr)); nparms++;
```

```
    }
  if (kids[0].sym.equals("QualifiedName")) {
    icode.addAll(kids[0].icode);
    icode.addAll(gen("PARM", kids[0].kids[0].addr));
    }
  else icode.addAll(gen("PARM", new address("self",0)));
  icode.addAll(gen("CALL", kids[0].addr,
                    new address("imm", nparms)));
}
```

What this section showed has probably convinced you that code generation for the calling side is more challenging than code generation for the return instruction, which you can examine in this chapter's code on GitHub. It is also worth mentioning that every method body's code might have a ret instruction appended, ensuring that code never executes past the end of a method body and into whatever comes after it.

Reviewing the generated intermediate code

You cannot run intermediate code, but you should check it carefully. Ensure that the logic looks correct on test cases for every feature that you care about. To check the generated code for a file such as hello.java, run the following command using either the Unicon (left-hand side) or Java implementation (right-hand side). As a reminder for Java, on Windows, you must execute something like set CLASSPATH=".;C:\byopl" first or the equivalent in your **Control Panel** or **Settings**. On Linux, it might look like export CLASSPATH=.:..:

```
j0 hello.java                    java ch9.j0 hello.java
```

The output should look similar to the following:

```
.string
L0:
        string  "hello, jzero!"
.global
        global  global:8,hello
        global  global:0,System
.code
proc    main,0,0
        ASIZE   loc:24,loc:8
        ASN     loc:16,loc:24
        ADD     loc:32,loc:16,imm:2
```

```
          ASN       loc:16,loc:32
  L138:
          BGT       L139,loc:16,imm:3
          GOTO      L140
  L139:
          PARM      strings:0
          PARM      loc:40
          CALL      PrintStream__println,imm:1
          SUB       loc:48,loc:16,imm:1
          ASN       loc:16,loc:48
          GOTO      L138
  L140:
          RET
  end
  no errors
```

Looking over intermediate code is when you start to realize that you may be able to finish this compiler and translate your source code down into machine code of some kind. If you are not excited, you should be. A lot of errors can be spotted at this point, such as omitted features, or branch statements that go to non-existent labels, so check it out before you rush ahead to generate the final code.

Summary

In this chapter, you learned how to generate intermediate code. Generating intermediate code is the first vital step in synthesizing the instructions that will eventually allow a machine to run the user's program. The skills you learned in this chapter build on the skills that are used in semantic analysis, such as how to add semantic attributes to the syntax tree nodes, and how to traverse syntax tree nodes in complex ways as needed.

One of the important features of this chapter was an example intermediate code instruction set that we used for the Jzero language. Since the code is abstract, you can add new instructions to this instruction set as needed for your language. Building lists of these instructions was easy using Unicon's list data type or Java's ArrayList type.

The chapter showed you how to generate code for straight-line expressions such as arithmetic calculations. Far more effort in this chapter went into the instructions for control flow, which often involve goto instructions whose target instructions must have labels. This entailed computing several attributes for labels, including inherited attributes, before building the lists of code instructions.

Now that you have generated intermediate code, you are ready to move on to the final code generation part. However, first, *Chapter 10, Syntax Coloring in an IDE*, will take you on a practical diversion, consisting of exploring how to use your knowledge to incorporate syntax coloring into an **integrated development environment (IDE)**.

Questions

1. Does it make sense to add three address instructions for doing input and output? Why or why not?

2. Explain the relationship between semantic rules, such as those shown in *Figure 9.4*, and tree traversal methods, such as genAddExpr().

3. Why might some compiler writers be insistent about computing the semantic attributes for instruction labels during the same tree traversal that generates code? Is that possible in a general case?

4. In this chapter, whenever a new local variable is required, the code generator just calls genlabel() to obtain one. In the absence of optimization, what effect might this have on the program?

Join our community on Discord

Join our community's Discord space for discussions with the authors and other readers:

https://discord.com/invite/zGVbWaxqbw

10

Syntax Coloring in an IDE

Creating a useful programming language requires more than just a compiler or interpreter that makes it possible to run programs—it requires an ecosystem of tools for developers. This ecosystem often includes debuggers, online help, or an **integrated development environment**, commonly called an **IDE**. An IDE can be broadly defined as any programming environment in which source code editing, compilation, linking steps (if any), and execution can all be performed within the same **user interface**. A good modern IDE typically includes many additional features, such as a graphical user interface builder and integrated debugger.

This chapter addresses some of the challenges of supporting your programming language in an IDE to provide syntax coloring and visual feedback about syntax errors. One reason that you want to learn how to do this is that many programmers will not take your language seriously unless it has an IDE with such features. We will start by adding support for Jzero in a mainstream IDE, Visual Studio Code. We will then show some example syntax-coloring IDE code written in Unicon. Unlike other chapters, there is no IDE that is implemented identically in Unicon and Java from which to draw a parallel dual-language example.

This chapter covers the following main topics:

- Writing your own IDE versus supporting an existing one
- Downloading the software used in this chapter
- Adding support for your language to Visual Studio Code
- Integrating a compiler into a programmer's editor
- Avoiding reparsing the entire file on every change
- Using lexical information to colorize tokens
- Highlighting errors using parse results

The skills to learn in this chapter revolve around software system communication and coordination. Primarily, by bundling the IDE and compiler into a single executable, high-performance communication is conducted by passing references to shared data, instead of resorting to file input/output or **inter-process communication**.

Writing your own IDE versus supporting an existing one

Writing an IDE is a large project and could be the subject of an entire book. If all you want is for your new language to be supported by a good IDE, just figure out how to support your language within an existing popular IDE. This is especially true if supporting your language requires nothing unusual that existing IDEs do not already do. It is a big job, bigger in some IDEs than in others, but most of this chapter provides an example of how to do it in one mainstream IDE.

On the other hand, there are several reasons that you might decide to write your own IDE. Writing an IDE puts you in control. Writing an IDE in your new language can be a convincing demonstration of how awesome your new language is, that it has better user interface capabilities than mainstream languages, or that it has achieved a level of maturity and usability by virtue of a suitably robust class library and system interface.

Unlike other chapters of this book where we present the compiler code from scratch, the discussion of adding syntax highlighting in an existing IDE involves no compiler code from our GitHub repository; instead, we will talk about configuration using various data files. The chapter includes a second section that describes how syntax coloring was added to the Unicon IDE, with a little bit of code that illustrates the ideas but is not intended for you to type in or run. The Unicon IDE was written by Clinton Jeffery and Nolan Clayton, with contributions from many other people since then. Luis Alvidres did the syntax coloring work as part of his master's degree project. Luis's project report can be found at `http://www.unicon.org/reports/alvidres.pdf`. The next section describes how to download the programs discussed in this chapter.

Downloading the software used in this chapter

In this chapter, we will be looking at one mainstream commercially supported IDE, plus a simpler IDE that illustrates some of the concepts presented. The first IDE is Visual Studio Code, a free IDE that you can download from `http://visualstudio.microsoft.com/downloads`. Since we are extending Visual Studio Code to know about your new language, additional tools are required. Microsoft's instructions say to install Git (which you have probably already installed in order to access the book's code on GitHub) and Node.js (from `https://nodejs.org/en/download/current`).

Node.js in turn installs many software components, including Python and some Visual Studio development tools. This requires a strong internet connection, many minutes of time, and requires quite a leap of blind faith on your part. Have fun. If the Node.js install went well, you should be able to install a tool called Yeoman with the command:

```
npm install -g yo generator-code
```

Installing Yeoman (yo) takes a few more minutes. After that, you can run yo from the command line, which we will do in order to add a new extension for Jzero.

After Visual Studio Code and its baggage, the second IDE discussed in this chapter is a program called ui, which stands for Unicon IDE. The ui program is included in the Unicon language distribution, where it can be found in a directory called uni/ide. The program consists of about 10,000 lines of code in 26 files, not counting code in library modules. The following screenshot shows the ui program:

Figure 10.1: The ui IDE

Before we explore syntax coloring within the ui program, let's learn what we can do easily to support a new language within the context of Visual Studio Code, a popular editor that many programmers already use.

Adding support for your language to Visual Studio Code

Microsoft Visual Studio Code (VS Code) is a popular IDE. It has been chosen to illustrate adding support for your language to an existing IDE because of its market share, but this is not an endorsement. For your language, you may find that another IDE is far easier to support, or far better for your users. Personally, I am an Emacs person, and I won't become interested in VS Code until someone demonstrates it running in Emacs mode. However, VS Code does have a lot of nice features.

VS Code knows how to display programs written in many programming languages, such as Java, out of the box. How it colors Java syntax depends on the selected color theme, but under the Black High Contrast, you get a very colorful display. It is unclear what colors mean in VS Code or how colors are assigned to different bits of the code, but hey—many pretty pastel shades of blue, green, pink, and orange are used. *Figure 10.2* shows a **"hello world"** Java program in VS Code with the Black High Contrast theme.

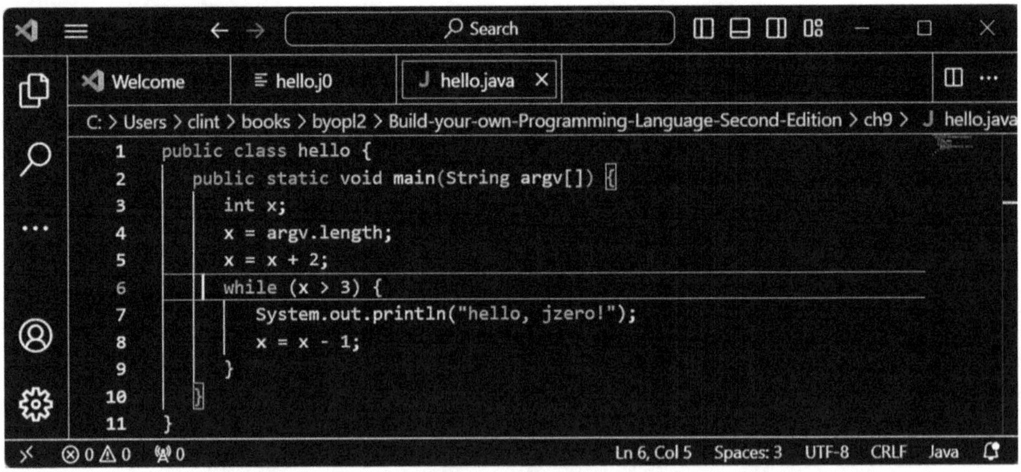

Figure 10.2: VS Code uses many colors to display a Java program

On the other hand, when you invent a new programming language, VS Code starts out knowing nothing about it. If you open a Jzero program named hello.j0 within VS Code, it will display with no syntax coloring. *Figure 10.3* shows the same program as *Figure 10.2* under the Light High Contrast theme, with the filename changed to hello.j0 instead of hello.java. Because VS Code doesn't know the .j0 extension, no coloring is applied under any theme, other than to highlight the current line that the cursor is on.

Figure 10.3: VS Code shows unknown languages without syntax coloring

Adding support for a new language within VS Code depends on the provisions for such extensions, which are discussed at code.visualstudio.com/api/language-extensions/overview. The next section presents a few key aspects of writing VS Code language extensions, a topic that is complex and goes well beyond the scope of this chapter. This chapter also uses ideas incorporated from the site https://macromates.com/manual/en/language_grammars.

Configuring Visual Studio Code to do Syntax Highlighting for Jzero

As downloaded, VS Code will know how to syntax highlight mainstream languages, especially Microsoft languages, such as C++, C#, and the like. It already knows Java, so to configure VS Code to work with Jzero as a brand new language, we use the file extension .j0 for Jzero files in order to create our own syntax highlighting rules instead of using those of Java.

The page at code.visualstudio.com/api/language-extensions/syntax-highlight-guide provides Microsoft's instructions on syntax highlighting new languages. This section is adapted from there and from https://code.visualstudio.com/api/get-started/your-first-extension.

To inform VSCode of the new language, run the Yeoman tool to generate a new language extension, and then see how to populate it. Run the command:

```
yo code
```

The Yeoman program asks you a series of questions, starting with what type of VS Code extension you want to create. For Jzero, select New Language Support. It will ask you for a grammar, but if you leave that blank, you start a new language extension with no grammar; we will discuss writing the grammar below. Yeoman then asks you for many fields of non-technical information about your language. It even offers to create a Git repository for you. If we answer these questions in the most basic way possible for Jzero, we will see the output shown in Figure 10.4 as the Yeoman program creates our new extension:

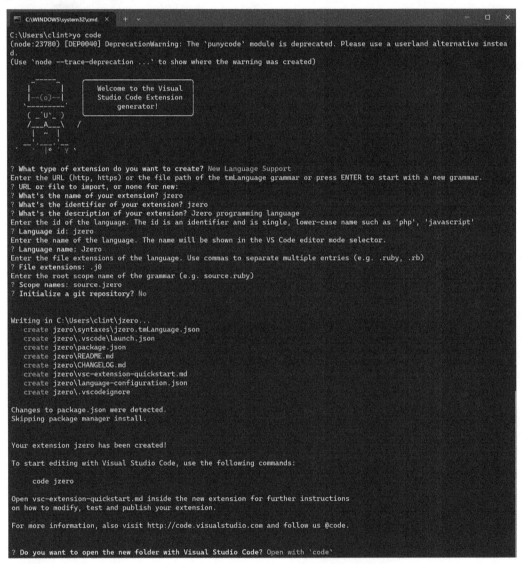

Figure 10.4: The output of running yo code *to create a language extension for Jzero*

Yeoman creates a directory for our new extension and populates it with several files. A quickstart file tells you to add your grammar to `syntaxes\jzero.tmLanguage.json` and add comment and bracket information to `language-configuration.json`. It also tells you to press F5 to open a new window with your extension loaded, and to use Control + R to reload when you make changes to your extension. When you have finished adding features and debugging your extension, copy it into the `.vscode/extensions` directory, where it will be used automatically whenever VS Code runs. There is a separate step for publishing your extension to the world when the time comes. At the time of writing, the link for it can be found at `https://marketplace.visualstudio.com/VSCode` and requires you to log in with a Microsoft account and password.

Visual Studio Code extensions using the JSON format

VS Code uses the **JavaScript Object Notation (JSON)** for many of its language extension configuration files. If you don't already know about JSON, you can learn all there is to know about it in around 5 minutes; no wonder it is so popular. JSON is described in detail at `json.org`, which uses railroad diagrams and a simple grammar to give a precise definition.

JSON is written in an ASCII text format. Its syntax comes from JavaScript, but it is similar to other popular languages like Python. There are six kinds of things that can appear in a JSON file: four kinds of simple atomic values, and two kinds of collections.

JSON atomic types

JSON allows four kinds of simple **atomic values**: string, number, boolean, or null. Strings are enclosed in double quotes and are pretty much the same as Java strings. JSON strings include obvious examples such as `"hello"`. Like Java, backslash is the escape character and can be followed by a double-quote mark, a backslash, a forward slash, `b` (backspace), `f` (form feed), `n` (newline), `r` (carriage return), `t` (tab), or a `u` followed by four hexadecimal symbols between `0-9` and `A-F` (or `a-f`).

JSON numbers include both integer and real numbers. No leading zero is allowed. Real numbers can have either or both of a decimal point followed by one or more digits, and an exponent (E or e) followed by an optional sign (+ or -) and one or more digits.

Booleans are the values `true` and `false`. The null type has only one value: `null`.

JSON collections

JSON has two kinds of **collections** that can contain multiple values: arrays and objects. An array is an ordered sequence of JSON values enclosed in square brackets and separated by commas. Array elements may themselves be other arrays or objects. Some example JSON arrays are `[1, 2]` and `["three", 4, 5]`. A more complex example is `[[1, 2, 3], [4, 5, 6], [7, 8, 9]]`.

A JSON object is an unordered collection of JSON values enclosed in curly brackets. Each element is preceded by an associated string key and a colon, and these key:element pairs are separated by commas. Consider the example {"hello": 3.14, "true": null}. It associates the key "hello" with the value 3.14 and the key "true" with the value null.

File organization for Visual Studio Code extensions

VS Code configuration files live in a .vscode\ directory under the home directory, as in c:\users\ clint\.vscode. Within that directory lives a directory named extensions. Each VS Code extension that you install gets its own subdirectory within .vscode\extensions; all the installed extensions are listed and described within a .vscode\extensions\extensions.json file. If you copy your new extension into the .vscode\extensions\ folder manually, you might need to modify the extensions.json file to tell it about your extension.

The extensions file

The extensions.json file is a JSON array with one element for each language extension. The format is shown below. To add your language as a new extension, if you are not the first installed extension, you add a comma after the last element already present, and then add the information for your new extension. Each extension is represented by one JSON object. The JSON format does not care if the contents are all given on one ridiculously long line, or spread across thousands. We provide this example on many lines for the sake of human readability:

```
[{"identifier":{
    "id":"cj.jzero",
    "uuid":"4cdff31d-3adb-47ad-a0eb-87489d90b470"},
  "version":"0.1.0",
  "location":{
      "$mid":1,
      "fsPath":"c:\\Users\\clint\\.vscode\\extensions\\cj.jzero-0.1.0",
      "_sep":1,
      "external":"file:///c%3A/Users/clint/.vscode/extensions/
cj.jzero-0.1.0",
      "path":"/c:/Users/clint/.vscode/extensions/cj.jzero-0.1.0",
      "scheme":"file"},
  "relativeLocation":"cj.jzero-0.1.0",
  "metadata":{
      "id":"4cdff31d-3adb-47ad-a0eb-87489d90b470",
      "publisherId":"aa569c26-c667-46dc-9b6d-0672063c01f8",
```

```
            "publisherDisplayName":"Clinton Jeffery",
            "targetPlatform":"undefined",
            "updated":false,
            "isPreReleaseVersion":false,
            "installedTimestamp":1702263162060,
            "preRelease":false}
    }]
```

Let's examine these sub-elements in detail. The `identifier` is a JSON object containing an `id` string in the format `publisher.extensionname`, and `uuid`, a unique user ID string. You can sort of make up your publisher and extension name; in this example, I use my initials and the language name. The UUID string consists of 128 bits in 32 hexadecimal digits, with hyphens after the 8th, 12th, 16th, and 20th digit. UUIDs can be generated by websites such as `uuidgenerator.net` and by library functions such as `java.util.UUID.randomUUID()`.

The `version` is a string of the format `major.minor.patch`. The `location` is a JSON object containing URI fields: `$mid`, `fsPath`, `_sep`, `external`, `path`, and `scheme`. For installed extensions, these URI fields can be set at installation time. You can use the location as-is, other than updating the three paths to show locations on your machine instead of mine. The `relativeLocation` is a string in the format `id-version`. The `metadata` is a JSON object containing `id`, `publisherId`, `publisherDisplayName`, `targetPlatform`, `update`, `isPreReleaseVersion`, `installedTimestamp`, and `preRelease`. It is mostly self-explanatory, other than to say that the extension installer normally generates this metadata.

The extension manifest

Within your extension directory, a file named `package.json` contains details about your language extension. When you created your extension with the Yeoman program, it created an initial version of this file, but you may want to review and edit it. The file `package.json` contains a single JSON object containing many fields. Different VS Code extensions will look different in this file, and this chapter only has room for a basic language extension for syntax coloring. We will break the `package.json` description into several pieces for exposition. The first section consists of seven simple string fields, identifying what the extension is and who provided it:

```
{
    "name": "jzero",
    "displayName": "Jzero Syntaxcolorer",
    "description": "Jzero Language",
    "author": "Clinton Jeffery",
```

```
        "license": "public domain",
        "version": "0.1.0",
        "publisher": "cj",
```

The next field is a JSON object that describes what the extension contributes to VS Code. In this case, we contribute a new language, which uses the extension .j0:

```
        "contributes": {
            "languages": [
                {
                    "id": "jzero",
                    "extensions": [
                        ".j0"
                    ],
                    "aliases": [
                        "Jzero",
                        "jzero"
                    ]
                }
            ]
        },
```

The "activationEvents" field is an array listing what events active this extension. For a language extension, this is almost always "onLanguage", which is concatenated with your language name and the event that occurs whenever a file with your language's extension is opened. There are a lot of other event types used for other kinds of extensions:

```
        "activationEvents": [
            "onLanguage:jzero"
        ],
```

The "keywords" field is an array listing up to five keywords describing your extension. The "categories" field lets you identify with the 17 kind(s) of extensions VS Code things it supports—in our case, "Programming Languages". The "engines" field specifies the minimum version(s) for the tools that use this extension—in our case, "vscode" version 1.85.0 or newer:

```
        "keywords": [
            "Jzero"
        ],
```

```
    "categories": [
        "Programming Languages"
    ],
    "engines": {
        "vscode": "^1.85.0"
    },
```

There may be other fields in the extension manifest. This section only described the most common ones. Now, it is time to look at the notation VS Code uses to specify token types and syntax rules: TextMate grammars.

Writing IDE tokenization rules using TextMate grammars

Syntax coloring can be provided by either a **TextMate grammar** or a more extensive language server. This section describes writing a TextMate grammar. TextMate grammars are similar to Flex+YACC specifications, only formatted in JSON. A TextMate grammar consists of two JSON files: a grammar contribution file, plus the grammar file itself. The grammar contribution file for Jzero looks like the following code. The grammar contribution file is optional, but recommended, organization, since "contributes" fields can appear directly in package.json files:

```
{
    "contributes": {
        "languages": [
            {
                "id": "jzero",
                "extensions": [".j0", ".jzero"]
            }
        ],
        "grammars": [
            {
                "language": "jzero",
                "scopeName": "source.jzero",
                "path": "jzero.tmGrammar.json"
            }
        ]
    }
}
```

A starting TextMate grammar file for Jzero, named `jzero.tmGrammar.json`, is shown below, divided into several sections for exposition purposes. A TextMate grammar encodes a subset of the same information depicted earlier by Flex and YACC in a longer and less readable format. A complete TextMate grammar for a mainstream language can run to well over a thousand lines, so we cannot present one in its entirety here.

The whole grammar is a single JSON object (a.k.a. a table or dictionary). In this example, there are four keys. Many additional keys are optional and you can look those up on reference websites. This top-level JSON object corresponds to the start symbol in the Jzero YACC grammar from *Chapter 4*. The first two keys in the grammar are the "name" and "scopename". The "name" is just a unique identifier, and each grammar sub-element will have a different one. The "scopename" here is the root of all VS Code scopes introduced by this grammar; scope names of sub-elements will be this root "scopename" concatenated into the front of their "name":

```
{
    "name": "Jzero",
    "scopeName": "source.jzero"
```

The third key gives the patterns, corresponding to production rules for this start symbol. Like this top-level JSON object, the various sub-elements that correspond to non-terminal symbols in your grammar must include a similar entry named "patterns", which is a list of the production rules for that non-terminal. The following Jzero patterns just say that Jzero programs have keywords, strings, and expressions in them and do not impose structure. The patterns entry here could contain actual patterns of what strings to match, but instead, this one just consists of forward references to the names of patterns to be found within the next key; pattern names are prefixed by a pound sign (#):

```
"patterns": [
    { "include": "#keywords" },
    { "include": "#strings" },
    { "include": "#expression" }
],
```

The fourth key gives the repository, which is the main body of the grammar and contains all the syntactic and lexical rules about your language that you care to encode. Each entry in the repository maps to a JSON object. The JSON object might consist of a regular expression for specific tokens, or references to sub-elements, which may be recursive:

```
"repository": {
```

```
        "keywords": {
            "patterns": [{
                "name": "keyword.control.jzero",
                "match": "\\b(if|while|for|return)\\b"
                }]
        },
        "strings": {
            "name": "string.quoted.double.jzero",
            "begin": "\"",
            "end": "\"",
            "patterns": [{
                "name": "constant.character.escape.jzero",
                "match": "\\\\."
                }]
        }
        "expression": {
            "patterns": [{ "include": "#paren-expression" }]
        },
        "paren-expression": {
            "begin": "\\(",
            "end": "\\)",
            "beginCaptures": {
                "0": { "name": "punctuation.paren.open" }
            },
            "endCaptures": {
                "0": { "name": "punctuation.paren.close" }
            },
            "name": "expression.group",
            "patterns": [{ "include": "#expression" }]
        }
    },
}
```

The sub-level in your grammar that corresponds to terminal symbols will often include an entry named "match" that gives the regular expression for that terminal symbol. Whether you match many keywords with one big pattern, as shown in the "keywords" element above, or instead write a separate rule for each keyword depends on your grammar and whether the syntax coloring should be different for different keywords.

Writing support for your language in VS Code can be as big a job as you have time to put into it. We have introduced this topic, but there is a much deeper end in this swimming pool, where you can write a whole external server to direct VS Code's handling of your language. Now, let's move on to a brief description of the Unicon IDE and how the Unicon compiler frontend code was integrated into that IDE for the purposes of syntax coloring.

Integrating a compiler into a programmer's editor

The front half of the Unicon compiler—loosely covered from *Chapter 2, Programming Language Design*, up to *Chapter 5, Syntax Trees*, in this book—was integrated into the Unicon IDE, known as ui. The Unicon frontend consists of three major components: a **preprocessor**, a **scanner** (also called a **lexical analyzer**), and a **parser**. While we discussed scanners and parsers in detail in *Chapters 3* and *4*, we have not discussed preprocessors, which implement symbolic macro substitutions and provide the ability to select platform-specific code at compile time. Preprocessors are a major subject in the next chapter.

In the Unicon translator, these components are called from a main() procedure. The translator opens, reads, and writes files in the filesystem to perform its I/O, providing feedback to the user by writing text to standard output or a standard error on a console or terminal window. In an IDE, the compiler components are called from behind the scenes while the user edits their code in a **graphical user interface (GUI)**. The source code is obtained directly from the memory in the IDE, and the compiler's output is obtained from the memory by the IDE and presented to the user. Altogether, seven files from the Unicon translator were modified to become library modules that can be linked in and used from other programs besides the Unicon compiler itself. An overview of this integration of compiler components into the Unicon IDE is shown in Figure 10.5. The compiler code is invoked from a method ReparseCode(), which is invoked whenever the cursor moves to a new line:

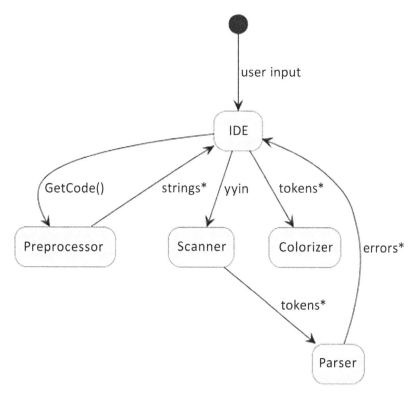

Figure 10.5: Overview of compiler integration into the IDE method ReparseCode()

The IDE method GetCode() is interesting in and of itself. It could just return the entire contents of the file as a list of strings, but in large files, this amount of redundant lexical and syntax analysis would be problematic. Instead, GetCode() looks forward and backward from the cursor to find the nearest global declaration boundaries, and it only reparses the current declaration (such as a procedure or class) where the user makes changes. GetCode() is described in more detail later in the section titled *Avoiding reparsing the entire file on every change.* The next section explores how source code in the IDE is fed into the compiler frontend. After that, we will consider how the compiler syntax results, including error messages, are fed into the IDE.

Analyzing source code from within the IDE

A compiler usually obtains its input by opening and reading from a named file. Compilers that feature preprocessors, such as C/C++ and Unicon, feed the named file a line at a time through a macro preprocessor that transforms the source code before it is input to the lexical analyzer. When a chunk of source code is selected for parsing in the IDE by GetCode( ), it is provided as a list of strings to the preprocessor, which in turn produces its results one line at a time.

The lex-compatible interface used by many compilers specifically designates that input comes from an opened file handle stored in a global variable named yyin. This is too slow for an IDE, which performs lexical and syntax analysis frequently and repeatedly as the user edits. Instead of reading from a file, the Unicon scanner was modified so that it could read source code that was already in main memory. The scanner code already read the entire file into a single large string and scanned that, so modifying it to accept a string as input was trivial. The code to go from the list of strings produced by GetCode( ) through the preprocessor and into the string format expected by Unicon's yylex( ) looks like the following. The function preprocessor( ) is a generator that produces a string for each line of source code. The lines are concatenated into one big string:

```
preproc_err_count := 0
yyin := ""
every yyin ||:= preprocessor(theCode, uni_predefs) do yyin ||:= "\n"
```

With yyin initialized to the part of the file that needs to have its syntax checked, the ReparseCode( ) method calls yyparse( ):

```
parsingErrors := []
rv := yyparse()
```

The list parsingErrors will be populated by yyparse( ) in the event that any syntax errors should occur. Now, let's look at how compiler messages are delivered to the GUI of the IDE.

Sending compiler output to the IDE

Instead of directly writing error output, the parser was modified to construct a list of error diagnostics. The regular compiler can then output these to the console, while the IDE displays messages in a sub-window or depicts them graphically. Consider a possible error message, such as the following:

```
hello.icn:5: '}' expected
```

Prior to integration, the compiler could have written that with the following line of code:

```
write(&errout, filename, ":", lineno, ": ", message)
```

To integrate such messages into the IDE, the compiler yyerror(s) function was modified to instead put the error message, along with the line number, into a ParseError object on a list named parsingErrors. The code in yyerror() is as follows:

```
/parsingErrors := []
errorObject := ParseError( yylineno, s )
put( parsingErrors, errorObject )
```

In the Unicon IDE, these parsing errors are displayed textually from within the ReparseCode() method. After the parser is invoked, if errors were encountered, the following lines execute:

```
every errorObject := !parsingErrors do {
  errorObject.lineNumber +:= lineNumberOffset
  if errorObject.lineNumber <= *contents then {
    SetErrorLineNumber(errorObject.lineNumber)
    uidlog.MsgBox.set_contents(
      [errorObject.lineNumber ||": " ||
        errorObject.errorMessage])
  }
}
```

The error message text is placed in a GUI component named MsgBox with a call to its set_contents() method. MsgBox is drawn below the source code. In addition to displaying the same output text that the compiler would show, in the event of an error, the IDE highlights the line on which the error occurs. This is discussed later in the *Highlighting errors using parse results* section.

This section on integrating a compiler into an IDE or programmer's editor discussed the nuts and bolts of how to combine these two large and complex pre-existing pieces of software. The Unicon compiler and IDE are maintained mostly independently. Keeping the connections between them simple reduces the likelihood of a change in one affecting the other. If you are writing a new IDE from scratch to go along with a new compiler, a more extensive integration might enable extra features or better performance, at a cost in complexity, maintainability, and portability. Now, let's look at how to invoke syntax checks without parsing a file constantly while the user edits code.

Avoiding reparsing the entire file on every change

The lexical and syntax analysis necessary to parse input and detect and report syntax errors presented in this book from *Chapter 2, Programming Language Design*, to *Chapter 8, Checking Types on Arrays, Method Calls, and Structure Accesses*, are substantial algorithms. Although the Flex and Yacc tools we've used are high-performance, if given a large input file, scanning and parsing become slow enough that users will not want to reparse the whole file each time a user modifies it in an IDE text editor. In testing, we found that reparsing the entire file became a problem on files larger than 1,000 lines.

Sophisticated incremental parsing algorithms that minimize the amount that must be reparsed after changes are the subject of Ph.D. dissertations and research articles. For the Unicon IDE, a simple approach is taken. Whenever the cursor moves away from a line that has been changed, a parsing unit is selected, starting with the changed line and extending above and below the boundaries of the nearest procedure, method, or another global declaration unit. That unit is reparsed.

In Unicon, this gives a very good performance. Luis Alvidres found that when an entire declaration unit is reparsed after a line is changed, 98% of the time the compiler reparses fewer than 100 lines of code. Most of the other 2% of cases—namely, procedures or methods larger than 100 lines—are still not a problem. Only the very largest procedure or method bodies result in slow reparsing. This is often machine-generated code, such as the output of Flex or Yacc, that a user seldom edits by hand. For this, the IDE disables syntax checking to avoid an unacceptable user response time.

The code to select a slice to reparse when a cursor moves off a line is in a method named `GetCode()`, this can be found in the `BuffEditableTextList` class, which is a subclass of Unicon's standard GUI editor component named `EditableTextList`. The `BuffEditableTextList` class lives in `uni/ide/buffertextlist.icn`. The `GetCode()` method is implemented as follows. First comes the method header and a set of local variable declarations:

```
method GetCode()
   local codeSubStringList,
         originalPositionY, currentPositionY, token,
         startPositionY := 0, endPositionY := 0,
         inClass := 0, inMethod := 0
```

Within the `GetCode()` method, these variables play the following roles:

- `codeSubStringList` is a list containing the line number to start error reporting on, followed by the strings to parse for code that could be affected by changes to the current line.
- `originalPositionY` is the text line where the text has been changed.

- currentPositionY is a variable used to walk up and down from the current line.
- Token is an integer category returned by yylex(), as seen in *Chapter 2, Programming Language Design*.
- startPositionY and endPositionY are the lines that identify the beginning and end of the current declaration.
- inClass and inMethod report whether the declaration is in a class or a method.

Initialization in the GetCode() method consists of resetting the parser and starting the position variables from the current cursor row, which indicates on which line the cursor is located. This is illustrated in the following code snippet:

```
reinitialize()
originalPositionY := currentPositionY := cursor_y
```

A primary loop in this procedure walks backward from the cursor location, using the compiler's yylex lexical analyzer function to look at the first token on each line and find the nearest previous line on which an enclosing declaration begins, as illustrated in the following code snippet:

```
while currentPositionY > 0 do {
    yyin := contents[currentPositionY]
    yylex_reinit()
    if (token := yylex()) ~=== EOFX then {
        if token = (PROCEDURE | METHOD | CLASS) then {
            if token=METHOD then inMethod := 1
            if token=CLASS then inClass := 1
            startPositionY := currentPositionY
        }
    }
    if startPositionY ~= 0 then break
    currentPositionY -:= 1
}
```

You can see that walking backward is achieved by decrementing the current line index held in the currentPositionY variable. The preceding while loop terminates when a line is found that begins with a procedure, method, or class reserved word. When this while loop terminates without finding an enclosing declaration, parsing starts from *line 1*. This is achieved with the following if statement:

```
if startPositionY = 0 then startPositionY := 1
```

The method then searches forward from the cursor to find the enclosing end token. Lexical features such as multiline continued string contents make this trickier than we might expect. The following while loop is long enough that it is split into multiple segments for explanation. The first segment shows that the while loop steps one line at a time through the code to be displayed, advancing currentPositionY on each line and fetching contents from the class member variable list of strings, named contents. In Unicon, unterminated string constants can span multiple lines that end in an underscore, which is handled by an inner while loop. In the unlikely event that we reach the end of a file while in a multiline string, the expression break break exits out of two while loops in one shot:

```
currentPositionY := cursor_y
while currentPositionY < *contents + 1 do {
    yyin := contents[ currentPositionY ]
    yylex_reinit()
    while countdoublequotes(yyin)%2=1 & yyin[-1]=="_" do {
        currentPositionY +:= 1
        if not (yyin ||:= contents[currentPositionY]) then {
            break break
            }
        }
    yylex_reinit()
```

The main task of the while loop given in the preceding code snippet is presented in what is the second half of the loop, shown next. This inner loop uses the compiler's lexical analyzer to identify tokens that would indicate the boundary of a compilable unit. The end token indicates the end of a unit that can be compiled, while class and procedure indicate the beginning of a subsequent unit:

```
while ( token := yylex() ) ~=== EOFX do {
    case token of {
    END: {
        endPositionY := currentPositionY
        break
        }
    CLASS | PROCEDURE: {
        if currentPositionY ~= startPositionY then {
            endPositionY := currentPositionY-1
            break
```

```
            }
        }
    default : break
    }
}
```

The method finishes by constructing a slice of the source code to reparse and returning it as a list of strings, prefixed by the line number immediately preceding the slice, as illustrated in the following code snippet:

```
if endPositionY = 0 then
    return codeSubStringList := [ 0 ] ||| contents
if startPositionY = 0 then startPositionY := 1
if inMethod = 1 then
    codeSubStringList := [ startPositionY,
        "class __Parse()" ] |||
        contents[ startPositionY : endPositionY+1 ] |||
        ["end"]
else if inClass = 1 then
    codeSubStringList := [ startPositionY ] |||
        contents[ startPositionY : endPositionY+1 ] |||
        ["end"]
else
    codeSubStringList := [ startPositionY ] |||
        contents[ startPositionY : endPositionY+1 ]
return codeSubStringList
```

A careful reader might worry about whether the GetCode() function as presented might sometimes miss a declaration boundary and grab too much code—for example, if the word procedure or end is not at the beginning of a line. This is true but non-fatal, since it just means that if the source code is written in a very strange and improbable manner, the syntax checker might reparse a larger amount of code than necessary. Now, let's look at how the source code is colorized.

Using lexical information to colorize tokens

Programmers need all the help they can get with reading, understanding, and debugging their programs. In *Figure 10.1*, the source code is presented in many different colors to enhance its readability. This coloring is based on the lexical categories of different elements of the text.

Although some people consider colored text as mere eye candy and others are not able to see colors at all, most programmers value it. Many forms of typos and text-editing bugs are spotted more quickly when a given piece of the source code is a different color than the programmer expected. For this reason, almost all modern programmer's editors and IDEs include this feature.

Extending the EditableTextList component to support color

EditableTextList is a Unicon GUI component that displays the visible portion of a list of strings using a single font and color selection. EditableTextList does not allow the setting of a font or foreground and background colors for individual letters or words. To support syntax coloring, the Unicon IDE extends a subclass of EditableTextList named BuffEditableTextList to present the user with source code. BuffEditableTextList is not a full rich-text widget. As with EditableTextList, it represents the source code as a list of strings, but BuffEditableTextList knows to apply syntax coloring (and highlight an error line, if any) on the fly when it draws the source code.

Coloring individual tokens as they are drawn

To color tokens, BuffEditableTextList calls yylex() to obtain the lexical category for each token when it is drawn. The following code, drawn from the left_string_unicon() method in the BuffEditableTextList class, sets the color, using a big case expression from five user-customizable colors specified in a preferences object. Most reserved words are drawn with a special color, designated as syntax_text_color in the preferences. Separate colors are used for global declarations, for the boundaries of procedures and methods, and for string and cset literals. This simple set of color designations could be extended by assigning different colors to a few other important lexical categories, such as comments or preprocessor directives:

```
while (token := yylex()) ~=== EOFX do {
    Fg(win, case token of {
        ABSTRACT | BREAK | BY | CASE | CREATE | DEFAULT |
        DO | ELSE | EVERY | FAIL | IF | INITIALLY |
        iconINITIAL | INVOCABLE | NEXT | NOT | OF | RECORD |
        REPEAT | RETURN | SUSPEND | THEN | TO | UNTIL |
            WHILE : prefs.syntax_text_color
        GLOBAL | LINK | STATIC |
            IMPORT | PACKAGE | LOCAL :
                prefs.glob_text_color
        PROCEDURE | CLASS |
```

```
        METHOD | END      : prefs.procedure_text_color
     STRINGLIT | CSETLIT : prefs.quote_text_color
     default              : prefs.default_text_color
     })
   new_s_Position := yytoken["column"] + *yytoken["s"]-1
   DrawString(win, x, y,
              s[ last_s_Position : (new_s_Position+1)])
   off := TextWidth(win,
              s[ last_s_Position : (new_s_Position + 1)])
   last_s_Position := new_s_Position + 1
   x +:= off
   }
```

As can be seen from the preceding code, after the foreground color is set from the token, the token itself is rendered by a call to `DrawString()`, and the pixel offset at which the subsequent text should be drawn is updated using a call to `TextWidth()`. All of this, when combined, allows different lexical categories of source code to be drawn in different colors in the IDE. The term used in the industry is *syntax coloring*, although the part of our compiler that we brought in was only the lexical analyzer, not the parser function that performs syntax analysis. Now, let's consider how to draw the user's attention to the line, should the parser determine that the edits that were made on a line leave the code with a syntax error.

Highlighting errors using parse results

In a `BuffEditableTextList` component, the `fire()` method is called whenever the content is changed, as well as whenever the cursor moves. When content is changed, it sets a flag named `doReparse`, indicating that the code should be syntax-checked. The check does not occur until the cursor is moved. The code for the `fire()` method is shown here:

```
method fire(type, param)
   self$Connectable.fire(type, param)
   if type === CONTENT_CHANGED_EVENT then
      doReparse := 1
   if type === CURSOR_MOVED_EVENT &
         old_cursor_y ~= cursor_y then
      ReparseCode()
end
```

In the preceding code, the ReparseCode() method is occasionally called in the Unicon IDE in response to a cursor move, in order to see whether editing has resulted in a syntax error. Only cursor moves that change the current line (old_cursor_y ~= cursor_y) trigger the ReparseCode() method, as shown here:

```
method ReparseCode ()
   local s, rv, x, errorObject, timeElapsed,
       lineNumberOffset
   if doReparse === 1 then {
     timeElapsed := &time
     SetErrorLineNumber ( 0 )
     uni_predefs := predefs()
     x := 1
     s := copy(GetCode()) | []
     lineNumberOffset := pop(s)
     preproc_err_count := 0
     yyin := ""
     every yyin ||:= preprocessor(s, uni_predefs) do
        yyin ||:= "\n"
     if preproc_err_count = 0 then {
        yylex_reinit()
        /yydebug := 0
        parsingErrors := []
        rv := yyparse()
     }
     if errors + (\yynerrs|0) + preproc_err_count > 0 then {
        # . . .every loop from Sending compiler output to the IDE here
     }
     else uidlog.MsgBox.set_contents(["(no errors)"])
   doReparse := 0
   }
end
```

The ReparseCode() method does nothing unless the code has changed, indicated by doReparse having the value 1. If the code has changed, ReparseCode() calls GetCode(), reinitializes the lexer and parser, calls yyparse(), and sends any error output to the IDE's message box. The actual line on which the error occurs is also highlighted when the code is redrawn as follows.

Within the draw_line() method in the BuffEditableTextList class, if the current line being drawn is the one found in the errorLineNumber variable, the foreground color is set to red:

```
if \errorLineNumber then {
    if i = errorLineNumber then {
        Fg(self.cbwin, "red")
        }
    }
```

You have now seen that setting different colors for different kinds of tokens, such as reserved words, is fairly easy and requires only the lexical analyzer to be involved, whereas checking for syntax errors in the background was a fair bit of work.

Summary

In this chapter, you learned how to use lexical and syntax information to provide the coloring of text in an IDE. Most of the coloring is based on relatively simple lexical analysis, and much of the work required involves modifying the compiler frontend to provide a memory-based interface, instead of relying on reading and writing files on disk. In this chapter, you picked up several skills. You learned how to color reserved words and other lexical categories in a programmer's editor, communicate information between the compiler code and the programmer's editor, and highlight syntax errors during editing.

Up to this point, this book has been about analyzing and using the information extracted from source code. The rest of this book is all about generating code and the runtime environments in which programs execute. The topic we will explore in the next chapter is bytecode interpreters.

Questions

1. A significant percentage of the population are partly or completely colorblind. How might you provide colorblind individuals with the same benefits that colors provide in this chapter?

2. Reparsing code to look for syntax errors whenever the user moves the cursor to a different line might be too often, or it might not be often enough. Can you suggest a better criterion for how often to reparse the code?

3. Suppose you want to add syntactic nesting information so that a user can visually tell which blocks of code are nested within which other blocks. Is there a way that you can add this information to the IDE in addition to the color-coding presented in this chapter? How might you go about that?

Join our community on Discord

Join our community's Discord space for discussions with the authors and other readers:

https://discord.com/invite/zGVbWaxqbw

Section III

Code Generation and Runtime Systems

After this section, you will finally be able to run programs written in your new programming language.

This section comprises the following chapters:

11

Preprocessors and Transpilers

This chapter returns us from our detour into IDEs back to the quest of generating output from our source program that can run. There are many ways to produce executable output from a programming language, and rather than pick just one in order to adhere to a rigid sequential narrative, this and the next couple chapters are a bit like a choose-your-own-adventure book that explores three ways of producing an executable: this chapter discusses translation to another high-level language, while *Chapter 12* presents translation to a lower-level software instruction set called a bytecode machine, and *Chapter 13* illustrates translation to native code that runs on the hardware's instruction set.

The ordering of these three chapters is intentional. The code generation for this chapter is easier to implement but offers slower performance than the strategy demonstrated in *Chapter 12*, which is easier but slower than the strategy of *Chapter 13*. You may want to start with the easiest implementation that will meet your requirements and migrate to a more challenging implementation strategy later if needed.

In the good old days, there were two primary means of implementing a programming language: either write an interpreter that implements that language directly or write a compiler that translates that language down to an assembler language or machine code. At this point, however, for many new programming languages, the quickest path to a first implementation is to generate source code for another more mainstream programming language that has an existing implementation. A language tool that generates code that is fed as input into another high-level language tool is commonly called a **preprocessor** or a **transpiler**.

The terms preprocessor and transpiler overlap but are not synonyms. When a new language extends an existing language or is limited to textual replacements, this is commonly called a preprocessor.

When the new language is substantially different from the target language, but both are considered high-level languages, the tool is called a transpiler. Using these definitions, the original AT&T implementation of C++, called Cfront and commonly referred to as a preprocessor, also may be called a transpiler using modern terminology. One possible way to distinguish between preprocessors and transpilers is by whether the tool fully parses the source language in order to generate its output.

This chapter introduces preprocessors and transpilers and aspects of their implementation. The sections in this chapter cover the following main topics:

- Understanding preprocessors
- Code generation in the Unicon preprocessor
- The difference between preprocessors and transpilers
- Transpiling Jzero code to Unicon

In this chapter, the skills learned include how to expand macros, generate output during a tree traversal, select different output operators depending on the source language type used, move around declarations to where the target language allows them, and how to manage data structures and semantic attributes for preprocessing and transpiling.

Understanding preprocessors

A preprocessor applies a transformation to source code. Some preprocessors are stand-alone tools, usable by and independent of any programming language tool. The most famous of these is probably the Unix m4 preprocessor. However, most preprocessors are tied to, associated directly with, and often integrated into a particular programming language and apply the transformation before the language compiler reads it in for lexical analysis. The output code from a preprocessor usually resembles its input with only a few changes, so you might wonder: why bother? Usually, the reason is that the judicious use of a preprocessor can make the code shorter and more readable.

A typical preprocessor transformation might be to replace all occurrences of some symbolic abbreviation such as PI with 3.1415. Another typical preprocessing example would be to expand some function-like syntax with parameters at compile-time, such as replacing occurrences of CUBED(X) with (X * X * X).

Symbols such as PI or CUBED that are replaced by preprocessors when they are encountered are called **macros**. Most preprocessors look for and apply these macro transformations to the source code one line at a time, although the replacement text of a macro may span several lines. Writing PI in place of 3.1415 merely shortens code and improves readability.

On the other hand, the notation CUBED(X) could be implemented by either a function call or a macro, and on many compilers, a macro at compile time will be faster than a function call performed at execution time. The next section illustrates preprocessor macros in a small real-world example.

A preprocessing example

Consider the symbol Key_Home used in the Unicon graphics facilities. It denotes the HOME key on the keyboard, which is returned as a different integer on different platforms. In the MS Windows implementation, a Unicon header file named keysyms.icn contains the following line:

```
$define Key_Home                36
```

This line defines Key_Home to be 36 on Windows. The event codes defined in keysyms.icn are returned by Unicon's Event() function, which reports the next key or mouse input event. Typical client code looks as follows:

```
case e := Event() of {
    Key_Home: do_home_key()
    other cases
  }
```

After preprocessing on an MS Windows platform, it looks like this:

```
case e := Event() of {
    36: do_home_key()
    other cases
  }
```

Other platforms such as Linux/X11 use different integer codes native to their platforms. The substitution of the symbol Key_Home for the appropriate code is performed in this manner on each platform. In addition to adding platform portability, code readability is improved by avoiding magic numbers like 36.

Macro preprocessors are important in low-level languages, particularly assembler languages in which much code is repetitive or otherwise cumbersome. In these low-level languages, programmers need all the help they can get. When used properly, macros make code more maintainable by reducing duplication; fixing a bug in a macro body may fix a bug in hundreds of locations where that macro is used. Of course, the flip side of this is true: if you write a macro and get it a little bit wrong, your one little bug may be introduced in hundreds of locations where that macro is used.

Macro preprocessors have been so popular in the past that entire languages and communities have revolved around advanced macro programming. For some personalities, macro coding may be addictive; some coders use macros whenever possible! It might be a joke, or it might be legitimate , to say that such programmers have adopted a "macro-oriented programming" paradigm.

On the other hand, macro preprocessors are also famous for being too powerful, resulting in obscure bugs. They have been over-used in some projects to the point where humans do not understand what their complex macros are doing, particularly when many macros are being used in combination. In such instances, macro bugs can be difficult to find and fix. For example, our macro CUBED(X) looks innocent enough, but what if you pass in a parameter that has side effects, such as CUBED(n++)? It would expand to (n++ * n++ * n++), which might not be what was intended.

The dangers of macros influenced the Java language designers, who feel they have improved upon C/C++ by omitting the preprocessor entirely. Icon and Unicon's more conservative reaction to these problems was to provide a preprocessor with macro parameters, such as the X in CUBED(X) intentionally omitted. This preserves the preprocessor's ability to provide symbolic names for magic constants and select different code to eliminate the most egregious preprocessor-induced bugs. Now it is time to look at other aspects of the Unicon preprocessor.

Identity preprocessors and pretty printers

Two special cases of preprocessing might be worth considering before looking at more complicated preprocessors. An **identity preprocessor** is a source code preprocessor that implements the identity transformation, generating identical output to its input. Identity preprocessors are not really a thing, but if you were starting a preprocessor project, it would make good sense to start with code that implements all the I/O correctly, before you started monkeying around with transformations of that code. To copy the input to the output character by character is easy and boring; to copy the input to the output after a full lexical analysis and possibly a full parse of the code might be a good test of a compiler frontend and its lexical attributes for line and column numbers, which are needed to support debugging tools.

A **pretty printer** is a bit like an identity preprocessor, except it reformats the whitespace around the code to improve readability. Pretty printers are at least as old as the LISP language, which is one of the oldest languages of all. Most languages are designed specifically to allow programmers to not worry about indentation, although bad indentation is known to result in occasional bugs and maintenance headaches.

Since a pretty printer may do a full parse of the input but applies a lexical transformation to the whitespace for increased human readability, it is arguably not a preprocessor or transpiler, but it does bring up the question of how closely the output of a preprocessor or transpiler should resemble the input, which may affect subsequent debugging efforts. If you were starting a new transpiler project from scratch, you might do very well to start from a pretty printer for the input language if one is available, and then gradually modify the pretty printer's output to be the target language instead. In any case, if you are using an IDE that reformats your code for you, the pretty printer transformation might be integrated into your IDE. Now let's consider a larger real-world preprocessor example.

The preprocessor within the Unicon preprocessor

This chapter discusses the Unicon translator as our example preprocessor. Unicon exhibits attributes of both a preprocessor and a transpiler. On the one hand, Unicon is a preprocessor since it translates Unicon, which has object-oriented constructs, into often similar-looking but non-object-oriented Icon code and then invokes (a modified version of) the Icon translator icont to generate VM bytecode. On the other hand, the Unicon translator does not just read lines and replace macro symbols with their bodies; it does a full parse of the input code. With a full syntax tree and symbol tables of the sort that a compiler would use, some of the Unicon translator output for constructs such as classes is not a simple text substitution and does not closely resemble anything in the input. Additionally, the presence of a complex multiple inheritance mechanism and packages goes far beyond the concerns of a normal preprocessor.

A second argument against calling Unicon a preprocessor is that Unicon contains an entire macro preprocessor as its first stage! This preprocessor within the "preprocessor" is described in this section. Space does not allow coverage of every aspect of the preprocessor; an emphasis is placed on the core functionality of defining symbols and subsequently looking for them and substituting their replacement text. *Figure 11.1* shows the Unicon translation pipeline with the preprocessor at the front, the Unicon translator as described above, the two phases of bytecode generation named itran and ilink that respectively translate to human-readable and binary bytecode, and the bytecode interpreter named iconx.

To make the code examples as clear as possible, a brief description of the preprocessor functionality is in order. The Unicon preprocessor is like the C/C++ preprocessor, with some major exceptions. Since the hash sign # used by the C/C++ preprocessor is Unicon's comment character, the preprocessor symbol in Unicon is the dollar sign $. So, directives like #if or #define in C/C++ are $if and $define in Unicon. Secondly, the Unicon preprocessor does not have macro parameters.

The ability to $define symbols and to include selected blocks of code via $ifdef...$endif are major features that were added to Icon and Unicon while avoiding the primary sources of problems introduced into C/C++ by their preprocessor.

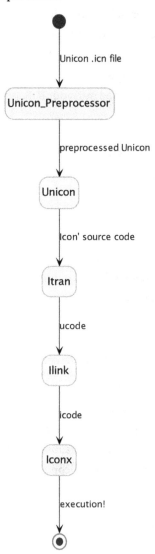

Figure 11.1: Preprocessing is the first stage in the Unicon translation process

The Unicon preprocessor is implemented by a 700-line module named preproce.icn that was written by Bob Alexander and previously used in Jcon, a Java-based Icon implementation. A generator function called preprocessor() works one line of source code at a time in a big while loop.

Each line is analyzed using the string scanning control structure, resulting in a sequence of output lines that are produced by the function preprocessor() and transmitted to the rest of the compiler as its input. The code for this outermost loop in the preprocessor looks like the following:

```
procedure preprocessor(fname, predefined_syms)
    # initialization
    while line := preproc_read() do line ? {
        # preprocess one line of code, suspending results
        }
    # finalization
end
```

Each iteration of the while loop analyzes the string corresponding to one line of source code. After skipping over whitespace, if a preprocessor directive is encountered, a helper function named preproc_scan_directive() is called, and otherwise, macro substitutions are performed on the line by a helper function named preproc_scan_text().

Let's look at these two functions, starting with preproc_scan_directive(). A lot of the logic of this function is dedicated toward maintaining a stack of conditions as to whether various $if blocks are enabled or disabled. The function starts by identifying which preprocessor directive is being invoked, from around twelve that are defined in the language. For the sake of time and space, we will confine our consideration to the $define directive that introduces macros and their definitions into the preprocessor. The $define directive doesn't do anything if it occurs inside a block that has been excluded by $if or $ifdef. In a live block, the symbol being defined is grabbed using the function prepoc_word(), followed by the text to substitute when that symbol is found in the code, which runs to the end of the line. It is an error to attempt to change the definition of a symbol that has already been defined, but new symbols are entered into a table named preproc_sym_table by the assignment preproc_sym_table[sym] := value. When symbols are encountered in regular source code lines, they are looked up in the table, and if a definition is present, the value will be used to replace the occurrence of the symbol:

```
procedure preproc_scan_directive()
    # initialization
    preproc_command := preproc_word()
    case preproc_command of {
        "define": {
            if /preproc_if_state then {
```

```
                if sym := preproc_word() then {
                    if value := preproc_scan_define_value() then {
                        if \(old := preproc_sym_table[sym]) ~=== value then {
                            preproc_error("redefinition of " || sym || " = " ||
                                          old)
                        }
                        else {
                            preproc_sym_table[sym] := value
                        }
                    }
                }
                else {
                    preproc_error()
                }
            }
        }
        # … other cases for other preprocessor directives, $undef, $ifdef …
    }
    # finalization
end
```

The procedure `preproc_scan_text()` processes a line of source code, looking for macro symbols and replacing them with their definitions. It is a long procedure so our discussion will skip large parts and focus on the macro search-and-replace code. It looks for possible macro symbols, skipping over comments and text inside single- and double-quote characters. For every identifier encountered by the preprocessor, if the identifier is found via `preproc_sym_table[ident]`, the value is substituted after first applying the `preproc_scan_text()` to it recursively, to expand macros within macros. Infinite and/or mutually recursive symbol expansions are prevented by a done_set of macros that are active at any given time:

```
procedure preproc_scan_text(done_set)
    initialization code, skip over code excluded by $if directives.
    result := ""
    while tab(upto(interesting_chars)) do {
        case move(1) of {
            "#": … skip over comments
```

```
                    "\"" | "'": … skip over quotes (no macros in constants)
                    default: { …possible start of macro symbol
                        move(-1)
                        p := &pos
                        ident := tab(many(preproc_word_chars))
                        if value := \preproc_sym_table[ident] then {
                            if /done_set |
                                    {type(done_set) == "string" &
                                        done_set := set([done_set])
                                    not member(done_set,ident)} then {
                                value ? value :=
                                    preproc_scan_text(insert(copy(\done_set),
                                                            ident) | ident)
                                result ||:= &subject[q:p] || value
                                q := &pos
                                }
                            }
                        }
                    }
                }
        finalization code, suspends results out to the caller
    end
```

Although Unicon's preprocessor is a lot smaller and simpler than that used in C/C++, it performs a similar task. It is line-oriented and primarily dedicated to macro-based rewrites of defined macro symbols to provide multi-platform portability and allow software to be more readable and configurable.

Code generation in the Unicon preprocessor

After the preprocessor, the rest of the Unicon translator takes Unicon input and outputs an extended dialect of Icon that has no name but is occasionally referred to as Icon' (Icon prime). Unicon is written in around 7,000 lines of Unicon code and another 1,100 lines of iyacc specification. It is tiny compared to a conventional compiler, but ten times the size of the preprocessor's preprocessor described in the previous section.

Transforming objects into classes

The output of Unicon frequently resembles the input closely enough to qualify as a preprocessor. Regular Icon code such as user-defined procedures with their statements and expressions pass through Unicon almost unmodified. However, the Unicon translator implements several language extensions by changing the source code. For example, packages are implemented via name mangling. You can see what Unicon does with a given input file foo.icn, if anything, by running unicon -E foo and looking at its output.

The primary transformation performed by this preprocessor turns each class definition into a pair of record declarations, a constructor procedure, and various initialization pieces. Consider the following simple class:

```
class myClass(a,b,c)
    method h()
        write("hello, world")
    end
  end
```

From this class, the Unicon translator writes out two record types: one for instances of the class and one for methods. Taken together, the field names in these two records comprise the namespace, and this namespace is resolved at runtime. The methods record for each class is a singleton; a single instance is created and then shared by all the class instances through an aliased pointer held by each instance in their __m field. The singleton is also accessible globally through a myClass__oprec variable that is only accessed within class constructor procedures:

```
record myClass_state(__s, __m, a, b, c)
record myClass__methods(h)
global myClass__oprec
```

Each method in a class is written out almost identically to how it appears in the source code, but name-mangled to introduce the class name, and a new parameter is inserted into the front – the self variable:

```
procedure myClass_h(self)
        write("hello, world")
  end
```

The self variable is used in several ways. Unless hidden by a local identifier, references to names in a class scope such as a, b, or c in class myClass above are textually replaced by self.a, self.b, or self.c. Similarly, the names of method calls are looked up in the methods vector; considerable magic occurs within the Unicon virtual machine to turn foo(…) into self.__m.foo(self, …) but only when needed.

There is one other fundamental part of generating a class, which is the class constructor procedure. The first time that a constructor for a given class is executed, its methods vector is instantiated and initialized by a procedure called classnameinitialize(). On that and all subsequent calls to the constructor, an instance is created:

```
procedure myClass(a,b,c)
local self,clone
initial {
    if /myClass__oprec then myClassinitialize()
    }
    self := myClass__state(&null,myClass__oprec,a,b,c)
    self.__s := self
    return self
end
```

This mapping of object-oriented constructs onto lower-level Icon procedures and records has a few historical warts but has been sufficient to inspire the successful creation of much larger projects in Unicon than have been developed in Icon. When classes use inheritance and define themselves in terms of other classes, an additional layer of complexity is introduced, which is omitted here for the sake of space. See the Unicon Implementation Compendium if you want to study more about inheritance. Now let's look at how Unicon generates the code described in this section.

Generating source code from the syntax tree

This section describes the code generation process within the Unicon translator. I will explain everything as best I can, but this code was not written for the pedagogical benefit of the learner. If you are not proficient in the Unicon language, you might skip this section, or skim it and avoid getting bogged down in the arcane code.

For a preprocessor or transpiler working from a syntax tree, code generation is a tree traversal in which most internal nodes just visit their children, and most leaf nodes that correspond to elements from the source code simply print themselves as they appeared in the original.

The Unicon syntax tree is heterogeneous. Most syntax tree nodes are of the type treenode from tree.icn. (Warning: Unicon's tree.icn is totally different from Jzero's tree.icn; do not confuse them!) The code for such nodes is printed by a procedure called yyprint(node). This procedure has a special-case code for many kinds of tree nodes, distinguished by their label field. The procedure yyprint() also incorporates other kinds of objects that occur in the tree, such as plain string literals, the token objects associated with leaves, and custom non-treenode objects for special entities, described below. The code for yyprint() starts with the following:

```
procedure yyprint(node)
    initialization
    case type(node) of {
        "treenode": {
            if node.label=="package" then …
            …other treenode types
        }
        other types of objects that appear in the tree
```

Source code tokens appear occasionally as leaves in the tree. Around such tokens with lexical line and column attributes, yyprint() emits #file and #line directives so that debuggers will report source code line numbers correctly. It is too long to present in its entirety, but some of the relevant parts of yyprint() include:

```
"token": {
    if \outfilename ~== \ (node.filename) | (outline > node.line)
    then {
        write(yyout,"\n#line ", node.line-1, " \"",
                node.filename, "\"")
        outline := node.line
        outcol := 1
        outfilename := node.filename
    }
    output spaces to get to the correct column, write the lexeme
```

In addition to tree nodes, tokens, and occasionally raw strings that should be output, several more complex classes are used for major declaration units with substantial code generation roles, including Class, Method, and about a dozen other types of special nodes. The syntax tree classes that represent major syntax constructs in Unicon know how to generate code for themselves.

These classes predate Unicon; they originated in Unicon's predecessor Idol, which was a more basic line-oriented preprocessor that did the job with no syntax tree. To this day, they live in a idol.icn file that originated in the Fall of 1989 as a university class project.

A class named declaration handles Icon and Unicon headers of the form:

```
tag name (field1, field2, …)
```

Where tag is one of several reserved words that use that syntax in Icon and Unicon, such as record and procedure. The class declaration provides two methods. Method Write(f) writes the header to file f, and method String() converts the declaration into a string.

Unicon's class Class is a subclass of class declaration. It provides around seventeen methods that perform semantic analysis tasks related to scope and inheritance calculations in addition to code generation. When yyprint() encounters a Class instance, it calls its Write() method via node.Write(yyout). The Write() method is long and will be presented in several parts. First, it writes out the method bodies for all the methods that are defined directly in the class:

```
method Write(f)
    nam := self.name
    yyprint("\n")
    writemethods(f)
```

The actual representation of the class includes a pair of record declarations, but before Write() can generate those, it must complete the calculation of what fields and what methods the class gains by means of inheritance. Inheritance is performed by a call to a method named resolve() that populates a list of inherited fields named ifields. The record that holds instances' state variables is written out; since it is not derived from any real location in the source code, a #line directive is emitted stating that the code comes from a fake location: line 1 of the imaginary file __faux.icn. A couple of fields in each instance contain pointers to the instance and to the methods record for the class. After that, the explicit fields of the class are written out by asking the fields object to produce its string representation via the method String(). The explicit fields are followed by the inherited fields that come from superclasses:

```
    if /self.ifields then self$resolve()
    write(f,"#line 1 \"__faux.icn\"")
    writes(f,"record ",nam,"__state(__s,__m") # reserved fields
    rv := ","
    rv ||:= self.fields$idTaque.String()                        # my fields
    if rv[-1] ~== "," then rv ||:= ","
```

```
     every ifi := (!self.ifields).ident do {
        if type(ifi) == "string" then
           rv := rv || ifi || "," # inherited fields
        else if type(ifi) == "treenode" & ifi.label == "arg3" then {
           rv := rv || (ifi.children[1].s) || ","
           }
        else stop("Write(): can't handle ", type(ifi))
        }
     yyprint(rv[1:-1] || ")\n")
```

After the instance record is written out, a record for the class fields is generated. The methods record is a singleton, providing a shared reference to the methods to all instances of the class:

```
     writes(f,"record ",nam,"__methods(")
     rv := ""
     every s := (((methods$foreach())$name()) | # my explicit methods
                 (!self.imethods).ident |        # my inherited methods
                 supers$foreach())                # super.method fields
        do rv := rv || s || ","
     if *rv>0 then rv[-1] := ""                   # trim trailing comma
     yyprint(rv||")\n")
```

Each class's methods record is stored in a global variable whose name is classname__oprec. The first instance of the class triggers the initialization of the methods record. This requires that all superclasses also be initialized, so references to their methods records are also declared:

```
     writes(f,"global ",nam,"__oprec")
     every writes(f,", ", supers$foreach(),"__oprec")
     yyprint("\n")
```

The constructor procedure for a class is the most complex piece of code to generate. It starts with writing out the procedure header, which is performed by the method writedecl(). The first time the constructor is called, an initial section calls the procedures that initialize the current class's and all superclasses' method records:

```
     self$writedecl(f,"procedure")
     yyprint("local self,clone\n")
     yyprint("initial {\n  if /"||nam||"__oprec then "||
             nam||"initialize()\n")
     if supers$size() > 0 then
```

```
                every (super <- supers$foreach()) ~== nam do
                    yyprint("   if /"||super||"__oprec then "||
                            super||"initialize()\n"||
                            "   "||nam||"__oprec."||super||" := "||
                            super||"__oprec\n")
            yyprint("   }\n")
```

Class constructor procedures can have default values for the various fields of the class. The default values are found to be present when the field is a treenode with the label arg3. In that case, the generated code for that field checks if the passed-in value is null using the unary null-check operator /, and if it is null, then an assignment of the default value is written out:

```
        every fld := fields$foreach() do {
            if type(fld) == "treenode" & fld.label == "arg3" then {
                writes(f,"/",fld.children[1].s, " := ")
                yyprint(fld.children[3])
                yyprint("\n")
                }
            }
        every ifi := (!(self.ifields)).ident do {
            if type(ifi) == "treenode" & ifi.label == "arg3" then {
                writes(f,"/", ifi.children[1].s, " := ")
                yyprint(ifi.children[3])
                yyprint("\n")
                }
            }
```

The constructor creates the actual instance and stores it in a variable named self:

```
        writes(f,"   self := ",nam,"__state(&null,",nam,"__oprec")
```

The fields of the instance are initialized from constructor parameters unless those parameters were passed into an initially method, in which case the initially method is responsible for any initialization of fields. An initially method is Unicon's way of supplying code to run whenever an object instance is created. Checking whether we have an initially method with fields is one heck of an if statement. The technique here is brute force. I was a student when I wrote this code, and it hasn't been broken enough to make us rewrite it:

```
    if (("initially" == (m := (methods$foreach()))$name()) & \(m.fields)) |
        (((mn := !(self.imethods)).ident == "initially") &
```

```
        (m := classes.lookup(mn.Class).methods.lookup("initially")) &
        \(m.fields))
    then {
```

If we do in fact have an initially method with fields, the next big wrinkle is whether the initially method takes variable arguments. If it does, we generate code to build the arguments as a list and then use the binary apply operator ! in the generated code to call the initially method from our constructor. The outermost else here shows the code is a lot simpler when the initially method is not a variable argument method. In that case, we just generate a code to call initially, inserting the instance followed by the arguments. In both cases, if the initially method fails, the constructor fails and the calling code gets no object to work with:

```
        yyprint(")\n  self.__s := self\n")
        if \ (m.fields.varg) then {
            m.fields.String()[1:-2] ? {
                if find(",") then {
                    writes(f,"  self.__m.initially!([self,")
                    while writes(f,tab(find(","))) do {
                        move(1)   # if last was nonfinal write it
                        if find(",") then writes(f,",")
                        }
                    write(f, "]|||", tab(0),") | fail")
                    }
                else {
                    write(f,"  self.__m.initially!(push(", tab(0),
                        ",self)) | fail")
                    }
                }
            }
        else {
            writes(f,"  self.__m.initially(self,")
            yyprint(m.fields)
            yyprint(") | fail\n")
            }
        }
```

If there was an initially section but no fields declared on it, the constructor still calls the initially method that is declared explicitly in the class or inherited from a superclass.

In this way, an object can initialize its fields entirely on its own, with no parameters. This is transitive with respect to inheritance; initially methods call superclasses' initially methods. This allows superclasses to initialize their own fields that the class may have inherited from those superclasses:

```
if ((methods$foreach())$name()| (!self.imethods).ident) == "initially"
then
    yyprint("  self.__m.initially(self) | fail\n")
    }
```

The return value from the constructor is the instance – the self object that has been constructed:

```
yyprint("  return self\nend\n\n")
```

After the constructor procedure, the remaining code to generate for a class consists of the methods record initializer procedure. Its code only executes the first time it is called:

```
yyprint("procedure "||nam||"initialize()\n")
 writes(f,"  initial ",nam,"__oprec := ",nam,"__methods")
 rv := "("
 every s := (methods$foreach())$name() do { # explicit methods
   if *rv>1 then rv ||:= ","
   rv := rv || nam || "_" || s
 }
 every l := !self.imethods do {              # inherited methods
   if *rv>1 then rv ||:= ","
   rv := rv || l.Class || "_" || l.ident
 }
 yyprint(rv||")\nend\n")
end
```

As you have seen, the code generation for Unicon classes is quite involved. We next present the code that Unicon executes in order to calculate what each class inherits from its superclasses, which is implemented in the resolve() method.

Closure-based inheritance in Unicon

Unicon's main inheritance mechanism is inspired by the concept of transitive closure. The basic idea is to start from those fields and methods that you have in your class and look in superclasses for things that can be added until nothing more can be found and added. Superclasses are visited from left to right in depth-first order.

One of the things you can find in a superclass is more superclasses, so this process repeats until no more superclasses can be found. Interestingly, because inheritance stops as soon as nothing more can be added, it is no problem for two or more classes to inherit from each other under these semantics. The code for all this is in the method resolve() within the class Class in the file idol.icn. The method resolve() builds lists whose elements are [class, ident] pairs.

The code below can be read in English as something like: for every superclass name, fetch its corresponding Class object. For every field in that superclass, see if we've got it, and if not, add it to our inherited fields. Then do the same for methods:

```
method resolve()
    self.imethods := []
    self.ifields := []
    ipublics := []
    addedfields := table()
    addedmethods := table()
    every sc := supers$foreach() do {
        if /(superclass := classes$lookup(sc)) then
            halt("class/resolve: couldn't find superclass ",sc)
        every superclassfield := superclass$foreachfield() do {
            if /self.fields$lookup(superclassfield) &
               /addedfields[superclassfield] then {
                addedfields[superclassfield] := superclassfield
                put ( self.ifields , classident(sc,superclassfield) )
                if superclass$ispublic(superclassfield) then
                    put( ipublics, classident(sc,superclassfield) )
            } else if \strict then {
                warn("class/resolve: '",sc,"' field '",superclassfield,
                    "' is redeclared in subclass ",self.name)
            }
        }
        every superclassmethod := (superclass$foreachmethod())$name() do {
            if /self.methods$lookup(superclassmethod) &
               /addedmethods[superclassmethod] then {
                addedmethods[superclassmethod] := superclassmethod
```

```
                        put ( self.imethods, classident(sc,superclassmethod) )
                }
        }
        every public := (!ipublics) do {
            if public.Class == sc then
                put (self.imethods, classident(sc,public.ident))
        }
    }
end
```

These core elements of Unicon have been around for 35 years and have not changed much in that time. Our discussion will now move to the creation of a more explicit transpiler, from our toy language Jzero to Unicon, but first, let's consider some of the differences between preprocessors and transpilers.

The difference between preprocessors and transpilers

The Unicon translator writes its output in Icon and then invokes an extended version of the Icon translator icont to generate bytecode. The use of a preprocessor and a full parser smells like things a transpiler would do, but the fact that Unicon's output looks almost identical to its input except when classes are involved makes Unicon feel like it is still a preprocessor in some respects.

Fundamentally, a transpiler writes out source code in a different high-level language than its input. Usually, there is in fact a difference in language level, and usually, the transpiler transpiles from a higher-level language down to a slightly lower-level language. The poster children for this process are the myriad of languages that are implemented by writing out C code and using a C compiler as their code generator. C is effectively used as a portable assembly language. The considerable optimization performed by most C compilers is cited as a great benefit, although if anything can defeat optimizers' assumptions, it is transpilers generating staggering amounts of machine-written C code.

When writing a transpiler, a higher-level language's control structures may need to be implemented using considerably different control structures in a lower-level or different-paradigm target language. In the extreme case, the higher-level language control structures may be turned into a data structure representation and the lower-level language may be implementing those control structures by a form of interpretation.

Similarly, the data values in the source language must be mapped by a transpiler onto data values in the target language. When the two languages are similar, substantial performance can be gained if the data representation is a 1:1 map from source values to target values, or conversely, a substantial performance loss will be incurred if the data representations chosen by the transpiler are inefficient.

Unfortunately, the values represented and the language semantics are different enough that it is usually easiest to map all source values onto objects in the target language's object-oriented facilities, and this mapping can then impose enough of a performance cost to pose substantial problems. A transpiler may need to incorporate substantial shortcuts or optimizations for common cases and fall back on a more expensive representation only when necessary.

Most of these considerations will seem like vague hand-wringing unless we can see how they impact a transpiler project in practice. For that, we will turn to a transpiler for the Jzero language.

Transpiling Jzero code to Unicon

All this talk about transpilers is all well and good, but seeing is believing. This section presents a transpiler implementation of the Jzero language, writing out Unicon code from the Jzero source code. The implementation is structured to resemble the tree traversal that is used to generate intermediate code in *Chapter 9*, so you might experience some déjà vu. But before we go there, consider what the transpiler code generator needs to do. As we've seen before, this involves a set of semantic attributes and a set of rules for how to compute them.

Semantic attributes for transpiling to Unicon

Our transpiler will introduce three semantic attributes. The first semantic attribute will be a representation of the output Unicon code. Like the `icode` attribute introduced for building a list of intermediate code instructions in *Chapter 9*, the transpiler implementation introduces an attribute named `icncode` that is used to build a list of output strings, generally one per output line. I say generally because, of course, there are times when different chunks of code need to be spliced together that must go on a single line, so some of the values in `icncode` may constitute parts of an output line, as little as a single token. In any case, the various tree nodes' `icncode` attributes will be concatenated together as we work our way up the tree so that at the root, we will have a gigantic list of code that we can write out and send to the Unicon compiler.

The other two semantic attributes are similar to `icncode` but serve a special purpose: they assist in moving code around to where Unicon requires it to be located. Java `static` declarations have substantially different semantics from Unicon statics. Static declarations must be pulled out to a different location in the generated code. For this reason, Jzero statics are placed into a separate semantic attribute analogous to `icncode`, named `staticcode`.

Similarly, Java local variables may be declared anywhere in a method body, while Unicon `local` declarations must be at the top of a procedure or method body. For this reason, local Jzero variable declarations are placed in an attribute named `vardecls` so that they can be moved to the top of a statement sequence. The Unicon version of the class tree is modified with the following additions:

```
class tree (id,sym,rule,nkids,tok,kids,isConst,stab,
            typ, icncode, staticcode, vardecls)
```

The corresponding Java additions to `tree.java` look like this:

```
class tree {
   . . .
   typeinfo typ;
   ArrayList<String> icncode, staticcode, vardecls;
```

Depending on the differences between the source language and transpiled target language, some of the attributes and algorithms developed for three-address code might still be useful for transpiler code generation. For Unicon, maybe we won't need them, but if you are transpiling between two very different languages, you may decide to have class tree track addresses and Boolean labels in semantic attributes, as developed in *Chapter 9*.

The attributes `icncode`, `staticcode`, and `vardecls` will be synthesized during a post-order tree traversal. We will show you the main transpiler traversal method soon, but first, let's consider the code generation model.

A code generation model for Jzero

Before presenting code to translate Jzero to Unicon, we need to design how Jzero code will be mapped to Unicon. For intermediate code, in *Chapter 9*, a set of semantic rules was presented that mapped source code production rules onto semantic attribute assignments. This achieved the one-to-many mapping of grammar rules onto much lower-level intermediate code addresses and instructions.

One might hope that the code generation model for transpiling to a much higher-level language will be a 1:1 mapping: a Java class will be mapped to a Unicon class, etc. How can we even tell what to include in this mapping? Let's start with a simple example and work out how to translate it, adding features to the model as we go.

The two languages both have classes, but they are different. For example, a Java class has two kinds of methods, static and non-static, while a Unicon class has only non-static methods. Because Jzero is a small subset of Java, our transpiler can use a simpler mapping than would be required if it had to handle the entire Java language.

This transpiler will map a Jzero class to a Unicon package containing a class. Consider the Hello World Java program. A class named `hello` containing a `public static main()` method will be mapped to a `package hello` containing a `procedure main()`, along with procedures for any other static methods in the class. If Jzero were extended to handle objects with non-static fields and methods, then the code generation would also map a Unicon `class hello` containing any non-static members, so let's stick one of those in just for fun. To summarize, we are mapping a simple Jzero program such as the following `hello.java` file:

```
public class hello {
    public static void main(String argv[]) {
        System.out.println("hello, jzero!");
    }
}
```

The corresponding output from our transpiler should generate a package, and within it, a class, as shown below, which might be written to a `hello.icn` file. The empty class is present to reflect the logical mapping of a Java class onto a package containing a class, in case we want to expand Jzero into a larger subset of Java in the future:

```
package hello
 procedure main(argv [ ])
 write("hello, jzero!")
 end
 class hello()
 end
```

If you attempt to compile and run the above example, you will notice that the preceding code is incomplete. In addition to the `package hello` code, the generated Unicon code needs an ordinary `main()` procedure that starts outside of `package hello` and invokes the `main()` procedure within the `package hello`. This `main()` will be placed in a `hello_helper.icn` module:

```
procedure main(args)
    hello__main ! args
end
```

If you are wondering about the curious syntax in the call from main() to package hello's main(), the argument to Unicon main() is a list of strings; the code here turns it into the actual parameters passed to the package hello's main() procedure using Unicon's apply operator, binary exclamation (!). This procedure has been declared to accept a variable number of arguments Unicon-style, by adding square brackets after the parameter name in the code above. It is one thing to see an example and another to write code for it.

The Jzero to Unicon transpiler code generation method

The main transpiler code generation algorithm is a bottom-up post-order traversal of the syntax tree. To present it in small chunks for exposition purposes, the traversal is broken into a primary method, genunicon(), and helper methods for each non-terminal. The main objective of the traversal is to assign the three semantic attributes icncode, staticode, and vardecls; in most tree nodes, the only attribute that is assigned is icncode. In Unicon, the genunicon() method in tree.icn looks as follows:

```
method genunicon()
    every (!\kids).genunicon()
    case sym of {
        "ClassDecl":                { genuniClassDecl() }
        "CondAndExpr"|"CondOrExpr"|
        "MulExpr"|"AddExpr":        { genuniBinaryExpr() }
        "RelExpr":                  { genuniRelExpr() }
        # ...
        "QualifiedName":            { genuniQualifiedName() }
        "token":                    { genunitoken() }
        "FormalParm":               { genuniFormalParm() }
        "MethodDeclarator":         { genuniMethodDeclarator() }
        default: {
            icncode := []
            every icncode |||:= (!\kids).icncode
            }
        }
    end
```

The preceding code resembles the method gencode() presented in *Chapter 9*. Tree nodes that have any custom role in code generation directly invoke a helper method. The default for non-terminals that do not have custom output semantics is to simply concatenate the code found in all the node's children. The corresponding Java code looks like this:

```
void genunicon() {
    if (kids != null) for(tree k : kids) k.genunicon();
    switch (sym)  {
    case "ClassDecl":        { genuniClassDecl(); break; }
    case "AddExpr":
    case "MulExpr":          { genuniBinaryExpr(); break; }
    case "RelExpr":          { genuniRelExpr(); break; }
    // ...
    case "QualifiedName":    { genuniQualifiedName(); break; }
    case "token":            { genunitoken(); break; }
    case "FormalParm":       { genuniFormalParm(); break; }
    case "MethodDeclarator": { genuniMethodDeclarator(); break; }
    default: {
        icncode := new ArrayList<String>();
        if (kids != null) for(tree k:kids)
            icncode.addAll(k.icncode);
        }
    }
}
```

The methods that are used to generate code for specific non-terminals must generate the Unicon source code that corresponds to the Java code represented in the production rule. We will go through the syntax tree and transpiler rules from the bottom up. For each kind of tree node, we will present a semantic rule if appropriate, followed by the code.

Transpiling the base cases: names and literals

The genunitoken() method generates Unicon code for terminal symbols. When the same tokens are legal in Jzero and in Unicon, the icncode attribute can be set to contain just the source code text. In a larger Java subset, or for more complex transpilers, appropriate variable naming or translations of literals will require more output construction here. For example, the rules for integer or string literals in Jzero and Unicon differ, so how much does a Java string have to be altered to turn it into a legal Unicon string? In Unicon, the genunitoken() method looks like this:

```
method genunitoken()
    icncode := [ tok.text ]
    case tok.cat of { # more elaborate name mangling or literal translation
      parser.IDENTIFIER: { }
      parser.INTLIT: { }
      parser.STRINGLIT: { }
    # ...
    }
  end
```

In Java, genunitoken() looks like this:

```
void genunitoken() {
    icncode = new ArrayList<String>();
    icncode.add( tok.text );
    switch (tok.cat) {
      case parser.IDENTIFIER: { break; }
      case parser.INTLIT: { break; }
    // ...
    }
}
```

From genunitoken(), the next larger entities to translate will be various kinds of expressions, starting with the highest precedence things such as complex names constructed using one or more periods, also known as dot characters: so-called qualified names.

Handling the dot operator

The period or dot operator plays two distinct but similar roles in both Java and Unicon. On the one hand, dot is used in so-called qualified names, where a package name is given first, then a dot, then a name within that package's namespace. A similar use of dot is found in Java when a class name is followed by a dot followed by a static name within the class. The line between packages and classes is a little blurry, and in both cases, most object-oriented languages resolve those uses of the dot operator at compile time. On the other hand, the dot operator is also used with objects, to pick out a member from an instance, and in the general case, this happens at runtime.

For Jzero, we have a simplified dot operator. The Jzero transpiler will just translate a dot on input into a dot on output, except for a few built-ins that have a more direct translation. If you choose to implement a more serious object-oriented language, expect to spend more time on the dot operator.

This is a reasonable location to implement translations of built-ins like turning Jzero's System. out.println() into Unicon's write(). It is also the place to handle method invocations on built-in class types in Jzero, such as implementing the .length pseudo-member for arrays. The Unicon code for qualified names in the Jzero transpiler is in the method genuniQualifiedName() shown below. The order in which System.out.println is detected appears reversed; this is a by-product of the shape of the syntax tree that was built. In an occurrence of System.out.println, println was attached to the tree last and is viewed from the topmost tree node, while System and out are viewed deeper in the tree on the left-hand side:

```
method genuniQualifiedName()
    if kids[2].icncode[1]=="println" & kids[1].kids[2].icncode[1]=="out" &
       kids[1].kids[1].sym == "token" &
      kids[1].kids[1].icncode[1] == "System" then {
         icncode := [ "write" ]
         }
    else {
      case kids[1].typ.basetype of {
      "array": {
         if kids[2].tok.text ==="length" then {
           icncode := [ "*" ||kids[1].icncode[1] ]
           }
         else stop("don't know how to generate code for array.",
                   kids[2].tok.text)
         }
      "class": {
        icncode := [ kids[1].icncode[1] || "." || kids[2].icncode[1]]
        }
      default: stop("don't know how to generate code for type ",
                  kids[1].typ.str())
        }
      }
    end
```

In this method, a regular dot operator on a Jzero class type is mapped to a Unicon dot operator. The corresponding Java implementation of genuniQualifiedName() is shown here:

```
void genuniQualifiedName() {
    if (kids[1].icncode.get(0).equals("println") &&
```

```
                kids[0].kids[1].icncode.get(0).equals("out") &&
                kids[0].kids[0].sym.equals("token") &&
                kids[0].kids[0].icncode.get(0).equals("System")) {
        icncode = new ArrayList<String>();
        icncode.add("write");
        }
      else {
        switch (kids[0].typ.basetype) {
        case "array": {
          if (kids[1].tok.text.equals("length")) {
            icncode = new ArrayList<String>();
            icncode.add( "*" + kids[0].icncode.get(0) );
            }
          else {
            System.err.println("don't know how to generate code for array." +
                                    kids[1].tok.text);
            System.exit(1);
            }
          }
        case "class": {
          icncode = new ArrayList<String>();
          icncode.add( kids[0].icncode.get(0) + "." +
                    kids[1].icncode.get(0) );
          }
        default: {
          System.err.println("don't know how to generate code for type " +
                                    kids[0].typ.str());
          }
        }
      }
    }
  }
```

Such a simple operator turned out to be a real can of worms. Hopefully, transpiling other simple expressions will not be quite so bad.

Mapping Java expressions to Unicon

Straight-line Java code will have a straightforward mapping to Unicon. For a transpiler, output code might or might not need to be formatted, but in some output languages, care may be required to follow rules for line-breaking or indentation. *Figure 11.2* shows the semantic rules for expressions.

Production	Semantic Rules
Assignment : IDENT '=' AddExpr	Assignment.icncode := [IDENT.text || " := "] ||| AddExpr.icncode
AddExpr : AddExpr$_1$ '+' MulExpr	AddExpr.icncode := AddExpr$_1$.icncode ||| ["+"] ||| MulExpr.icncode
AddExpr : AddExpr$_1$ '-' MulExpr	AddExpr.icncode := AddExpr$_1$.icncode ||| ["-"] ||| MulExpr.icncode
MulExpr : MulExpr$_1$ '*' UnaryExpr	MulExpr.icncode := MulExpr$_1$.icncode ||| ["*"] ||| UnaryExpr.icncode
MulExpr : MulExpr$_1$ '/' UnaryExpr	MulExpr.icncode := MulExpr$_1$.icncode ||| ["/"] ||| UnaryExpr.icncode
UnaryExpr : '-' UnaryExpr$_1$	UnaryExpr.icncode := ["-"] ||| UnaryExpr$_1$.icncode
UnaryExpr : '(' AddExpr ')'	UnaryExpr.icncode := ["("] ||| AddExpr.icncode ||| [")"]
UnaryExpr : IDENT	UnaryExpr.icncode := [IDENT.text]

Figure 11.2: Semantic rules mapping Jzero class to Unicon

The Unicon code for binary arithmetic expressions in the helper function genuniBinaryExpr() is shown here. Boolean AND and OR are included as two more binary operators with straight-forward mappings to Unicon, although the output token for those operators is slightly different than in Java, with && turning into & and || turning into |. Fully parenthesizing the output reduces the potential for surprise. Although the code for AND and OR are given here along with binary operators for many other non-terminals, their semantic rules are presented later, along with those for relational expressions:

```
method genuniBinaryExpr()
    case rule of {
    1320: if typ.str()=="String" then op := "||" else op := "+"
    1321: op := "-"
    1310: op := "*"
    1311: op := "/"
    1312: op := "%"
    1350: op := "&"
    1360: op := "|"
    }
    icncode := ["( "||kids[1].icncode[1]||" "||op||
               " "||kids[2].icncode[1]||")"]
end
```

The corresponding Java code is shown below:

```
void genuniBinaryExpr() {
      String op;
      switch (rule) {
      case 1320:
        if (typ.str().equals("String")) op = "||"; else op = "+";
        break;
      case 1321: op = "-"; break;
      case 1310: op = "*"; break;
      case 1311: op = "/"; break;
      case 1312: op = "%"; break;
      case 1350: op = "&"; break;
      case 1360: op = "|"; break;
      }
      icncode = new ArrayList<String>();
      icncode.add("( " + kids[0].icncode.get(0) + " " + op +
                  " " + kids[1].icncode.get(0) + ")");
   }
```

Besides these binary expressions, the code must consider unary expressions such as -x. While a unary minus transpiles as itself, a unary NOT (!) in Jzero translates to a tilde (~) in Unicon:

```
method genuniUnaryExpr()
    case rule of {
    1300: op := "-"
    1301: op := "~"
    }
    icncode := ["( "||op||" "||kids[2].icncode[1]||")"]
  end
```

The corresponding Java code is shown below:

```
void genuniUnaryExpr() {
      String op;
      switch (rule) {
      case 1300: op = "-"; break;
      case 1301: op = "~"; break;
      }
```

```
        icncode = new ArrayList<String>();
        icncode.add("( " + op + " " + kids[1].icncode.get(0) + ")");
    }
```

Other than the wrinkle with the plus operator, binary operators are easy to transpile, an identity transformation or, at most, syntactic sugar. Now let's consider method calls.

Transpiler code for method calls

Mapping a Java method call onto a call to a Unicon procedure or method is straightforward. Calls to built-in functions might be handled specially with calls to the runtime system, but for the most part, the call syntax uses parentheses in the same manner, handling parameters and return values similarly. There are two production rules for method calls, one for names (possibly qualified by dots) to the left of the parentheses, and the other for when a more complex expression produces the object upon which a dot operator specifies the method. *Figure 11.3* shows the semantic rules for method calls. The argument list in both rules is optional, so the code should not fail in the event of a null value for ArgList.

Production	Semantic Rules
MethodCall : Name (ArgList)	MethodCall.icncode := Name.icncode ||| ["("] ||| ArgList.icncode ||| [")"]
MethodCall : Primary . IDENT (ArgList)	MethodCall.icncode := Name.icncode ||| ["."] || IDENT.text || "(") ||| ArgList.icncode ||| [")"]

Figure 11.3: Semantic rules for method calls

Within the method genuniMethodCall(), which production rule is to be used could be determined either by directly checking the treenode's rule field, or by checking how many children are present: 2 or 3. The Unicon implementation of genuniMethodCall() is shown here:

```
  method genuniMethodCall()
        if rule=1290 then {
           icncode := kids[1].icncode ||| ["("] |||
                      kids[2].icncode ||| [")"]
        }
        else { # rule 1291
           icncode := kids[1].icncode ||| [kids[2].text || "("] |||
                 kids[3].icncode ||| [ ")" ]
        }
  end
```

The corresponding Java code from tree.java is shown below:

```
void genuniMethodCall() {
    icncode = new ArrayList<String>();
    if (rule==1290) {
        icncode.addAll( kids[0].icncode );
        icncode.add( "(" );
        icncode.addAll( kids[1].icncode);
        icncode.add( ")" );
    }
    else { // rule 1291
        icncode.addAll( kids[0].icncode );
        icncode.add( kids[1].text || "(" );
        icncode.addAll( kids[2].icncode );
        icncode.add( ")" );
    }
}
```

Are we done transpiling method calls? Not quite. There is the question of how parameters are chained together to form argument lists. Just because it is straightforward doesn't mean we can overlook it. The Unicon code for this just consists of list concatenations that place the comma between arguments:

```
method genuniArgList()
    icncode := kids[1].icncode ||| [","] ||| kids[2].icncode
end
```

The corresponding Java implementation of genuniArgList() is shown here:

```
void genuniArgList() {
    icncode = new ArrayList<String>();
    icncode.addAll(kids[0].icncode);
    icncode.add(",");
    icncode.addAll(kids[1].icncode);
}
```

There is another kind of expression that we must transpile, and those are assignments, including augmented assignments.

Assignments

A basic level of syntactic sugar is needed to convert Jzero assignments using the = operator into Unicon assignments, which use the := operator that is used in languages such as Pascal. The Unicon code to perform such a substitution is shown here:

```
method genuniAssignment()
        op := kids[2].icncode[1]
        op[find("=", op)] := ":="
        icncode := [kids[1].icncode[1] || op || kids[3].icncode[1]]
    end
```

The Java code is similar:

```
void genuniAssignment() {
    op = kids[1].icncode.get(0);
    if (op.equals("=")) op = ":=";
    else op = op.substring(0,op.size-1) + ":=";
    icncode = new ArrayList<String>();
    icncode.add( kids[0].icncode.get(0) + op + kids[2].icncode.get(0));
}
```

At this point, we have presented the semantic rules for the expressions. It is time to consider the larger statement-level parts of the grammar, such as control structures.

Transpiler code for control structures

The mapping of common control structures in Jzero and Unicon is kept as simple as possible. The underlying language control structures take care of determining code locations where control is to be transferred. *Figure 11.4* shows semantic rules for some common control structures.

Production	Semantic Rules
IfThenStmt : if '(' Expr ')' Stmt	`IfThenStmt.icncode := [ "if " ] ||| Expr.icncode |||` `                       [" then "] ||| Stmt.icncode`

IfThenElseStmt :
if '(' Expr ')' Stmt₁ else Stmt₂
	`IfThenElseStmt.icncode := [ "if " ] ||| Expr.icncode |||`
	`                       [" then "] ||| Stmt₁.icncode |||`
	`                       [" else "] ||| Stmt₂.icncode`

WhileStmt : while '(' Expr ')' Stmt	`WhileStmt.icncode := [ "while " ] ||| Expr.icncode |||` `                       [" do "] ||| Stmt.icncode`

Figure 11.4: Semantic rules for transpiling Jzero control structures

Even with such simple semantic rules, there are at least two tricky parts of which to be aware. One is that the output language Unicon is line-sensitive and inserts semi-colons in ways that may break the translation if we are not careful. For example, because the reserved word do is optional in while loops and can be the start of a new do-while control structure, the language might insert a semi-colon after the condition and before a do reserved word on the next line. This issue is resolved by concatenating the strings at the ends of certain sublists with the strings at the start of the next sublist. The Unicon code corresponding to the IfThenStmt is shown in method genuniIfThenStmt():

```
method genuniIfThenStmt()
    if kids[2].icncode[1] === "{" then {
        icncode := ["if " || kids[1].icncode[1] || " then {"] |||
                    kids[2].icncode[2:0]
    }
    else
        icncode := ["if " || kids[1].icncode[1] || " then"] |||
                    kids[2].icncode
end
```

The corresponding Java code is as follows:

```
void genuniIfThenStmt() {
    icncode = new ArrayList<String>();
    if (kids[1].icncode.get(0).equals("{")) {
        icncode.add("if " + kids[0].icncode.get(0) + " then {";
        for(int i = 1; i < kids[1].icncode.size(); i++)
            icncode.add( kids[1].icncode.get(i) );
    }
    else {
        icncode.add( "if " + kids[0].icncode.get(0) + " then");
        icncode.addAll( kids[1].icncode );
    }
}
```

The code for genuniIfThenElseStmt() and genuniWhileStmt() are similar.

A second tricky part of transpiling Jzero control structures to Unicon is that the semantics of Jzero conditional expressions compute Booleans and perform C-style short-circuit evaluation, while the semantics of Unicon conditionals are goal-directed and are evaluated for success or failure. Despite this, it is easy to map Boolean conditions in Jzero to corresponding Unicon expressions. Boolean && and || in Java turn into & and | in Unicon. The not operator, !, becomes the tilde ~ in Unicon. The semantic rules are shown in *Figure 11.5*.

Production	Semantic Rules
RelExpr : RelExpr$_1$ RelOp AddExpr	`RelExpr.icncode := RelExpr`$_1$`.icncode |||` `                      [RelOp.text] ||| AddExpr.icncode`
AndExpr : AndExpr$_1$ && EqExpr	`AndExpr.icncode := AndExpr`$_1$`.icncode |||` `                      ["&"] ||| EqExpr.icncode`
OrExpr : OrExpr$_1$ || AndExpr	`OrExpr.icncode := OrExpr`$_1$`.icncode |||` `                      ["|"] ||| AndExpr.icncode`
UnaryExpr : ! UnaryExpr$_1$	`UnaryExpr.icncode := ["~"] ||| UnaryExpr`$_1$`.icncode`

Figure 11.5: Semantic rules for transpiling Jzero Boolean and relational operators

Here is the Unicon implementation of the `genuniRelExpr()` method, which is called from `genunicon()` to generate code for relational expressions. It depends on the fact that the relational operators <, <=, >, and >= are the same in Java and Unicon, and thus the source operator can just be copied to the output:

```
method genuniRelExpr()
        op := kids[2].tok.text
        icncode := [kids[1].icncode[1] || " " || op || " " ||
                kids[3].icncode[1] ]
    end
```

This code is fragile as written; it assumes the left and right sides of the relational operator icncode attributes are lists of length 1. We may wish to rewrite it to be more robust. The corresponding Java code is here:

```
void genuniRelExpr() {
        String op = kids[1].tok.text;
    icncode = new ArrayList<String>();
        icncode.add(kids[0].icncode.get(0) + " " + op + " " +
                kids[2].icncode.get(0));
    }
```

Earlier, for the plus operator, we saw that it was special-cased to turn into different operators in Unicon depending on whether the operands are numeric or strings. Unicon has separate relational operators for strings as well, but we do not have a similar special casing required here because Java instead uses methods for those operations. We do, however, need to consider equality testing. When Java code says ==, what does it mean? For numbers, it checks if they are bitwise identical. For strings and other objects, it checks if they are the same reference. The closest mapping onto Unicon is the === operator, although its string semantics are different:

```
method genuniEqExpr()
        if rule=1340 then op := "===" else op := "~==="
        icncode := [kids[1].icncode[1] || " " || op || " " ||
                kids[2].icncode[1] ]
end
```

The corresponding Java code is below:

```
void genuniEqExpr() {
    String op;
    if (rule==1340) op = "===" else op = "~===";
    icncode = new ArrayList<String>();
    icncode.add(kids[0].icncode.get(0) + " " + op + " " +
            kids[1].icncode.get(0));
}
```

Besides relational expressions such as x < y, conditions may use Boolean expressions containing AND, OR, and NOT whose code was shown earlier with binary and unary expressions. There are two more productions that arise.

Transpiling Jzero declarations

This section considers two forms of Jzero declarations: first, we will look at methods and then variables. A method consists of a header followed by a code body. We can stitch these two non-terminals' code together by simply concatenating their icncode attributes, as usual. Transpiling the code body is discussed later in the *Transpiling Jzero block statements* section, where we talk about grouping sequences of statements together. In this section, we are just worried about the header, and the overall location of the method.

It is a simple matter to transform the header line from Java syntax to Unicon. One aspect of transpiling the header line itself is that if it is a static method, indicated by production rule 1070, it is moved out of the class and called a procedure, while non-static methods in Unicon use the reserved word method.

We'll look at constructing the modified header line first, and then the more challenging aspect of moving static methods up, out of their class. The Unicon code for transpiling a method header is shown in the following method genuniMethodHeader():

```
method genuniMethodHeader()
   procname := kids[2].kids[1].icncode[1]
   if rule === 1070 then resword := "procedure "
   else resword := "method "
   icncode := [resword || procname || "(" || kids[2].icncode[1] || ")" ]
end
```

The corresponding Java implementation of genuniMethodHeader() is shown here. The code again depends mainly on whether the production rule for static methods was used:

```
void genuniMethodHeader() {
   String procname = kids[1].kids[0].icncode.get(0);
   String resword;
   if (rule == 1070) resword = "procedure ";
   else resword = "method ";
   icncode = new ArrayList<String>();
   icncode.add( resword + procname + "(" + kids[1].icncode.get(0) + ")" );
}
```

The one line of Unicon code needed for method headers is the least of our concerns. We have already mentioned two of the prominent differences between Java and Unicon that both involve moving declarations around. In Java, static methods are declared inside classes and local variables may be declared almost anywhere within a code block. In Unicon, static methods are called procedures; procedures are not declared inside classes. Also, in Unicon, local variable declarations must appear at the top of the procedure or method before other executable statements. We introduced two semantic attributes to help with this. First, let's see how the attribute staticcode is assigned and used.

In the Jzero grammar, the non-terminal MethodDecl connects the header and body. If the production rule for a MethodDecl includes the static keyword (production rule 1070), then the method should be placed on the staticcode list; otherwise, it is placed as usual on the icncode list. The Unicon code for this originates with the method genuniMethodDecl() as shown here:

```
method genuniMethodDecl()
        staticcode := []
        icncode := []
```

```
            if kids[1].rule === 1070 then { # public static
               staticcode := kids[1].icncode ||| kids[2].icncode[2:-1] |||
                           ["end"]
            }
         else {
            every icncode |||:= \ ((!\kids).icncode)
            put ( icncode, "end")
         }
   end
```

The corresponding Java code for genuniMethodDecl() is as follows:

```
void genuniMethodDecl() {
    staticcode = new ArrayList<String>();
    icncode = new ArrayList<String>();
    if (kids[0].rule == 1070) { // production rule # for public static
      staticcode.addAll(kids[0].icncode);
      for(int i = 1; i < kids[1].icncode.size()-1; i++)
        staticcode.add( kids[1].icncode[i] );
      staticcode.add( "end" );
      }
    else {
      for(tree k : kids)
        icncode.addAll( k.icncode );
      icncode.add ( "end" );
      }
  }
```

Before we can move on, we must consider how the static code gets from this origin point in method declarations up to the location in the tree where it can be placed outside the generated class. It turns out that it does not have to go far, logically. A MethodDecl is one of three kinds of ClassBodyDecl, which are all chained together into a list of ClassBodyDecls within a ClassBody. The transpiler must chain together all the static code found within the ClassBodyDecls. The Unicon implementation of this is shown below:

```
method genuniClassBodyDecls()
        staticcode := kids[1].staticcode ||| kids[2].staticcode
        icncode := kids[1].icncode ||| kids[2].icncode
  end
```

The Java implementation of genuniClassBodyDecls() is similarly straightforward:

```
void genuniClassBodyDecls() {
    staticcode = new ArrayList<String>();
    icncode = new ArrayList<String>();
    staticcode.addAll( kids[0].staticcode );
    staticcode.addAll( kids[1].staticcode );
    icncode.addAll( kids[0].icncode );
    icncode.addAll( kids[1].icncode );
}
```

Having looked at how to move statics outside of the class, it is time to consider how local variable declarations are moved up to the top of a procedure or method. It is a bit tempting to not bother; Unicon does not require that local variables be declared at all. However, local variable declarations prevent accidental collisions with the global namespace. The data types indicated in local variables and parameters, held in kids[1] in the syntax tree, are discarded and replaced by the reserved word local. The Unicon code for this is as follows:

```
method genuniLocalVarDecl()
        vardecls := [ "local " ] ||| kids[2].icncode
        icncode := []
    end
```

The corresponding Java code is shown here:

```
void genuniLocalVarDecl() {
    vardecls = new ArrayList<String>();
    vardecls.add( "local " );
    vardecls.addAll( kids[1].icncode );
    icncode = new ArrayList<String>();
}
```

The second child of a local variable declaration is a list of variables being declared. The code for chaining together a list of variables separated by commas looks similar to what was shown earlier for chaining together lists of arguments to a method in genuniArgList(). We can skip the code without sacrificing anything.

In this context of declarations, variable names in Jzero may be decorated with additional type information, such as specifying that something is an array by suffixing it with square brackets, as in int x []. When we write out the corresponding Unicon code, we throw away type information; in addition to changing int to local, the square brackets should just disappear.

This is accomplished by simply copying up the identifier leaf node's icncode into the VarDeclarator node without appending any brackets. The code for genuniVarDeclarator() is shown below:

```
method genuniVarDeclarator()
        icncode := kids[1].icncode
  end
```

The corresponding Java code looks like this:

```
void genuniVarDeclarator() {
   icncode = kids[0].icncode;
}
```

You may be wondering at this point: do we have to initialize empty lists of strings for staticcode, and for vardecls, in every node in the entire syntax tree? One answer is that it would be safe to do so, but in most nodes, those attributes would go to waste. For the most part, local variable declarations only must be percolated through the BlockStmts nodes up to the top of the Block.

Transpiling Jzero block statements

The BlockStmts non-terminal constitutes a linked list of the statements in a given block. We might write a genuniBlockStmts() method as follows:

```
method genuniBlockStmts()
        icncode := kids[1].icncode ||| kids[2].icncode
        vardecls := (\ (kids[1].vardecls)|[]) |||
                    (\ (kids[2].vardecls)|[])
    end
```

with the parallel Java code reading like this:

```
void genuniBlockStmts() {
    icncode = new ArrayList<String>();
    icncode.addAll( kids[0].icncode );
    icncode.addAll( kids[1].icncode );
    icncode = new ArrayList<String>();
    vardecls.addAll( kids[0].vardecls );
    vardecls.addAll( kids[1].vardecls );
}
```

Percolating the vardecls attribute up to the Block allows all the vardecls to be explicitly placed into the icncode before all the regular output code for executable statements. The Block is naturally enclosed in curly brackets:

```
method genuniBlock()
        icncode := ["{"]
        every icncode |||:= \ ((!\kids).vardecls) # move decls up front
        every icncode |||:= (!\kids).icncode
        icncode |||:= ["}"]
end
```

The Java version of this code is as follows:

```
void genuniBlock() {
    icncode = new ArrayList<String>();
    icncode.add( "{" );
    for (tree k : kids)
      icncode.addAll( k.vardecls ); // move decls up front
    for (tree k : kids)
        icncode.addAll(k.icncode);
    icncode.add( "}" );
}
```

Have you picked up on the one crucial omission from this discussion of moving variable declarations up to the front of the block? What do we do about variables declared inside nested blocks? For that matter, what do we do about variables declared inside for loop headers? Unicon certainly does not support these features.

A crude answer is to just disallow these features in Jzero. A more satisfying answer might be to go into all helper functions for production rules that contain a Block on their righthand side, and copy up their vardecls attributes so that they are found in the enclosing Block. If you go beyond Jzero in this way to support a larger subset of Java, be prepared to also rename, or perform name-mangling on, all such variables declared in nested blocks. The propagating from nested blocks to enclosing blocks can be illustrated for if statements by adding a line to our genuniIfThenStatement() method:

```
method genuniIfThenStmt()
    if kids[2].icncode[1] === "{" then {
      icncode := ["if " || kids[1].icncode[1] || " then {"] |||
                    kids[2].icncode[2:0]
      }
```

```
        else
            icncode := ["if " || kids[1].icncode[1] || " then"] |||
                        kids[2].icncode
        vardecls := kids[2].vardecls
    end
```

The corresponding Java code is as follows:

```
    void genuniIfThenStmt() {
        icncode = new ArrayList<String>();
        if (kids[1].icncode.get(0).equals("{")) {
            icncode.add("if " + kids[0].icncode.get(0) + " then {";
            for(int i = 1; i < kids[1].icncode.size(); i++)
                icncode.add( kids[1].icncode.get(i) );
            }
        else {
            icncode.add( "if " + kids[0].icncode.get(0) + " then");
            icncode.addAll( kids[1].icncode );
        }
        vardecls = kids[1].vardecls;
    }
```

Now it is time to look at how methods are assembled to form classes, and how classes are transpiled into Unicon packages.

Transpiling a Jzero class into a Unicon package that contains a class

For the sake of consistency, we will present the code generation model via semantic rules specifying the values of semantic attributes, as we did in *Chapter 9*. The semantic rules for the top-level Jzero code generation that maps a Java class to a Unicon package with a class inside are shown in *Figure 11.6*.

Production	Semantic Rules

ClassDecl : public class name ClassBody	ClassDecl.icncode := ["package " || name.text] |||
	     ClassBody.staticcode |||
	     ["class " || name.text || "()"] |||
	     ClassBody.icncode
	 ClassDecl.helper := ["procedure main(argv)",
	     " " || name.text || "__main ! argv",
	     "end"]

Figure 11.6: Semantic rules mapping a Jzero class to Unicon

The code to implement this transpilation of Java classes happens at the very end of our bottom-up traversal of the syntax tree. The part that builds the complete translation in the root's icncode attribute is located in a method genuniClassDecl(). The Unicon version of this code is shown here:

```
# generate public static methods outside the class, then generate the
class
 method genuniClassDecl()
    icncode := ["package " || kids[1].icncode[1]]
    icncode |||:= kids[-1].staticcode
    icncode |||:= ["class " || kids[1].icncode[1] || "()"]
    icncode |||:= kids[-1].icncode
 end
```

The corresponding Java code is given below:

```
void genuniClassDecl() {
    icncode = new ArrayList<String>();
    icncode.add("package " + kids[1].icncode.get(1));
    icncode.addAll(kids[kids.length-1].staticcode);
    icncode.add("class " + kids[1].icncode.get(1) + "()");
    icncode.addAll(kids[kids.length-1].icncode);
}
```

The actual writing out of the code takes place in a class j0 method genunicon(root). Its name is deliberately similar to the class tree's genunicon() method, since it calls the root's genunicon() to do the tree traversal, and then writes the results to a file. Here is the Unicon version of genunicon(root) from the j0.icn file:

```
method genunicon(root)
    root.genunicon()
    basenam := yyfilename[1:find(".java", yyfilename)]
    outfileh := basenam || "_helper.icn"
    if fout := open(outfileh, "w") then {
        write(fout, "procedure main(argv)")
        write(fout, "   ", basenam, "__main ! argv")
        write(fout, "end")
        close(fout)
    }
    outfilename := basenam || ".icn"
    if fout := open(outfilename, "w") then {
```

```
            every s := !(root.icncode) do {
                write(fout, s)
                write(s)
                }
            close(fout)
            system("unicon -s " || outfilename || " " || outfileh)
            }
    end
```

The corresponding Java code shown here is from j0.java. It required a bunch of imports, added to the top of j0.java, for classes FileWriter, BufferedWriter, and PrintWriter, because Java loves imports:

```
void genunicon(parserVal r) {
    tree root = (tree)(r.obj);
    root.genunicon();
    int len = yyfilename.length();
    String basenam = yyfilename.substring(0,length-5);
    String outfileh = basenam + "_helper.icn";
    PrintWriter fout;
    try {
        fout = new PrintWriter(new BufferedWriter(new FileWriter(outfileh)));
        fout.printf("procedure main(argv)\n");
        fout.printf("    " + basenam + "__main ! argv\n");
        fout.printf("end\n");
        fout.close();
        }
    catch (java.io.IOException ioException) {
        System.err.println("can't write "+outfileh);
        System.exit(1);
        }
    String outfilename = basenam + ".icn";
    try {
        fout = new PrintWriter(new BufferedWriter(
                            new FileWriter(outfilename)));
        for(int i=0; i<root.icncode.size(); i++) {
            String s = root.icncode.get(i);
            fout.printf(s + "\n");
```

```
        System.out.println(s);
      }
    fout.close();
```

After writing the output, j0.java's genunicon() continues by invoking the Unicon compiler. This was a one-liner in Unicon and would be so in most high-level languages, but Java culture must be different:

```
    String cmd = "unicon -s " + outfilename + " " + outfileh;
    Process p = Runtime.getRuntime().exec(cmd);
    int eValue=0;
    try {
      eValue = p.waitFor();
    } catch (InterruptedException iException) {
      System.err.println("unicon interrupted");
    }
    System.out.println("unicon exited with status " +
                       String.valueOf(eValue));
    }
  catch (java.io.IOException ioException) {
    System.err.println("can't write "+outfilename);
    System.exit(1);
    }
  }
```

Class j0's genunicon(root) is invoked from yyparse() after the root node has been constructed and the semantic analysis is performed on it. The code below that launches transpilation in j0gram.y is shared by Unicon and Java implementations:

```
ClassDecl: PUBLIC CLASS IDENTIFIER ClassBody {
    $$=j0.node("ClassDecl",1000,$3,$4);
    j0.semantic($$);
    j0.genunicon($$);
  } ;
```

This transpiler turns out to be a lot simpler than generating bytecode or native code instructions; for high-level languages from similar language paradigms, a transpiler is probably simpler than generating intermediate code. Hopefully, this code will be helpful should you find yourself writing a transpiler in the future.

Summary

In this chapter, you learned the basics of preprocessors and transpilers. While a preprocessor can be thought of as a transformation that alters code slightly from its input to its output, a transpiler turns source code from one high-level language into another high-level language that is substantially different.

For any new programming language project that you undertake, you may want to first consider the implementation techniques shown in this chapter, either to produce a prototype or as your primary implementation. If you write a transpiler, you will get things working fast, and you will be able to try it out and determine whether your language semantics work the way that you intend. Later on, you may want to write a bytecode interpreter or native code implementation, but even if you do, having the transpiler around will be helpful. You can reuse its frontend code, while switching the backend over to the intermediate code generator from *Chapter 9*, followed by one of the code generators that are presented next, in *Chapters 12* and *13*.

Questions

1. Java was designed to avoid having a preprocessor, despite resembling C/C++, which features a powerful preprocessor that is heavily used. What are some of the disadvantages of preprocessors that might have motivated the designers of Java to omit a preprocessor from their language?

2. The promise of transpilers is to get a new language working very quickly without requiring any machine code. What are some of the limitations and pitfalls of this implementation approach?

Join our community on Discord

Join our community's Discord space for discussions with the authors and other readers:

`https://discord.com/invite/zGVbWaxqbw`

12

Bytecode Interpreters

A new programming language may include novel features that are not supported directly by mainstream CPUs. The most practical way to generate code for many programming languages is to generate bytecode for an abstract machine whose instruction set directly supports the language's intended domain. This is important because it sets your language free from the constraints of what current hardware CPUs know how to do. It also allows the generation of code that is tied more closely to the types of problems that you want to solve. If you create your own bytecode instruction set, you can execute programs by writing a virtual machine that knows how to interpret that instruction set. This chapter covers how to design an instruction set and an interpreter that executes bytecode. Because this chapter is tightly connected to *Chapter 13*, *Generating Bytecode*, you may want to read them both before you dive into the code.

This chapter covers the following main topics:

- Understanding what bytecode is
- Comparing bytecode with intermediate code
- Building a bytecode instruction set for Jzero
- Implementing a bytecode interpreter
- Examining iconx, the Unicon bytecode interpreter

Technical requirements

The code for this chapter is available on GitHub: `https://github.com/PacktPublishing/Build-Your-Own-Programming-Language-Second-Edition/tree/master/ch12`

The Code in Action video for the chapter can be found here: `https://bit.ly/327bZWn`

A bytecode interpreter is a piece of software that executes an abstract machine instruction set. We are going to learn about bytecode interpreters by looking at a simple bytecode machine for Jzero and taking a quick peek at the Unicon virtual machine. But first, let's explore what we mean by **bytecode**.

Understanding what bytecode is

Bytecode is a sequence of machine instructions encoded in a binary format and written not for a CPU to execute, but instead for an abstract (or virtual) machine instruction set that embodies the semantics of a given programming language. Although many bytecode instruction sets for languages such as Java use a byte as the smallest instruction size, almost all of them include longer instructions. Such longer instructions have one or more operands. Since many kinds of operands must be aligned at a word boundary with an address that is a multiple of four or eight, a better name for many forms of bytecode might be wordcode. The term bytecode is commonly used for such abstract machines, regardless of the instruction's size.

The languages that are directly responsible for popularizing bytecode are Pascal and SmallTalk. These languages adopted bytecode for different reasons that remain important considerations for programming languages that are defined in terms of their bytecode. Java took this idea and made it more widely known throughout the computer industry.

For Pascal, bytecode is used to improve the portability of a language implementation across different hardware and operating systems. It is much easier to port a bytecode interpreter to a new platform than to write a new compiler code generator for that platform. If most of a language is written in that language itself, the bytecode interpreter may be the only part that has to be ported to a new machine.

SmallTalk popularized bytecode for a different reason: to create a layer of abstraction upon which to implement novel features that were far removed from the hardware at the time. A bytecode interpreter allows a language developer to design new instructions as needed, as well as to define runtime system semantics that are present for all the implementations of that language.

To explain what bytecode is, consider the bytecode that's generated from the following Unicon code:

```
write("2 + 2 is ", 2+2)
```

Bytecode breaks down the execution of this expression into individual machine instructions. The human-readable representation of the bytecode for this expression might look like the following Unicon bytecode, called **ucode**:

```
      mark    L1
      var     0
      str     0
      pnull
      int     1
      int     1
      plus
      invoke    2
      unmark
  lab L1
```

Going line by line, the mark instruction designates the destination label where the execution should proceed if any instruction *fails*. In Unicon, control flow is mostly determined by failure, rather than by Boolean conditions and explicit goto instructions. The var instruction pushes a reference to variable #0 (write) onto an evaluation stack. Similarly, the str instruction pushes a reference to string constant #0 ("2 + 2 is "). The pnull instruction is pushed to provide a space on the evaluation stack where the result of an operator (+) may be placed. The int instruction pushes a reference to the integer constant in constant region location #1, which is the value 2; this is done twice for the two operands of the addition. The plus instruction pops the top two stack elements and adds them, placing the result on the top of the stack. The invoke instruction performs a call with two arguments. The function to be called was specified on the stack prior to the two arguments; in this case, it was the function write that was supplied way back in the var instruction. When invoke comes back, the arguments will have been popped, and the top of the stack, where the reference to the write function had been pushed, will hold the function's return value.

From the preceding example, you can see that bytecode somewhat resembles intermediate code, and that is intentional. So, what is the difference?

Comparing bytecode with intermediate code

In *Chapter 9, Intermediate Code Generation*, we generated machine-independent intermediate code using abstract three-address instructions. Bytecode instruction sets are in between the three-address intermediate code and a real hardware instruction set in their complexity. A single three-address instruction may map to multiple bytecode instructions. This refers to both the direct translation of any instance of a three-address instruction, as well as to the fact that there may be several bytecode instruction opcodes that handle various special cases of a given three-address opcode. Bytecode is generally more involved than intermediate code, even if it manages to avoid the complexities of operand addressing modes found on a lot of CPUs.

Many or most bytecode instruction sets explicitly or implicitly use (virtual, logical) registers, although bytecode machines are usually far simpler than CPU hardware in terms of the number of registers and the register allocation that the compiler must perform to generate code.

Bytecode is generally a binary file format. Binary formats are very difficult for humans to read. When talking about bytecode in this chapter, we will provide examples in an assembler-like format, but the bytecode itself is all ones and zeros.

Comparing a hello world program in intermediate code and bytecode might give you some idea of their similarities and differences. We will use the following hello.java program as an example. It just prints a message if you give it command-line arguments, but it contains arithmetic as well as control flow instructions:

```java
public class hello {
    public static void main(String argv[]) {
        int x = argv.length;
        x = x + 2;
        if (x > 3) {
            System.out.println("hello, jzero!");
        }
    }
}
```

The Jzero three-address code for this program looks as follows. Its operands include several kinds of memory references, ranging from local variables to code region labels. The main() function consists of 11 instructions and 20 operands, averaging almost two operands per instruction:

```
.string
L0:       string  "\"hello, jzero!\""
.global
          global  global:8,hello
          global  global:0,System
.code
proc      main,0,0
          ASIZE   loc:24,loc:8
          ASN     loc:16,loc:24
          ADD     loc:32,loc:16,imm:2
          ASN     loc:16,loc:32
L75:      BGT     L76,loc:16,imm:3
```

```
        GOTO    L77
L76:    PARM    strings:0
        FIELD   loc:40,global:0,class:0
        PARM    loc:40
        CALL    PrintStream__println,1
L77:    RET
end
```

The JVM bytecode for this program, as produced by compiling the program with javac and then disassembling it with the javap -c command, is shown here (comments have been removed). The main() function consists of 14 instructions with four operands, which equates to less than a third of an operand per instruction:

```
public class hello {
  public hello();
    Code:
        0: aload_0
        1: invokespecial #1
        4: return
  public static void main(java.lang.String[]);
    Code:
        0: aload_0
        1: arraylength
        2: istore_1
        3: iload_1
        4: iconst_2
        5: iadd
        6: istore_1
        7: iload_1
        8: iconst_3
        9: if_icmple     20
       12: getstatic     #2
       15: ldc           #3
       17: invokevirtual #4
       20: return
}
```

The instructions in this `main()` method illustrate some characteristics of their underlying Java bytecode interpreter virtual machine. It is a stack machine. The `load` and `store` families of instructions push and pop a variable between a numbered slot in the main memory region and the top of the stack, where expressions are evaluated. This instruction set is typed, with mnemonic prefixes for each of the built-in scalar atomic types of the Java language (`i` for integer, `f` for float, `a` for array, and so on). It has object-oriented conventions, such as specifying that local variable #0 is the current object, so the instruction `aload_0` pushes a reference to the current object (known as `this` or `self`). It has built-in instructions for special purposes such as the `arraylength` instruction for returning the length of an array. Seven integers from -1 through 5 have opcodes that push those constants: `iconst_1`, `iconst_2`, `iconst_3`, etc. An instruction such as `iadd` pops two values, adds them, and then pushes the result.

We will present a simpler bytecode instruction set in this chapter, but it is nice to know what the most brilliant minds in the industry are churning out. Now, let's look at a simpler bytecode instruction set that's suitable for Jzero.

Building a bytecode instruction set for Jzero

This section describes a simple file format and instruction set for Jzero code, generated from three-address intermediate code. This is very much a toy instruction set. For the language that you create, you instead might decide to use (possibly a subset of) a real instruction set such as the Java bytecode instruction set. Java bytecode is a complicated format; if it wasn't, we wouldn't be going to the trouble of presenting something simpler. The instruction set presented here is slightly more capable than Jzero uses, to allow for common extensions.

Defining the Jzero bytecode file format

The Jzero bytecode format consists of a header, followed by a data section, followed by a sequence of instructions. Jzero bytecode files are interpreted as a sequence of 8-byte words in little-endian format. The header consists of an optional self-execution script, a magic word, a version number, and the word offset of the first instruction, relative to the magic word. A self-execution script is a set of commands written in some platform-dependent language that invokes the interpreter, feeding the Jzero file to it as a command-line argument. If present, the self-execution script must be padded if necessary to comprise a multiple of 8 bytes. The magic word is 8 bytes containing the "Jzero!!\0" string. The version number is another 8 bytes containing a version such as 1.0 padded with NUL bytes, as in "1.0\0\0\0\0\0". The word offset of the first instruction would, at its smallest, be 3; this number is relative to the magic word. A word offset of 3 indicates an empty constant section of 0 words. After the magic word, the version word, and the word offset, execution starts at the instruction whose offset is given in the third word. *Figure 12.1* illustrates the memory layout for the shortest Jzero program. Each row denotes one word of eight bytes, and each rectangle contains one byte, with the byte offset shown in a small font in the upper-left corner, and the byte contents printed in the center. In this figure, after the three-word header and the empty data section, the fourth word is a HALT instruction, whose first byte is the code for HALT: "\x01".

0x00 J	0x01 z	0x02 e	0x03 r	0x04 o	0x05 !	0x06 !	0x07 "\x00"
0x08 1	0x09 .	0x0a 0	0x0b "\x00"	0x0c "\x00"	0x0d "\x00"	0x0e "\x00"	0x0f "\x00"
0x10 "\x00"	0x11 "\x00"	0x12 "\x00"	0x13 "\x00"	0x14 "\x00"	0x15 "\x00"	0x16 "\x00"	0x17 "\x03"
0x18 "\x01"	0x19 "\x00"	0x1a "\x00"	0x1b "\x00"	0x1c "\x00"	0x1d "\x00"	0x1e "\x00"	0x1f "\x00"

Figure 12.1: Memory layout for the shortest Jzero program

After the header, there is an optional static data section, which, in Jzero, includes static variables as well as constants, including string literals. In a more serious production language, there might be several kinds of static data sections. For example, there might be one subsection for read-only data, one for data that starts uninitialized and doesn't need to physically occupy space in the file on disk, and a third for statically initialized (non-zero) data. For Jzero, we will just allow one section on disk for all of that.

After the data section, the rest of the file consists of instructions. Every instruction in Jzero format is a single 64-bit word containing an opcode (8 bits), an operand region (8 bits), and an operand (48 bits). The operand region and operand are not used in all opcodes. *Table 12.1* shows the opcodes that are defined in the Jzero instruction set:

Opcode	Mnemonic	Description
1	HALT	Halt
2	NOOP	Do nothing
3	ADD	Add the top two integers on the stack, push the sum
4	SUB	Subtract the top two integers on the stack, push the difference
5	MUL	Multiply the top two integers on the stack, push the product
6	DIV	Divide the top two integers on the stack, push the quotient
7	MOD	Divide the top two integers on the stack, push the remainder
8	NEG	Negate the integer at the top of the stack
9	PUSH	Push a value from memory to the top of the stack
10	POP	Pop a value from the top of the stack and place it in memory
11	CALL	Call a function with n parameters on the stack
12	RETURN	Return to the caller with a return value of x
13	GOTO	Set the instruction pointer to location L
14	BIF	Pop the stack; if it is non-zero, set the instruction pointer to L
15	LT	Pop two values, compare, push 1 if less than, else 0
16	LE	Pop two values, compare, push 1 if less or equal, else 0
17	GT	Pop two values, compare, push 1 if greater than, else 0
18	GE	Pop two values, compare, push 1 if greater or equal, else 0
19	EQ	Pop two values, compare, push 1 if equal, else 0
20	NEQ	Pop two values, compare, push 1 if not equal, else 0
21	LOCAL	Allocate n words on the stack
22	LOAD	Indirect push; reads through a pointer
23	STORE	Indirect pop; writes through a pointer

Table 12.1: The Jzero instruction set

Compare the Jzero instruction set with the set of instructions defined for intermediate code in *Chapter 9*, *Figure 9.2*. The intermediate code instruction set is higher-level, allowing up to three operands. The Jzero instruction set is lower-level and instructions have zero or one operand.

The operand region byte is treated as a signed 8-bit value. For non-negative values, the Jzero format defines the following operand regions:

- region 0 == no operand (R_NONE).
- region 1 == absolute (R_ABS): The operand is a word offset relative to the magic word.
- region 2 == immediate (R_IMM): The operand is the value.
- region 3 == stack (R_STACK): The operand is a word offset relative to the current stack pointer.
- region 4 == heap (R_HEAP): The operand is a word offset relative to the current heap pointer.

The bytecode interpreter source code needs to be able to refer to these opcodes and operand regions by name. In Unicon, a set of $define symbols could be used, but instead, a set of constants in a singleton class called Op is used to keep the code similar in Unicon and Java. The Op.icn file, which contains the Unicon implementation, is shown here:

```
class Op(HALT, NOOP, ADD, SUB, MUL, DIV, MOD, NEG, PUSH,
  POP,
  CALL, RETURN, GOTO, BIF, LT, LE, GT, GE, EQ, NEQ, LOCAL,
  LOAD, STORE, R_NONE, R_ABS, R_IMM, R_STACK, R_HEAP)
initially
  HALT := 1;  NOOP := 2; ADD := 3; SUB := 4; MUL := 5
  DIV := 6; MOD := 7; NEG := 8; PUSH := 9; POP := 10
  CALL := 11; RETURN := 12; GOTO := 13; BIF := 14; LT := 15
  LE := 16; GT := 17; GE := 18; EQ := 19; NEQ := 20
  LOCAL := 21; LOAD := 22; STORE := 23
  R_NONE := 0; R_ABS := 1; R_IMM := 2
  R_STACK := 3; R_HEAP := 4
  Op := self
end
```

The corresponding Java class looks like this:

```
public class Op {
  public final static short HALT=1, NOOP=2, ADD=3, SUB=4,
    MUL=5, DIV=6, MOD=7, NEG=8, PUSH=9, POP=10, CALL=11,
```

```
      RETURN=12, GOTO=13, BIF=14, LT=15, LE=16, GT=17, GE=18,
      EQ=19, NEQ=20, LOCAL=21, LOAD=22, STORE=23;
   public final static short R_NONE=0, R_ABS=1, R_IMM=2,
      R_STACK=3, R_HEAP=4;
}
```

Having a set of opcodes is all well and good, but the more interesting differences between the
three-address code and bytecode lie in the semantics of the instructions. We will discuss this later
in the *Executing instructions* section. Before we get to that, you need to know more about how a
stack machine operates, as well as a few other implementation details.

Understanding the basics of stack machine operation

Like Unicon and Java, the Jzero bytecode machine uses a stack machine architecture. Most of the
instructions implicitly read or write values to or from the stack. Popping to read, and pushing to
write values is an expected side-effect in a stack machine. For example, consider the ADD instruction. To add two numbers, you push them onto the stack and execute an ADD instruction. The
ADD instruction itself takes no operands; it pops two numbers, adds them, and pushes the result.

Now, consider a function call with n parameters whose syntax looks like this:

```
      arg0 (arg1, …, argN)
```

On a stack machine, this can be implemented by the sequence of instructions shown here:

```
      push reference to function arg0
      evaluate (compute and push) arg1
      . . .
      evaluate (compute and push) argN
      call n
```

The function call will use its operand (n) to locate arg0, the address of the function to be called.
When the function call returns, all the arguments will be popped and the function return value
will be on the top of the stack, in the location that previously held arg0. Now, let's consider some
other aspects of how to implement a bytecode interpreter.

Implementing a bytecode interpreter

A bytecode interpreter runs the following algorithm, which implements a *fetch-decode-execute
loop* in software. Most bytecode interpreters use at least two registers almost continuously: an
instruction pointer and a **stack pointer**. The Jzero machine also includes a **base pointer register**
to track function call frames and a **heap pointer register** that holds a reference to a current object.

While the instruction pointer is referenced explicitly in the following fetch-decode-execute loop pseudocode, the stack pointer is used almost as frequently, but it's more often used implicitly as a byproduct of the instruction semantics of most opcodes:

```
load the bytecode into memory
initialize interpreter state
repeat {
    fetch the next instruction, advance the instruction pointer
    decode the instruction
    execute the instruction
}
```

Bytecode interpreters are usually implemented in a low-level systems programming language such as C, rather than a high-level applications language such as Java or Unicon. The sample implementations will perhaps feel somewhat iconoclastic to hardened systems programmers for this reason. Everything in Java is object-oriented, so the bytecode interpreter is implemented in a class named bytecode. The most native representation of a raw sequence of bytes in Unicon is a string, while in Java, the most native representation is an array of bytes.

To implement the bytecode interpreter algorithm, this section presents each of the pieces of the algorithm in separate subsections. First, let's consider how to load bytecode into memory.

Loading bytecode into memory

To load bytecode into memory, the bytecode interpreter must obtain the bytecode via an input/output of some kind. Typically, this will be done by opening and reading from a named local file. When executable headers are used, a launched program opens itself and reads itself in as a data file. The Jzero bytecode is defined as a sequence of 64-bit binary integers, but this representation is more native in some languages than in others.

In Unicon, loading a file might look like this. The loaded file contents are returned as the return value from method loadbytecode(), or the method fails:

```
class j0machine(code, ip, stack, data, finstr, sp, bp, hp, op, opr, opnd)
  method loadbytecode(filename)
    sz := stat(filename).st_size
    f := open(filename) | stop("cannot open ", filename)
    s := reads(f, sz) | stop("cannot read from ", filename)
    close(f)
    s ? {
```

```
          if tab(find("Jzero!!\01.0\0\0\0\0\0\0")) then {
             return tab(0)
             }
          else stop("file ", filename, " is not a Jzero file")
          }
      end
      ...
   end
```

The call to reads() in this example reads the entire bytecode file into a single contiguous sequence of bytes. In Unicon, this is represented as a string. The corresponding Java uses an array
of bytes, with a ByteBuffer wrapper to provide easy access to the words within the code. The
loadbytecode() method within j0machine.java looks like this. The method returns whether
the codebuf was loaded or not:

```
import java.io.IOException;
import java.nio.file.Files;
import java.nio.file.Paths;
import java.nio.charset.StandardCharsets;
import java.nio.ByteBuffer;
public class j0machine {
   public static byte[] code, stack;
   public static ByteBuffer codebuf, stackbuf;
   . . .
   public static boolean loadbytecode(String filename)
      throws IOException {
         code = Files.readAllBytes(Paths.get(filename));
         byte[] magstr = "Jzero!!\01.0\0\0\0\0\0\0".getBytes(
                               StandardCharsets.US_ASCII);
         int i = find(magstr, code);
         if (i>=0) {
            codebuf = ByteBuffer.wrap(code);
            return true;
         }
         else return false;
      }
}
```

Finding the magic string within a Java byte array requires the following helper method:

```
public static int find(byte[]needle, byte[]haystack) {
    int i=0;
    for( ; i < haystack.length - needle.length+1; ++i) {
        boolean found = true;
        for(int j = 0; j < needle.length; ++j) {
            if (haystack[i+j] != needle[j]) {
                found = false;
                break;
            }
        }
        if (found) return i;
    }
    return -1;
}
```

In addition to loading bytecode into memory and before starting execution, the bytecode interpreter must initialize its registers.

Initializing the interpreter state

The bytecode interpreter state includes the memory regions, instruction and stack pointers, and a small amount of constant or static data used by the interpreter. The init() method allocates and initializes the code region by calling the loadbytecode() method and allocates a stack region. The init() method sets the instruction register to 0, indicating that execution will start at the first instruction in the code region. The stack is initialized to be empty.

In Unicon, initialization consists of the following init() method within the j0machine class. For static variables, Unicon must allocate a separate static data region because the string type that is used to load the bytecode is immutable. Both it and the bytecode interpretation stack are implemented as lists of integers; this exploits the fact that Unicon version 13 and higher implements lists of integers in a contiguous block of memory:

```
method init(filename)
    if not (code := loadbytecode(filename) then
        stop("cannot open ", filename)
    Op()
    ip := 16
```

```
      ip := finstr := 8*getOpnd()
      data := Data(code[25:ip+1])
      stack := list()
   end
```

The corresponding Java code is as follows. The allocation of a 100,000-word stack is somewhat arbitrary:

```
   public static void init(String filename)
      throws IOexception {
      ip = sp = 0;
      if (! loadbytecode(filename)) {
         System.err.println("cannot open ", filename);
         System.exit(1);
      }
      stack = new byte[800000];
      stackbuf = ByteBuffer.wrap(stack);
   }
```

Program executions in Jzero start with the execution of a function named main(). This is a function in the Jzero bytecode, not in the Java implementation of the bytecode interpreter.

When the Jzero main() function runs, it expects to have a normal activation record on the stack, where parameters can be accessed. The easiest way to provide this is to initialize the instruction pointer to a short sequence of bytecode instructions that call main(), and exit after it returns. To set this up, you can initialize the stack, and point the instruction pointer at a CALL instruction that calls main, followed by a HALT instruction.

In the case of Jzero, main() has no parameters and the start sequence is always:

```
   PUSH main
   CALL   0
   HALT
```

Since the startup sequence is the same for every program, it would be possible to embed this bytecode sequence into the virtual machine interpreter code itself, and some bytecode machines do this. The catch is that the code offset (address) of main() will vary from program to program unless it is hardwired, and the linker is forced to always place main() in the same location.

In the case of Jzero, it is sufficient and acceptable for the startup sequence to always begin the code section, at the word offset specified in the header. Now, let's consider how the interpreter fetches the next instruction.

Fetching instructions and advancing the instruction pointer

A register ip, called the **instruction pointer**, holds the location of the current instruction. Bytecode interpreters can represent this as a variable that denotes a pointer into the code, or an integer index, viewing the code as an array. In Jzero, it is a byte offset from the magic word. An instruction fetch in bytecode is an operation that reads the next instruction in the code. This includes the opcode that must be read, as well as any additional bytes or words that have operands for some instructions. In Unicon, this fetch() method is located in the j0machine class. It looks as follows:

```
method fetch()
    op := ord(code[1+ip])
    opr := ord(code[2+ip])
    if opr ~= 0 then opnd := getOpnd()
    ip +:= 8
end
```

The corresponding Java version of the fetch() method looks like this:

```
public static void fetch() {
    op = code[ip];
    opr = code[ip+1];
    if (opr != 0) { opnd = getOpnd(); }
    ip += 8;
}
```

The fetch() method depends on the getOpnd() method, which reads the next word from the code. In Unicon, the getOpnd() method might be implemented as follows:

```
method getOpnd()
    return signed(reverse(code[ip+3+:6]))
end
```

The corresponding Java implementation of the getOpnd() method does a lot of bit shifting and OR operations:

```
public static long getOpnd() {
    long i=0;
```

```
    if (codebuf.get(ip+7) < 0) i = -1;
    for(int j=7;j>1;j--) i = (i<<8) | codebuf.get(ip+j);
    return i;
}
```

Now that we've looked at instruction fetching, let's look at how instruction decoding is performed.

Instruction decoding

The decoding step is a big deal in hardware CPUs; in a bytecode interpreter, it is no big deal, but it needs to be fast. You do not want a long chain of *if-else-if* statements in the main loop that is going to execute extremely frequently. You want decoding to take a small constant amount of time, regardless of the number of opcodes in your instruction set, so usually, you should implement it with either a table lookup or a switch or case control structure. A Unicon implementation of instruction decoding can be seen in the case expression in the following interp() method, which implements the fetch-decode-execute loop:

```
method interp()
  repeat {
    fetch()
    case (op) of {
        Op.HALT: { stop("Execution complete.") }
        Op.NOOP: { . . . }

        . . .

        default: { stop("Illegal opcode " + op) }
        }
    }
end
```

The corresponding Java code looks like this:

```
public static void interp() {
  for(;;) {
    fetch();
    switch (op) {
        case Op.HALT: { stop("Execution complete."); break; }
        case Op.NOOP: { break; }

        . . .

        default: { stop("Illegal opcode " + op); }
```

```
            }
          }
        }
```

The key pieces of the interpreter loop that remain to be shown are the implementation of the various instructions. A couple of examples have been given here that depend on the stop() method to implement the execution of the HALT instruction. In Unicon, stop() is a built-in, but in Java, it can be implemented as follows:

```
    public static void stop(String s) {
      System.err.println(s);
      System.exit(1);
    }
```

The next section describes the rest of the execute portion of the fetch-decode-execute cycle.

Executing instructions

For each of the Jzero instructions, their execution consists of filling in the body of the corresponding case. In Unicon, the ADD instruction might look like this case branch:

```
  Op.ADD: {
      val1 := pop(stack); val2 := pop(stack)
      push(stack, val1 + val2)
  }
```

The corresponding Java implementation is as follows:

```
  case Op.ADD: {
      long val1 = stackbuf.getLong(sp--);
      long val2 = stackbuf.getLong(sp--);
      stackbuf.putLong(sp++, val1 + val2);
      break;
  }
```

Similar code applies for SUB, MUL, DIV, MOD, LT, and LE.

The PUSH instruction takes a memory operand and pushes it onto the stack. The challenging part of this (in Unicon and Java, where pointers are being faked) is the interpretation of the operand to fetch a value from memory. This is performed by a separate dereferencing method. Internal helper functions such as deref() are part of the runtime system and will be covered in the *Writing a runtime system for Jzero* section.

The Unicon implementation of the PUSH instruction is as follows:

```
Op.PUSH: {
    val := deref(opr, opnd)
    push(stack, val)
}
```

The equivalent Java code looks like this:

```
case Op.PUSH: {
    long val = deref(opr, opnd);
    push(val);
    break;
}
```

The POP instruction removes a value from the stack and stores it in a memory location designated by a memory operand. The Unicon implementation of the POP instruction is as follows:

```
Op.POP: {
    val := pop(stack)
    assign(opnd, val)
}
```

The equivalent Java code looks like this:

```
case Op.POP: {
    long val = pop();
    assign(opnd, val);
    break;
}
```

The GOTO instruction sets the instruction pointer register to a new location. In Unicon, this is just as straightforward as you would expect:

```
Op.GOTO: {
    ip := opnd
}
```

The equivalent Java code looks like this:

```
case Op.GOTO: {
    ip = (int)opnd;
    break;
}
```

The conditional branch instruction, BIF (branch-if), pops the top of the stack. If it is non-zero, then it sets the instruction pointer register to a new location, such as a GOTO instruction. In Unicon, the implementation is as follows:

```
Op.BIF: {
    if pop(stack)~=0 then
        ip := opnd
}
```

The equivalent Java code looks like this:

```
case Op.BIF: {
    if (pop() != 0)
        ip = (int)opnd;
    break;
}
```

The call instruction is also like GOTO. It saves an address indicating where execution should resume after a return instruction. The function to call is given in an address just before the n parameters on the top of the stack. A non-negative address in the function slot is the location where the instruction pointer must be set. If the function is negative, it is a call to runtime system function number -n. This is shown in the following Unicon implementation of the CALL instruction:

```
Op.CALL: {
    f := stack[1+opnd]
    if f >= 0 then {
        push(stack, ip) # save old ip
        push( stack, bp) # save old bp
        bp := *stack      # set new bp
        ip := f
        }
    else if f = -1 then do_println()re
}
```

The equivalent Java code looks like this:

```
case Op.CALL: {
    long f;
    f = stackbuf.getLong(sp-8-(int)(8*opnd));
```

```
    if (f >= 0) {
        push( ip);
        push( bp);
        bp = sp;
        ip = (int)f;
        }
    else if (f == -1) do_println();
    else { stop("no CALL defined for " + f); }
    break;
}
```

The return instruction is also a GOTO, except it goes to a location that was previously stored on the stack:

```
Op.RETURN: {
    while *stack > bp do pop(stack)
    bp := pop(stack)
    ip := pop( stack )
}
```

The equivalent Java code looks like this:

```
case Op.RETURN: {
    sp = bp;
    bp = (int)pop();
    ip = (int)pop();
    break;
}
```

The Jzero interpreter's execute operation is pretty short and sweet. Some bytecode interpreters would have additional instructions for input/output, but we are delegating those tasks to a small set of functions that can be called from the generated code. We'll cover those runtime functions shortly, but first, we'll look at the main() method, which starts the Jzero interpreter from the command line.

Starting up the Jzero interpreter

The main() function that launches the Jzero interpreter lives in a module named j0x. This launcher is short and sweet. The Unicon code looks like this, and it can be found in j0x.icn:

```
procedure main(argv)
```

```
   if not (filename := argv[1]) then
     stop("usage: j0x file[.j0]")
   if not (filename[-3:0] == ".j0") then argv[1] ||:= ".j0"
   j0machine := j0machine()
   j0machine.init(filename)
   j0machine.interp()
 end
```

The corresponding Java code in j0x.java looks like this:

```java
public class j0x {
  public static void main(String[] argv) {
    if (argv.length < 1) {
      System.err.println("usage: j0x file[.j0]");
      System.exit(1);
    }
    String filename = argv[0];
    if (! filename.endsWith(".j0"))
      filename = filename + ".j0";
    try {
      j0machine.init(filename);
    } catch(Exception ex) {
      System.err.println("Can't initialize. Exiting.");
      System.exit(1);
    }
    j0machine.interp();
  }
}
```

We will see how well this interpreter runs shortly. But first, let's look at how built-in functions are incorporated into the Jzero runtime system.

Writing a runtime system for Jzero

In a programming language implementation, the runtime system is the code that is included to provide basic functionalities needed for the generated code to run. Generally, the more high-level the language is and the greater its distance from the underlying hardware, the larger the runtime system. Since Jzero is a toy language, its toy runtime system is incomplete and only supports a few internal helper functions such as deref() and some basic functions for input and output.

These functions are written in the implementation language (in our case, Unicon or Java), not the Jzero language. Here is the deref() method in Unicon. Additional runtime system functions are left as exercises for the reader:

```
method deref(reg, od)
  case reg of {
    Op.R_ABS: {
      if od < finstr then return data.word(od)
      else return code[od]
      }
    Op.R_IMM: { return od }
    Op.R_STACK: { return stack[bp+od] }
    default: { stop("deref region ", reg) }
  }
end
```

Each region has different dereferencing code that is appropriate to how that region is stored. The corresponding Java implementation of deref() looks like this:

```
public static long deref(int reg, long od) {
switch(reg) {
case Op.R_ABS: { return codebuf.getLong((int)od); }
case Op.R_IMM: { return od; }
case Op.R_STACK: { return stackbuf.getLong(bp+(int)od); }
default: { stop("deref region " + reg); }
}
return 0;
}
```

In the case of built-in functions, we must be able to call them from the generated Jzero code. The implementation of built-in functions such as System.out.println() and how they are called from the bytecode interpreter will be covered in *Chapter 14, Implementing Operators and Built-In Functions*. Now, it is finally time to look at how to run the Jzero bytecode interpreter.

Running a Jzero program

At this point, we need to be able to test our bytecode interpreter, but we haven't presented the code generator that generates this bytecode yet! For this reason, most of the testing for this chapter's bytecode interpreter will have to wait until the next chapter, where we will present the code generator. For now, here is a hello world program. The source code is as follows:

```
public class hello {
    public static main(String argv[]) {
        System.out.println("hello");
    }
}
```

The corresponding Jzero bytecode might look something like this. One word is shown per line; the lines in hexadecimal show each byte as two hex digits. The opcode is in the leftmost byte, then the operand region byte, and then the operand in the remaining 6 bytes:

```
"Jzero!!\0"
"1.0\0\0\0\0\0"
0x0000040000000000
"hello\0\0\0"
0x0902380000000000                push main
0x0B02000000000000                call 0
0x0100000000000000                halt
0x0902FFFFFFFFFFFF                push -1 (println)
0x0902180000000000                push "hello"
0x0B02010000000000                call 1
0x0C02000000000000                return 0
```

If this is written in binary to a file called hello.j0, then executing the j0x hello command (or running java j0x hello for the Java version) will write out hello, as expected. This tiny but concrete example should whet your appetite for the much more interesting examples that we will generate in the next chapter. In the meantime, compare the simplicity of Jzero with some of the more interesting features that can be found by examining the Unicon bytecode interpreter.

Examining iconx, the Unicon bytecode interpreter

The Unicon language and its predecessor, Icon, share a common architecture and implementation in the form of a bytecode interpreter and runtime system program named iconx. Compared to the Jzero bytecode interpreter in the previous section, iconx is large and complex and has the benefit of real-world use over a sustained period. Compared to the Java virtual machine, iconx is small and simple, and it's relatively accessible for studying. A thorough description of iconx can be found in *The Implementation of Icon and Unicon: a Compendium*. This section can be viewed as a brief introduction to that work.

Understanding goal-directed bytecode

Unicon has an unusual bytecode. A brief example was provided earlier in this chapter in the *Understanding what bytecode is* section. The language is goal-directed. All expressions succeed or fail. Many expressions, called **generators**, can produce additional results on demand when a surrounding expression fails. **Backtracking** is built into the bytecode interpreter to save the state of such generator expressions and resume them later on if needed.

Under the covers, goal-directed expression evaluation can be implemented in many ways, but Unicon's bytecode instruction set, which it inherits largely from Icon, has very unusual semantics that mirror the goal direction found in the source language. Chunks of instructions are marked with information to tell them where to go if they fail. Within such chunks of instructions, the state of generators is saved on a spaghetti stack, and if an expression fails, the most recently suspended generator is resumed.

Leaving type information in at runtime

In Unicon, variables can hold any type of value, and values know what type they are. This contributes to the flexibility of the language and supports polymorphic code, at the cost of slower execution that requires more memory to run. In the C implementation, all variables, including ones stored in structures such as lists or records, are represented in a *descriptor*, declared to be of the struct descrip type. struct descrip contains two words: a dword or descriptor word that mainly holds type information and a vword or value word that holds the value or a pointer to the value. The C implementation of this struct is shown here:

```
struct descrip {
    word dword;
    union {
        word integr;
        double realval;
        char *sptr;
        union block *bptr;
        dptr descptr;
        } vword;
    };
```

Strings are special-cased in dword; for a string, the dword contains the string length; the sign bit of that word is a flag that indicates whether the value is a non-string, which is to say whether a type information code is present. Numbers are special-cased in vword of a descriptor; for integers and real numbers, the value word contains the value; for all other types, the value word is a pointer to the value. Three different kinds of pointers are used, and the pointer to a union block can point at any of a couple dozen or so different Unicon data types. Which field of the vword union to use is decided in all cases by inspecting the dword.

Fetching, decoding, and executing instructions

In the Unicon bytecode interpreter, the fetch-decode-execute loop lives in a C function named interp(). Consistent with this chapter, this consists of an infinite loop with a switch statement inside it. One difference between Unicon instructions and Jzero, as described in this chapter, is that Unicon opcodes are generally a half-word in size, and if they contain an operand, it is generally a full word following that half-word opcode. Since many instructions have no operand, this may make the code more compact, and since operands are full words, they can contain a native C pointer rather than an offset relative to a base pointer for a given memory region. Unicon bytecode is computed by the compiler and stored in the executable on disk using offsets, and when they first execute, the offsets are converted into pointers and the opcode is modified to indicate that they now contain pointers. This clever self-modifying code poses extra pain for thread safety, and it means bytecode cannot be executed from constant or read-only memory.

Crafting the rest of the runtime system

Another difference between iconx and the Jzero interpreter presented in this chapter is that the Unicon bytecode interpreter has an enormous runtime system consisting of numerous sophisticated capabilities, such as high-level graphics and networking. Where the Jzero bytecode interpreter might be 80% of the code, with 20% left to the runtime system, the interp() function at Unicon's core might be only 5% of the code, with the other 95% being the implementation of the many built-in functions and operators. This runtime system is written in a language called **Runtime Language (RTL)**. RTL is a kind of superset of C with special features to support the Unicon type system, type inferencing, and automatic type conversion rules.

This section presented a brief introduction to the Unicon bytecode interpreter implementation. You saw that programming language bytecode interpreters are often a lot more interesting and complex than the Jzero interpreter. They may involve novel control structures, high-level and/or domain-specific data types, and more.

Summary

This chapter presented the essential elements of bytecode interpreters. Knowing how to implement a bytecode interpreter frees you to generate flexible code, without having to worry about hardware instruction sets, registers, or addressing modes.

First, you learned that the definition of an instruction set includes the opcodes and rules for processing any operands in those instructions. You also learned how to implement generic stack machine semantics, as well as bytecode instructions that correspond to domain-specific language features. Then, you learned how to read and execute bytecode files, including interchangeably working with sequences of bytes and words in both Unicon and Java.

Given the existence of a bytecode interpreter, in the next chapter, we will discuss generating bytecode from intermediate code so that we can run programs that are compiled using our compiler!

Questions

1. A bytecode interpreter could use an instruction set with up to three addresses (operands) per instruction, such as three-address code. Instead, the Jzero interpreter uses zero or one operands per instruction. What are the pros and cons of using three-address code in the bytecode interpreter, just as it is used in intermediate code?

2. On real CPUs and in many C-based bytecode interpreters, bytecode addresses are represented by literal machine addresses. However, the bytecode interpreters that were shown in this chapter implement bytecode addresses as positions or offsets within allocated blocks of memory. Is a programming language that does not have a pointer data type at a fatal disadvantage in implementing a bytecode interpreter, compared to a language that does support pointer data types?

3. If code is represented in memory as an immutable string value, what constraints does that impose on the implementation of a bytecode interpreter?

Join our community on Discord

Join our community's Discord space for discussions with the authors and other readers:

```
https://discord.com/invite/zGVbWaxqbw
```

13

Generating Bytecode

In this chapter, we continue with code generation, taking the intermediate code from *Chapter 9, Intermediate Code Generation*, and generating bytecode from it. When you translate from intermediate code into a format that will run, you are generating **final code**. Conventionally this happens at compile time, but it could occur later—at link time, load time, or runtime. We will generate bytecode in the usual way at compile time. This chapter and the following chapter on generating native code present you with two additional forms of final code that you can choose besides transpiling to another high-level language as described in *Chapter 11*.

Translation from intermediate code to bytecode is performed by walking through a list of intermediate instructions, translating each intermediate code instruction into one or more bytecode instructions. A straightforward loop is used to traverse the list, with a different chunk of code for each intermediate code instruction. Although the loop used in this chapter is simple, generating the final code remains very important as the culminating essential skill you must acquire to bring your new programming language to life.

This chapter covers the following main topics:

- Converting intermediate code to Jzero bytecode
- Comparing bytecode assembler with binary formats
- Linking, loading, and including the runtime system
- Unicon example: bytecode generation in i cont

With the functionality that we build in this chapter, we will be able to generate code that runs on the bytecode interpreter presented in the previous chapter.

Technical requirements

The code for this chapter is available on GitHub: `https://github.com/PacktPublishing/Build-Your-Own-Programming-Language-Second-Edition/tree/master/ch13`

The Code in Action video for the chapter can be found here: `https://bit.ly/3oR6zGt`

Converting intermediate code to Jzero bytecode

The Jzero intermediate code generator from *Chapter 9, Intermediate Code Generation*, traversed a tree and created a list of intermediate code as a synthesized attribute in each tree node, named `icode`. The intermediate code for the whole program is the `icode` attribute in the root node of the syntax tree. In this section, we will use this list to produce our output bytecode. To generate byte-code, the `gencode()` method in the `j0` class calls a new method in this class, named `bytecode()`, and passes it the intermediate code in `root.icode` as its input. The Unicon `gencode()` method that invokes this functionality in `j0.icn` looks like the following code block. The two highlight-ed lines at the end of the following code snippet are added for bytecode generation, verified by simple text output:

```
method gencode(root)
    root.genfirst()
    root.genfollow()
    root.gentargets()
    root.gencode()
    labeltable := table()
    bcode := bytecode(root.icode)
    if \textoutput then
        every (! (\bcode)).print()
    else
        genbytecode()
end
```

The `bytecode()` method takes in an `icode` list, and its return value is a list whose elements are objects that represent bytecode instructions. These elements are instances of the `byc` class, which stands for bytecode. If the `textoutput` global flag has been set, the bytecode is printed out in text form; a binary format is output by default. The corresponding Java code for the `gencode()` method is shown in the following code snippet. Output generation performed in the `if` statement is a little more convoluted in this case:

```
public static void gencode(parserVal r) {
```

```
      tree root = (tree)(r.obj);
      root.genfirst();
      root.genfollow();
      root.gentargets();
      root.gencode();
      labeltable = new HashMap<>();
      methodAddrPushed = false;
      ArrayList<byc> bcode = bytecode(root.icode);
      if (textoutput) {
        if (bcode != null) {
          for (int i = 0; i < bcode.size(); i++)
            bcode.get(i).print();
        }
      }
      else genbytecode(bcode);
  }
```

Now, let's examine the code for the byc class.

Adding a class for bytecode instructions

We could represent our bytecode literally, using a 64-bit word in the same format presented in *Chapter 12, Bytecode Interpreters*. Representing bytecode instructions as objects instead of 64-bit words facilitates output in both human-readable text and binary form. The list-of-objects representation also makes analysis for final code optimization more convenient.

The byc class resembles the tac class from *Chapter 9, Intermediate Code Generation*, but instead of an **operation code** (opcode) and fields for up to three operands, it just represents an opcode, an operand region, and—if present—an operand, as described in *Chapter 12, Bytecode Interpreters*. The class also contains several methods, including ones for printing in text and binary forms. The print() and printb() methods will be presented in the section titled *Comparing bytecode assembler with binary formats*. Here is an outline of the byc Unicon class from byc.icn:

```
class byc(op, opreg, opnd)
   method print() … end
   method printb() … end
   method addr(a) … end
initially(o, a)
   op := o; addr(a)
end
```

The corresponding Java class in byc.java looks like this:

```
public class byc {
    int op, opreg;
    long opnd;
    public byc(int o, address a) {
        op=o; addr(a);
    }
    public void print() { … }
    public void printb() { … }
    public void addr(address a) { … }
}
```

As a part of this byc class, we need a method named addr() that provides a mapping from three-address code addresses to bytecode addresses. Let's examine this next.

Mapping intermediate code addresses to bytecode addresses

Although the instruction sets are quite different, the addresses in the intermediate and final code denote approximately the same thing. Since we design both the intermediate code and bytecode, we can define addresses in bytecode to be a lot closer to intermediate code addresses than will be the case when we are mapping from intermediate code to native code in the next chapter. In any case, the region and offset from the address class from *Chapter 9, Intermediate Code Generation*, must be mapped onto opreg and opnd in the byc class. This is handled by an addr() method in the byc class that takes an instance of the address class as a parameter and sets opreg and opnd. The Unicon code in byc.icn looks like this:

```
method addr(a)
    if /a then opreg := Op.R_NONE
    else case a.region of {
    "loc": { opreg := Op.R_STACK; opnd := a.offset }
    "glob": { opreg := Op.R_ABS; opnd := a.offset }
    "const": { opreg := Op.R_ABS; opnd := a.offset }
    "lab": { opreg := Op.R_ABS; opnd := a.offset }
    "obj": { opreg := Op.R_HEAP; opnd := a.offset }
    "imm": { opreg := Op.R_IMM; opnd := a.offset }
    }
end
```

The corresponding Java method in byc.java is shown here:

```
public void addr(address a) {
    if (a == null) opreg = Op.R_NONE;
    else switch (a.region) {
    case "loc": { opreg = Op.R_STACK; opnd = a.offset; break; }
    case "glob": { opreg = Op.R_ABS; opnd = a.offset;  break; }
    case "const": { opreg = Op.R_ABS; opnd = a.offset; break; }
    case "lab": { opreg = Op.R_ABS; opnd = a.offset;    break; }
    case "obj": { opreg = Op.R_HEAP; opnd = a.offset;   break; }
    case "imm": { opreg = Op.R_IMM; opnd = a.offset;    break; }
    }
}
```

Given the byc class, one more helper function is needed in order to formulate the bytecode() code generator method. We need a convenient factory method for generating bytecode instructions and attaching them to the bcode list. We will call this method bgen().

The bgen() method in the j0 class is similar to gen() from the tree class; it produces a one-element list containing a byc instance. The Unicon code looks like this:

```
method bgen(o, a)
    return [byc(o, a)]
end
The corresponding Java implementation looks like this:
public ArrayList<byc> bgen(int o, address a) {
    ArrayList<byc> L = new ArrayList<byc>();
    byc b = new byc(o, a);
    L.add(b);
    return L;
}
```

Now, finally, it's time to present the bytecode generator.

Implementing the bytecode generator method

The Unicon implementation of the bytecode() method in the j0 class is shown next. The implementation must fill in one case branch for each opcode in the three-address instruction set given in *Chapter 9, Intermediate Code Generation*. There will be a lot of cases, so we present each one separately. The outline of the entire method is as follows.

The case bodies for the various instructions make use of a **singleton** class named Op that provides names for the opcodes, such as Op.ADD. The Op class constructor is replaced by its singleton instance at the top of the bytecode() method, after which bytecode translations of instructions are placed onto an output list named rv that is bytecode()'s return value:

```
method bytecode(icode)
    if type(Op)=="procedure" then Op()
    rv := []
    every i := 1 to *\icode do {
        instr := icode[i]
        case instr.op of {
            "ADD": { ... append translation of ADD to return val }
            "SUB": { ... append translation of SUB to return val }

            ...
        }
    }
    return rv
end
```

The Java implementation of bytecode() is shown here:

```
public static ArrayList<byc> bytecode(ArrayList<tac> icode)
{
    ArrayList<byc> rv = new ArrayList<byc>();
    for(int i=0; i<icode.size(); i++) {
        tac instr = icode.get(i);
        switch(instr.op) {
        case "ADD": { ... append translation of ADD to rv }
        case "SUB": { ... append translation of SUB to rv }

        ...
        }
    }
    return rv;
}
```

Within the framework of this bytecode() method, we now get to provide translations for each of the three-address instructions. We will start with simple expressions.

Generating bytecode for simple expressions

The different cases for each three-address opcode have many elements in common, such as the pushing of values from memory onto the evaluation stack. The case for addition perhaps shows the most common translation pattern. In Unicon, addition is handled like this:

```
"ADD": {
    bcode |||:= j0.bgen(Op.PUSH, instr.op2) |||
        j0.bgen(Op.PUSH, instr.op3) ||| j0.bgen(Op.ADD) |||
        j0.bgen(Op.POP, instr.op1)
}
```

This code reads operand 2 and operand 3 from memory and pushes them onto the stack. The actual ADD instruction works entirely from the stack. The result is then popped off the stack and placed into operand 3. In Java, the implementation of addition consists of the following code:

```
case "ADD": {
    rv.addAll(j0.bgen(Op.PUSH, instr.op2));
    rv.addAll(j0.bgen(Op.PUSH, instr.op3));
    rv.addAll(j0.bgen(Op.ADD, null));
    rv.addAll(j0.bgen(Op.POP, instr.op1));
    break;
}
```

The intermediate code instruction set presented in *Chapter 9, Intermediate Code Generation*, defines 19 three-address instructions that must be translated to final code. The final code generation pattern illustrated by the preceding ADD instruction is used for the other arithmetic instructions. For a unary operator such as NEG, the pattern is slightly simplified, as we can see here:

```
"NEG": {
    bcode |||:= j0.bgen(Op.PUSH, instr.op2) |||
        j0.bgen(Op.NEG) ||| j0.bgen(Op.POP, instr.op1)
}
```

In Java, the implementation of negation consists of the following code:

```
case "NEG": {
    rv.addAll(j0.bgen(Op.PUSH, instr.op2));
    rv.addAll(j0.bgen(Op.NEG, null));
    rv.addAll(j0.bgen(Op.POP, instr.op1));
```

```
        break;
    }
```

An even simpler instruction such as ASN may be worth special-casing when you design the instruction set of your bytecode machine, but for a stack machine you can stick with the same script
and simplify the preceding pattern further, as illustrated in the following code snippet:

```
"ASN": {
    bcode |||:= j0.bgen(Op.PUSH, instr.op2) |||
      j0.bgen(Op.POP, instr.op1)
}
```

In Java, the implementation of assignment might look like this:

```
case "ASN": {
    rv.addAll(j0.bgen(Op.PUSH, instr.op2));
    rv.addAll(j0.bgen(Op.POP, instr.op1));
    break;
}
```

Code consisting of arithmetic expressions and assignments is the core of most programming languages. Now, it is time to look at code generation for some other intermediate code instructions,
starting with the ones used for manipulating pointers.

Generating code for pointer manipulation

Three of the intermediate code three-address instructions defined in *Chapter 9, Intermediate
Code Generation*, pertain to the use of pointers: ADDR, LCON, and SCON. The ADDR instruction turns
an address in memory into a piece of data that can be manipulated to perform operations such
as pointer arithmetic. It translates to a bytecode instruction named LOAD, as illustrated in the
following code snippet:

```
"ADDR": {
    bcode |||:= j0.bgen(Op.LOAD, instr.op1)
}
```

In Java, the implementation of the ADDR instruction consists of this code:

```
case "ADDR": {
    rv.addAll(j0.bgen(Op.LOAD, instr.op1));
    break;
}
```

The LCON instruction reads from memory pointed at by other memory, as illustrated here:

```
"LCON": {
    bcode |||:= j0.bgen(Op.LOAD, instr.op2)
    bcode |||:= j0.bgen(Op.POP, instr.op1)
}
```

In Java, the implementation of the LCON instruction consists of the following code:

```
case "LCON": {
    rv.addAll(j0.bgen(Op.LOAD, instr.op2));
    rv.addAll(j0.bgen(Op.POP, instr.op1));
    break;
}
```

The SCON instruction writes to memory pointed at by other memory, as illustrated here:

```
"SCON": {
    bcode |||:= j0.bgen(Op. STORE, instr.op2) |||
                j0.bgen(Op.POP, instr.op1)
}
```

In Java, the implementation of the SCON instruction consists of the following code:

```
case "SCON": {
    rv.addAll(j0.bgen(Op.STORE, instr.op2));
    rv.addAll(j0.bgen(Op.POP, instr.op1));
    break;
}
```

These instructions are important for supporting structured data types such as arrays. Now, let's consider bytecode code generation for control flow, starting with the GOTO family of instructions.

Generating bytecode for branches and conditional branches

Seven of the intermediate code instructions pertain to conditional and unconditional branch instructions. The simplest of these is the unconditional branch or GOTO instruction. The GOTO instruction assigns a new value to the instruction pointer register. It should be no surprise that the GOTO bytecode is the implementation of the intermediate code GOTO instruction. The Unicon code for translating GOTO intermediate code into GOTO bytecode is shown here:

```
"GOTO": {
```

```
        bcode |||:= j0.bgen(Op.GOTO, instr.op1)
   }
```

In Java, the implementation of the GOTO instruction consists of the following code:

```
case "GOTO": {
    rv.addAll(j0.bgen(Op.GOTO, instr.op1));
    break;
}
```

The conditional branch instructions in the three-address code are translated down into simpler final code instructions. For the instruction set bytecode presented in the previous chapter, this means pushing operands onto the stack prior to the conditional branch instruction bytecode. The Unicon implementation of the BLT instruction looks like this:

```
"BLT": {
    bcode |||:= j0.bgen(Op.PUSH, instr.op2) |||
        j0.bgen(Op.PUSH, instr.op3) ||| j0.bgen(Op.LT) |||
        j0.bgen(Op.BIF, instr.op1)
}
```

In Java, the generation of bytecode for the BLT instruction consists of the following code:

```
case "BLT": {
    bcode.addAll(j0.bgen(Op.PUSH, instr.op2));
    bcode.addAll(j0.bgen(Op.PUSH, instr.op3));
    bcode.addAll(j0.bgen(Op.LT, null));
    bcode.addAll(j0.bgen(Op.BIF, instr.op1));
    break;
}
```

This pattern is employed for several of the three-address instructions, with slightly simpler code used for BIF and BNIF. Now, let's consider the more challenging forms of control flow transfer that relate to method calls and returns.

Generating code for method calls and returns

Three of the three-address instructions handle the very important topic of function and method calls and returns. A sequence of zero or more PARM instructions pushes values onto the stack, the CALL instruction performs a method call, and the RET instruction returns from a method to the caller.

But this three-address code calling convention must be mapped down onto the underlying instruction set, which in this chapter is a bytecode stack machine instruction set that requires the address of the procedure to be called to be pushed (in a stack slot where the return value will be found), prior to pushing other parameters. We could go back and modify our three-address code to fit the stack machine better, but then it would not fit so well for x86_64 native code.

The PARM instruction is a simple push, except when it is the first parameter and the procedure address is needed, as illustrated in the following code snippet:

```
"PARM": {
    if /methodAddrPushed then {
        every j := i+1 to *icode do
            if icode[j].op == "CALL" then {
                if icode[j].op1 === "PrintStream__println" then {
                    bcode |||:= j0.bgen(Op.PUSH, address("imm", -1))
                else {
                    bcode |||:= j0.bgen(Op.PUSH, icode[j].op1)
                }
                break
            }
        methodAddrPushed := 1
        }
    bcode |||:= j0.bgen(Op.PUSH, instr.op1)
}
```

The every loop looks for the nearest CALL instruction and pushes its method address. In Java, the implementation of the PARM instruction is similar, as we can see here:

```
case "PARM": {
    if (methodAddrPushed == false) {
        for(int j = i+1; j<icode.size(); j++) {
            tac callinstr = icode.get(j);
            if (callinstr.op.equals("CALL")) {
                if (callinstr.op1.str().equals("PrintStream__println:0")) {
                    rv.addAll(j0.bgen(Op.PUSH, new address("imm", -1)));
                } else {
                    rv.addAll(j0.bgen(Op.PUSH, callinstr.op2));
                }
                break;
```

```
        }
    methodAddrPushed = true;
        }
    }
    rv.addAll(j0.bgen(Op.PUSH, instr.op1));
    break;
}
```

Having pushed the method address ahead of time, the CALL instruction is straightforward. After the call, the op1 destination in the three-address code is popped from the stack, as with other expressions. The op2 source field is the method address that was used prior to the first PARM instruction. The op3 source field gives the number of parameters, which is used as is as the operand in CALL: bytecode, as illustrated in the following code snippet:

```
"CALL": {
    bcode |||:= j0.bgen(Op.CALL, instr.op2)
    methodAddrPushed := &null
}
```

In Java, the implementation of the CALL instruction consists of the following code:

```
case "CALL": {
    rv.addAll(j0.bgen(Op.CALL, instr.op2));
    methodAddrPushed = false;
    break;
}
```

The Unicon implementation of the RETURN instruction looks like the following code snippet. This code does not distinguish between a void return with no return value and a non-void return. A bytecode interpreter designer could have a separate instruction for returning with no return value, but perhaps most folks would just return a 0 in that case, and the caller would just not use the bogus return value:

```
"RETURN": {
    bcode |||:= j0.bgen(Op.RETURN, instr.op1)
}
```

In Java, the implementation of the RETURN instruction consists of the following code:

```
case "RETURN": {
    rv.addAll(j0.bgen(Op.RETURN, instr.op1));
```

```
    break;
 }
```

Generating code for method calls and returns is not too difficult. Now, let's consider how to handle pseudo-instructions in the three-address code.

Handling labels and other pseudo-instructions in intermediate code

Pseudo-instructions do not translate into code, but they are present in the linked list of three-address instructions and require consideration in the final code. The most common and obvious pseudo-instruction is label. If the final code is being generated in a human-readable assembler format, labels can be generated as part of the output in whatever format the target requires. A hash table named labeltable associates all the label names with byte offsets in the code. During the traversal of the list of instructions, label names obtained from the three-address code are mapped to how far (in bytes) into the generated code we have reached at this point. In Unicon, this is expressed in this way:

```
"LAB": {
    labeltable[instr.op1.offset] := *bcode * 8
}
```

This is the corresponding code in Java:

```
case "LAB": {
    labeltable.put("L"+String.valueOf(instr.op1.offset), rv.size() * 8);
    break;
}
```

For final code generated in a binary format, labels require some additional handling since they must be replaced by corresponding byte offsets or addresses.

Since a label is really a name or alias for the address of a particular instruction, in a binary byte-code format it is typically replaced by byte offsets in some form. As the final code is generated, a table containing the mapping between labels and offsets is constructed.

The past several sections produced a data structure containing a representation of the bytecode and then showed how various three-address instructions are translated. Now, let's move on to producing the code in textual and binary formats.

Comparing bytecode assembler with binary formats

Bytecode machines tend to use simpler formats than native code, where binary object files are the norm. Some bytecode machines, such as Python, hide their bytecode format entirely or make it optional. Others, such as Unicon, use a human-readable assembler-like text format for compiled modules. In the case of Java, they seem to have gone out of their way to avoid providing an assembler, to make it more difficult for other languages to generate code for the Java **virtual machine (VM)**.

In the case of Jzero and its machine, space limits motivate us to keep things as simple as possible. The byc class defines two output methods: print() for human-friendly text format and printb() for machine-friendly binary format. You can decide for yourself which one you prefer for any given application.

Printing bytecode in assembler format

The print() method in the byc class is similar to the one used in the tac class. One line of output is produced for each element in the list. The Unicon implementation of the print() method in the byc class is shown here. The f parameter, which defaults to the standard output, specifies the name:

```
method print(f:&output)
   write(f, "\t", nameof(), " ", addrof()) |
      write(&errout, "can't print ", image(self), " op ", image(op))
end
```

The corresponding Java implementation is shown here. Method overloading is used to make the parameter optional:

```
public void print(PrintStream f) {
   f.println("\t" + nameof() + " " + addrof());
}
public void print() { print(System.out); }
```

The text-based print() methods just punt off most of the work to helper methods that produce human-readable representations of the opcode and the operand. The Unicon code for the nameof() method that maps opcodes back to strings is shown in the following example. A table is stored as a static variable inside the nameof() method; the table is initialized the first time the nameof() method is called:

```
method nameof()
```

```
    static opnames
    initial opnames := table(Op.HALT, "halt", Op.NOOP, "noop",
        Op.ADD, "add", Op.SUB, "sub", Op.MUL, "mul",
        Op.DIV, "div", Op.MOD, "mod", Op.NEG, "neg",
        Op.PUSH, "push", Op.POP, "pop", Op.CALL, "call",
        Op.RETURN, "return", Op.GOTO, "goto", Op.BIF, "bif",
        Op.LT, "lt", Op.LE, "le", Op.GT, "gt", Op.GE, "ge",
        Op.EQ, "eq", Op.NEQ, "neq", Op.LOCAL, "local",
        Op.LOAD, "load", Op.STORE, "store")
    return opnames[op]
end
```

The corresponding Java code from the byc.java file shown here uses a HashMap. In contrast with the Unicon implementation, in Java, the static variable and the static code block that initializes it are placed outside of the nameof() method within the byc class:

```
static HashMap<Short,String> ops;
static { ops = new HashMap<>();
  ops.put(Op.HALT,"halt"); ops.put(Op.NOOP,"noop");
  ops.put(Op.ADD,"add"); ops.put(Op.SUB,"sub");
  ops.put(Op.MUL,"mul"); ops.put(Op.DIV, "div");
  ops.put(Op.MOD,"mod"); ops.put(Op.NEG, "neg");
  ops.put(Op.PUSH,"push"); ops.put(Op.POP, "pop");
  ops.put(Op.CALL, "call"); ops.put(Op.RETURN, "return");
  ops.put(Op.GOTO, "goto"); ops.put(Op.BIF, "bif");
  ops.put(Op.LT, "lt"); ops.put(Op.LE, "le");
  ops.put(Op.GT, "gt"); ops.put(Op.GE, "ge");
  ops.put(Op.EQ, "eq"); ops.put(Op.NEQ, "neq");
  ops.put(Op.LOCAL, "local"); ops.put(Op.LOAD, "load");
  ops.put(Op.STORE, "store");
}
public String nameof() {
    return ops.get(op);
}
```

Another helper function called from the print() method is the addrof() method, which prints a human-readable representation of an address based on the operand region and operand fields. Its Unicon implementation is shown here:

```
method addrof()
```

```
      case opreg of {
         Op.R_NONE | &null: return ""
            Op.R_ABS : return "@" || sprint("%x", opnd)
         Op.R_IMM: return string(opnd)
         Op.R_STACK: return "stack:" || opnd
         Op.R_HEAP: return "heap:" || opnd
         default: return string(opreg) || ":" || opnd
         }
   end
```

The corresponding Java code for addrof() is shown here:

```
public String addrof() {
   switch (opreg) {
   case Op.R_NONE: return "";
   case Op.R_ABS: return "@"+ java.lang.Long.toHexString(opnd);
   case Op.R_IMM: return String.valueOf(opnd);
   case Op.R_STACK: return "stack:" + String.valueOf(opnd);
   case Op.R_HEAP: return "heap:" + String.valueOf(opnd);
   }
   return String.valueOf(opreg)+":"+String.valueOf(opnd);
}
```

Now, let's look at the corresponding binary output.

Printing bytecode in binary format

The printb() methods are organized similarly, but where print() needs names of things, printb() needs to put all the bits in a row and output a binary word. Its Unicon implementation is shown below. Unicon output functions write strings. The built-in char(i) function takes an integer and returns a string encoding of that integer. For example, char(65) returns "A":

```
method printb(f:&output)
   writes(f, char(op), char(\opreg|0))
   x := (\opnd | 0)
   every !6 do {
      writes(f, char(iand(x, 255)))
      x := ishift(x, -8)
      }
end
```

The corresponding Java implementation of `printb()` is shown here. The code originally cast values to (byte) but I found, counterintuitively, that I had to modify the code to use (char):

```java
public void printb(PrintStream f) {
    long x = opnd;
    f.print((char)op);
    f.print((char)opreg);
    for(int i = 0; i < 6; i++) {
        f.print((char)(x & 0xff));
        x = x>>8;
    }
}
public void printb() { printb(System.out); }
```

In this section, we considered how to output our code in binary format to external storage. The contrast between text and binary formats was stark, with binary formats being more work, at least from a human perspective. Now, let's look at other issues necessary for program execution beyond the generated code. This includes linking generated code with other code, especially the runtime system.

Linking, loading, and including the runtime system

In a separately compiled native code language, the output binary format from the compile step is not usually executable. Machine code is output in an object file that must be linked together with other modules, and addresses between them resolved, to form an executable. The runtime system is included at this point, by linking in object files that come with the compiler, not just other modules written by the user. In the old days, loading the resulting executable was a trivial operation. In modern systems, it is more complex due to things such as shared object libraries.

A bytecode implementation often has substantial differences from the traditional model just described. Java performs no link step, or perhaps you can say that it links code in at load time. The Java runtime system might be considered sharply divided between a large amount of functionality that is built into the **Java VM (JVM)** interpreter and an also-large amount of functionality that must be loaded, both bytecode and native code, to run various parts of the standard Java language. From an outsider's perspective, one of the surprising things in Java is the enormous number of `import` statements a developer must place at the top of every file that uses things in Java's standard libraries. Almost all Java classes of non-trivial size and complexity use a lot of functionality from Java's standard libraries, so using many `import` statements is typical.

In the case of Jzero, severe limitations keep all this as simple as possible. There is no separate compilation or linking. Loading is kept extremely simple and was covered in the previous chapter. The runtime system is built into the bytecode interpreter, another dodge to enable the language to avoid linking. Now, let's look at bytecode generation in Unicon, another real-world bytecode implementation that does things far differently than either Java or Jzero.

Unicon example – bytecode generation in icont

Unicon's bytecode translator outputs human-readable text in **ucode** files. The ucode format serves as both an assembler and object file format in the Unicon ecosystem. Such ucode files are initially generated, and then linked and converted into binary icode format by a C program named icont that is invoked by the Unicon translator. The icont program plays the role of code generator for Unicon. Its back-end functions as an assembler and linker to form a complete bytecode program in binary format. Here are some of the details.

A C function about 400 lines long named gencode() in icont's lcode.c module reads lines of ucode text and turns them into binary format following the code outline shown below. For each line, an opcode is read using the getopc() function. After that, a gigantic switch statement emits different binary code appropriate to different instruction opcodes.

It is no accident that there is an interesting similarity between this pseudo-code and the fetch-decode-execute loop used in the bytecode interpreter. Here, we are fetching text bytecode from input, decoding the opcode, and writing binary bytecode with slight differences in format depending on the bytecode. Although the Unicon bytecode instruction set has close to 125 different opcodes, the switch statement emits code in the same format for similar opcodes and contains only about 25 distinct code blocks for different types of opcodes. Around 67 opcodes share the same code emission pattern as Op_Plus in the example below:

```
void gencode() {
    while ((op = getopc(&name)) != EOF) {
        switch(op) {
        ...
        case Op_Plus:
            newline();
            lemit(op, name);
            break;
        ...
        }
```

```
        }
    }
```

The lemit() function and about seven related functions with the lemit*() prefix are used to append bytecode within a contiguous array of bytes in binary format. Labels associated with instructions are turned into byte offsets. Forward references to labels that have not been encountered yet are placed in linked lists and backpatched later when the target label is encountered. The C code for lemitl() emits an instruction with a label, as shown here:

```
    static void lemitl(int op, int lab, char *name)
    {
    misalign();
    if (lab >= maxlabels)
        labels  = (word *) trealloc(labels, NULL, &maxlabels,
            sizeof(word), lab - maxlabels + 1, "labels");
    outop(op);
    if (labels[lab] <= 0) {        /* forward reference */
        outword(labels[lab]);
        labels[lab] = WordSize - pc;
        }
    else outword(labels[lab] - (pc + WordSize));
    }
```

Opcodes are generally short; in Unicon bytecode, opcodes can easily fit in a 16- or 32-bit value and could be made to fit in a byte if so desired. In the absence of operands, multiple instructions might fit in a single word. On the other hand, operands are big. Many CPU architectures process larger values such as 64-bit integers only when those values are word-aligned. As you may guess from its name, the misalign() function generates no-op instructions as needed in order to ensure that instructions with operands are emitted at offsets where their operand will start on a word boundary. The first if statement grows the array table if needed. The second if statement handles a label that is a forward reference to an instruction that does not exist yet, by inserting it onto the front of a linked list of instructions that will have to be backpatched when the instructions are all present.

The guts of the binary code layout are done by outop() and outword(), to output an opcode and an operand that are of integer and word length, respectively. These macros may be defined differently on different platforms, but on most machines, they simply call functions named intout() and wordout(). Note in the following code snippet that the binary code is in machine-native format and is different on **central processing units (CPUs)** with different word sizes or endianness.

This gives good performance at the cost of bytecode portability. Some popular languages, such as Java, make the opposite design decision and preserve portability at all costs:

```
static void intout(int oint)
    {
    int i;
    union {
        int i;
        char c[IntBits/ByteBits];
        } u;
    CodeCheck(IntBits/ByteBits);
    u.i = oint;
    for (i = 0; i < IntBits/ByteBits; i++)
        codep[i] = u.c[i];
    codep += IntBits/ByteBits;
    pc += IntBits/ByteBits;
    }
```

After all this glorious example C code, you will probably be glad to get back to Unicon and Java. But C really does make lower-level binary manipulation somewhat simpler than it is in Unicon or Java. The moral of the story is *learn and use the right tools for each kind of job.*

Summary

This chapter showed you how to generate bytecode for software bytecode interpreters. The skills you learned include how to traverse a linked list of intermediate code and, for each intermediate code opcode and pseudo-instruction, how to convert it into instructions in a bytecode instruction set. There were big differences between the semantics of the three-address machine and the byte-code machine. Many intermediate code instructions were converted into three or more bytecode machine instructions. The handling of CALL instructions was a bit hairy, but it is important for you to perform function calls in the manner required by the underlying machine. While learning all this, you also learned how to write out bytecode in text and binary formats.

The next chapter presents an alternative that is more attractive for some languages: generating native code for a mainstream CPU.

Questions

1. Describe how intermediate code instructions with up to three addresses are converted into a sequence of stack machine instructions that contain at most one address.

2. If a particular instruction (say it is instruction 15, at byte offset 120) is targeted by five different labels (for example, L2, L3, L5, L8, and L13), how are the labels processed when generating binary bytecode?

3. In intermediate code, a method call consists of a sequence of PARM instructions followed by a CALL instruction. Does the described bytecode for doing a method call in bytecode match up well with the intermediate code? What is similar and what is different?

4. CALL instructions in **object-oriented (OO)** languages such as Jzero are always preceded by a reference to the object (self or this) on which the methods are being invoked... or are they? Explain a situation in which the CALL method instruction may have no object reference, and how the code generator described in this chapter should handle that situation.

5. Our code for pushing method addresses at the first PARM instruction assumed that no nested PARM...CALL sequences occur inside a surrounding PARM...CALL sequence. Can we guarantee that to be the case for examples such as f(0, g(1), 2)?

6. The Java language popularized the idea that running identical bytecode regardless of the CPU word sizes or endianness was a top priority: portability trumps performance. The Unicon bytecode machine takes the opposite view. Which is correct? Is it possible to have the best of both worlds? How?

Join our community on Discord

Join our community's Discord space for discussions with the authors and other readers:

`https://discord.com/invite/zGVbWaxqbw`

14

Native Code Generation

This chapter shows you how to take the intermediate code from *Chapter 9, Intermediate Code Generation*, and generate **native code**. The term *native* refers to whatever instruction set is provided by hardware on a given machine. This chapter presents a simple native code generator for x64, a dominant architecture on laptops and desktops.

This chapter covers the following main topics:

- Deciding whether to generate native code
- Introducing the x64 instruction set
- Using registers
- Converting intermediate code to x64 code
- Generating x64 output

The skills developed here include basic register allocation, instruction selection, writing assembler files, and invoking the assembler and linker to produce a native executable. The functionality built into this chapter generates code that runs natively on typical computers.

Technical requirements

The code for this chapter is available on GitHub: https://github.com/PacktPublishing/Build-Your-Own-Programming-Language-Second-Edition/tree/master/ch14

The Code in Action video for the chapter can be found here: https://bit.ly/2Zdky0I

Deciding whether to generate native code

Generating native code is more work than bytecode but enables faster execution. Native code may also use less memory or electricity. Native code pays for itself if end users save time or money. However, targeting a specific **central processing unit** (**CPU**) sacrifices portability. You may want to implement bytecode first, and only generate native code if the language becomes popular enough to justify the effort. However, there are other reasons to generate native code. You may be able to write your runtime system using the facilities provided for another compiler. For example, our Jzero x64 runtime system is built using the **GNU's Not Unix** (**GNU**) C library. Now, let's look at some of the specifics of the x64 architecture.

Introducing the x64 instruction set

This section provides a brief overview of the x64 instruction set. You are encouraged to consult **Advanced Micro Devices** (**AMD**) or **Intel**'s architecture programmer's manuals for more information. Douglas Thain's book *Introduction to Compilers and Language Design*, available at http://compilerbook.org, has helpful x64 material.

x64 is a complex instruction set with many backward-compatibility features. This chapter covers the subset of x64 that is used to build a basic Jzero code generator. We are using **AT&T assembler syntax** so that our generated output can be converted into the binary object file format by the GNU assembler. This is for the sake of multiplatform portability.

x64 has hundreds of instructions with names such as ADD for addition or MOV to move (copy) a value to a new location. When an instruction has two operands, at most, one may be a reference to main memory. x64 instructions can have a suffix to indicate how many bytes are being read or written, although the name of a **register** in the instruction often makes the suffix redundant. Jzero primarily uses the x64 instruction suffix Q (or q) for 64-bit **quadword** operations. A 64-bit word is a **quad** based on the late-1970s Intel 16-bit instruction set. We use the following instructions and pseudo-instructions in this chapter. Some of the instructions use the stack pointer register %rsp. The names of the other x64 registers are given in the next section.

Instruction	Description
`addq`	Add a 64-bit into another 64-bit value
`call`	Store a return address to (%rsp), decrement %rsp, goto function
`cmpq`	Compare two values and set condition code bits
`goto`	Jump to a new location in the code
`jle`	Jump if less than or equal
`leaq`	Compute an address
`movq`	Move a 64-bit value from source to destination
`negq`	Negate a 64-bit value
`popq`	Fetch a value from (%rsp) and increment %rsp
`pushq`	Store a value to (%rsp) and decrement %rsp
`ret`	Fetch a value from (%rsp), increment %rsp and goto the address
`.global`	This symbol should be visible from other modules
`.text`	Place the bytes to follow in the code region
`.type`	This symbol is the following type

Table 14.1: Instructions for the examples in this chapter

Now, it's time to define a class to represent these instructions in memory.

Adding a class for x64 instructions

The **x64** class represents **operation code** (**opcode**) and operands as allowed in x64. An operand may be a register or reference to a value in memory. You can see an illustration of this class in the following Unicon code snippet from x64.icn:

```
class x64(op, opnd1, opnd2)
    method print() ... end
initially(o, src, dest)
    op := o; opnd1 := src; opnd2 := dest
end
```

The corresponding Java class in x64.java looks like this. The source and destination operands opnd1 and opnd2 are of type x64loc, a class for representing native x64 locations in memory and/ or registers. Class x64loc is described later in this chapter in the section titled *Mapping intermediate code address to x64 locations*:

```
public class x64 {
    String op;
    x64loc opnd1, opnd2;
    public x64(String o, Object src, Object dst) {
        op=o; opnd1 = loc(src); opnd2 = loc(dst); }
    public x64(String o, Object opnd) {
        op=o; opnd1 = loc(opnd); }
    public x64(String o) { op=o; }
    public void print() { ... }
}
```

This x64 class is where we map three-address code addresses to x64 addresses.

Mapping memory regions to x64 register-based address modes

To implement the code, global/static, stack, and heap memory regions on x64, we decide how to access memory in each memory region. x64 instructions allow operands to be either a register or a memory address. Many address modes are supported in x64, some of which are for legacy compatibility or arcane specialized uses. Jzero keeps things as simple as possible and typically just adds an offset constant to a register to compute the address indirectly, as illustrated here:

Access Mode	Description
$k	Immediate mode, value given in the instruction
k(r)	Indirect mode, fetch memory k bytes relative to register r

Table 14.2: Memory access modes used in this chapter

In immediate mode, the value is in the instruction. There are limits to how large a value can be represented in immediate mode, but on a 64-bit processor, these limits can be large. In indirect mode, main memory is relative to an x64 register. The various memory regions are accessed as offsets relative to different registers. Global and static memory are accessed relative to the instruction pointer, locals are accessed relative to the base pointer, and heap memory is accessed relative to a heap pointer register. Let's look more broadly at how registers are used.

Using registers

Main memory access is slow. Performance is heavily impacted by how registers are used. Optimal register allocation is **nondeterministic polynomial-time complete** (**NP-complete**) – very difficult. Optimizing compilers expend great effort on register allocation to make the generated code very efficient; the generated code might or might not be optimal. That is beyond the scope of this book.

x64 has 16 general-purpose registers, illustrated in the following table. Many registers have a special role. Arithmetic is performed on an accumulator register, rax. Several of the x64 registers have multiple names for accessing from 8 bits to all 64 bits of the register. This appears mainly to be for backward compatibility with legacy x86 code. In any case, Jzero only uses the 64-bit versions of registers, plus whichever 8-bit registers are necessary for strings. In AT&T syntax, register names are preceded by a percentage sign, as in %rax:

Register	Description/Role
rip	Instruction pointer.
rax	Accumulator. Also: function return value.
rbx	A secondary accumulator.
rbp	Frame pointer. Local variables are relative to this pointer.
rsp	Stack pointer. Memory between rbp and rsp is the local region.
rdi	Destination index. Holds parameter #1.
rsi	Source index. Holds parameter #2.
rdx	A secondary accumulator. Holds parameter #3.
rcx	Holds parameter #4.
r8	Holds parameter #5.
r9	Holds parameter #6.
r10-r15	Open registers usable for any purpose.

Table 14.3: x64 registers

Many registers are saved as part of a call instruction. The more registers there are, the slower it is to perform function calls. These issues are determined by the **calling conventions** of the compiler. Jzero only saves modified registers prior to a given call. Before we get to the actual code generator, let's consider a bit further how native code uses registers.

Starting from a null strategy

The minimal register strategy is the null strategy, which maps intermediate code addresses down to x64 addresses. Values are loaded into the rax accumulator register to perform operations on them. Results go immediately back to main memory.

The rbp base pointer and the rsp stack pointer manage activation records, which are also called frames. The current activation record revolves around the rbp base pointer register. The current local region on the stack lies between the base pointer and the stack pointer. *Figure 14.1* shows an x64 stack layout. The stack grows downward from higher addresses to lower ones by subtracting from the stack pointer. Parameters and locals are referenced as offsets relative to the base pointer; parameters are positive offsets while local variables are found at negative offsets.

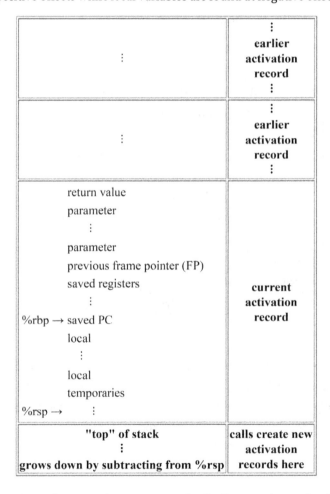

Figure 14.1: x64 stack, managed as a sequence of activation records, growing downward

x64 tweaks the classical stack layout slightly. Six registers, rdi through r9 are used to pass the first six parameters. The null strategy stores parameters to memory when a function call starts. The div instruction uses the rdx register, so besides being used to pass parameter #3, rdx is also required for div instructions. The null strategy is not affected by this design quirk.

Assigning registers to speed up the local region

Jzero maps registers rdi-r14 onto the first 88 bytes of the local region. As it walks the three-address instructions, the code generator tracks for each register if a value is loaded and if it was modified from the corresponding main memory location. The code generator uses the value in the register until that register is used for something else.

Here is a class named RegUse that tracks the main memory locations' corresponding register, if any, and whether its value has been modified since it was last loaded in main memory. The Unicon implementation of RegUse in RegUse.icn is shown here:

```
class RegUse (reg, offset, loaded, dirty)
    method load()
        if \loaded then fail
        loaded := 1
        return j0.xgen("movq", offset||"(%rbp)", reg)
    end
    method save()
        if /dirty then fail
        dirty := &null
        return j0.xgen("movq", reg, offset||"(%rbp)")
    end
end
```

The reg field denotes the string register name; offset is the byte offset relative to the base pointer. The loaded and dirty Boolean flags track whether the register contains the value and whether it has been modified, respectively. The load() and save() methods do not load and save; they generate instructions to load and save the register and set the loaded and dirty flags accordingly. The corresponding Java code looks like this:

```
public class RegUse {
    public String reg;
    int offset;
    public boolean loaded, dirty;
    public RegUse(String s, int i) {
```

```
            reg = s; offset=i; loaded=dirty=false;
        }
        public ArrayList<x64> load() {
            if (loaded) return null;
            loaded = true;
            return j0.xgen("movq", offset+"(%rbp)", reg);
        }
        public ArrayList<x64> save() {
            if (!dirty) return null;
            dirty = false;
            return j0.xgen("movq", reg, offset+"(%rbp)");
        }
    }
```

A list of instances of the RegUse class is held in a variable named regs in the j0 class so that for
each of the first words in the local region, the corresponding register is used appropriately. The
list is constructed in Unicon, as follows:

```
off := 0
regs := [: RegUse("%rdi"|"%rsi"|"%rdx"|"%rcx"|"%r8"|
        "%r9"|"%r10"|"%r11"|"%r12"|"%r13"|"%r14", off-:=8) :]
```

This Unicon code is showing off a bit. The | alternator produces all the register names for sep-
arate calls to RegUse(), triggered and captured by the [: :] list comprehension operator. One
x64 tricky bit is that the offsets are all negative integers because the stack grows downward. In
Java, this initialization is performed in RegUse.java, as shown here:

```
RegUse [] regs = new RegUse[]{ new RegUse("%rdi", -8),
    new RegUse("%rsi", -16), new RegUse("%rdx", -24),
    new RegUse("%rcx", -32), new RegUse("%r8", -40),
    new RegUse("%r9",-48), new RegUse("%r10", -56),
    new RegUse("%r11", -64), new RegUse("%r12", -72),
    new RegUse("%r13", -80), new RegUse("%r14", -88) };
```

The data structure operates on **basic block** boundaries, storing modified registers in memory
and clearing loaded flags whenever a label or a branch instruction occurs. At the top of a called
function, the loaded and dirty flags of parameters are set to true, indicating values that must
be saved to the local region before that register can be reused. Now, it's time to look at how each
intermediate code element is converted into x64 code.

Converting intermediate code to x64 code

The intermediate code generator from *Chapter 9, Intermediate Code Generation*, placed the intermediate code for the whole program in the icode attribute at the root of the syntax tree. A Boolean
named isNative says whether to generate bytecode, as shown in the previous chapter, or native
x64 code. To generate x64 code, the gencode() method in the j0 class calls a new method in this
class, named x64code(), and passes it the intermediate code in root.icode as its input. Output
x64 code is placed in a j0 list variable named xcode. The Unicon gencode() method that invokes
this functionality in j0.icn looks like this:

```
method gencode(root)
    root.genfirst()
    root.genfollow()
    root.gentargets()
    root.gencode()

    labeltable := table()
    methodAddrPushed := &null
    if \isNative then {
        xcode := x64code(root.icode)
        genx64code()
    }
    else {
        bcode := bytecode(root.icode)
        genbytecode(bcode)
    }
end
```

The new highlighted code layers the native alternative around the previous generation of bytecode, which is still available from the command-line option. The x64code() method takes in
an icode list, and its return value is a list of x64 class objects. In this example, the resulting x64
code is printed out in textual form; we let an assembler do the work for us to produce a binary
format. The corresponding Java code for the gencode() method found in j0.java is shown here:

```
public static ArrayList<x64> xcode;
public static void gencode(parserVal r) {
    tree root = (tree)(r.obj);
    root.genfirst();
    root.genfollow();
```

```
        root.gentargets();
        root.gencode();
        labeltable = new HashMap<>();
        methodAddrPushed = false;
        if (isNative) {
            xcode = x64code(root.icode);
            genx64code();
        } else {
            ArrayList<byc> bcode = bytecode(root.icode);
            if (bcode != null) {
                genbytecode(bcode);
            }
        }
    }
}
```

Now, let's examine how intermediate code addresses become x64 memory references.

Mapping intermediate code addresses to x64 locations

Addresses in intermediate code are abstract (region, offset) pairs represented in the address class from *Chapter 9, Intermediate Code Generation*. The corresponding x64loc class represents x64 locations that include addressing mode information or a register to use. The Unicon implementation in x64loc.icn looks like this:

```
class x64loc(reg, offset, mode)
initially(x,y,z)
    if \z then { reg := x; offset := y; mode := z }
    else if \y then {
        if x === "imm" then { offset := y; mode := 5 }
        else if x === "lab" then { offset := y; mode := 6 }
        else {
            reg := x; offset := y
            if integer(y) then mode := 3 else mode := 4
        }
    }
    else {
        if integer(x) then { offset := x; mode := 2 }
        else if string(x) then { reg := x; mode := 1 }
        else stop("bad x64loc ", image(x))
```

```
    }
  end
```

The reg field is the string register name, if there is one. The offset field is either an integer offset or a string name from which the offset is calculated. mode is 1 for a register, 2 for an absolute address, and 3 for a register and an integer offset. Mode 4 is for a register and a string offset name, 5 is for an immediate value, and 6 is for a label. The Java implementation in x64loc.java looks like this:

```java
public class x64loc {
  public String reg;  Object offset;
  public int mode;
  public x64loc(String r) { reg = r; mode = 1; }
  public x64loc(int i) { offset=(Object)Integer(i); mode=2; }
  public x64loc(String r, int off) {
    if (r.equals("imm")) {
      offset=(Object)Integer(off); mode = 5; }
    else if (r.equals("lab")) {
      offset=(Object)Integer(off); mode = 6; }
    else { reg = r; offset = (Object)Integer(off); mode = 3; }
  }
  public x64loc(String r, String s) {
    reg = r; offset = (Object)s; mode=4;
  }
}
```

The Java code has constructors for different memory types. The region and offset of the address class must be mapped onto an instance of the x64loc class that is an operand in the x64 class. This is done by a loc() method in the j0 class that takes an address as a parameter and returns an x64loc instance. The Unicon code for loc() in j0.icn looks like this:

```
method loc(a)
  if /a then return
  case a.region of {
  "loc": { if a.offset <= 88 then return loadreg(a)
             else return x64loc("rbp", -a.offset) }
  "glob": { return x64loc("rip", a.offset) }
  "const": { return x64loc("imm", a.offset) }
  "lab": { return x64loc("lab", a.offset) }
  "obj": { return x64loc("r15", a.offset) }
```

```
          "imm": { return x64loc("imm", a.offset) }
      }
  end
```

As the code converts an address to an x64loc instance, local region offsets are converted into negative values, since the stack grows downward. The Java methods in j0.java are shown here:

```
  public static x64loc loc(String s) { return new x64loc(s);}
  public static x64loc loc(Object o) {
      if (o instanceof String) return loc((String)o);
      if (o instanceof address) return loc((address)o);
      return null;
  }
  public static x64loc loc(address a) {
      switch (a.region) {
      case "loc": { if (a.offset <= 88) return loadreg(a);
                      else return x64loc("rbp", -a.offset); }
      case "glob": { return x64loc("rip", a.offset); }
      case "const": { return x64loc("imm", a.offset); }
      case "lab": { return x64loc("lab", a.offset); }
      case "obj": { return x64loc("r15", a.offset); }
      case "imm": { return x64loc("imm", a.offset); }
      default: { semErr("x64loc unknown region"); return null; }
      }
  }
```

A loadreg() helper method is used for local offsets in the first 88 bytes. If the value is not already present in its designated register, a movq instruction is emitted to place it there, as illustrated in the following code snippet:

```
  method loadreg(a)
    r := a.offset/8 + 1
    if / (regs[r].loaded) then {
      every put(xcode,
                 !xgen("movq",(-a.offset)||"(%rbp)",regs[r].reg))
      regs[r].loaded := "true"
      }
    return x64loc(regs[a.offset/8+1].reg)
  end
```

The Java implementation of loadreg() is shown here:

```
public static x64loc loadreg(address a) {
  long r = a.offset/8;
  if (!regs[r].loaded) {
    xcode.addAll(xgen("movq",
             String.valueOf(-a.offset)+"(%rbp)", regs[r].reg));
    regs[r].loaded = true;
  }
  return x64loc(regs[a.offset/8+1].reg);
}
```

Given the x64 class, one more helper function is needed in order to formulate the x64code() code generator method. We need a convenient factory method to generate x64 instructions. This xgen() method converts source and destination operands into x64loc instances, which may add movq instructions to load values into registers. The Unicon code looks like this:

```
method xgen(o, src, dst)
   return [x64(o, loc(src), loc(dst))]
end
```

There are many versions of the corresponding Java implementation shown here, handling cases where the source or destination are addresses or the string names of registers:

```
public static ArrayList<x64> l64(x64 x) {
    return new ArrayList<x64>(Arrays.asList(x)); }
public static ArrayList<x64> xgen(String o){ return l64(new x64(o)); }
public static ArrayList<x64> xgen(String o, address src, address dst) {
    return l64(new x64(o, loc(src), loc(dst))); }
public static ArrayList<x64> xgen(String o, address opnd) {
    return l64(new x64(o, loc(opnd))); }
public static ArrayList<x64> xgen(String o, address src, String dst) {
    return l64(new x64(o, loc(src), loc(dst))); }
public static ArrayList<x64> xgen(String o, String src, address dst) {
    return l64(new x64(o,loc(src),loc(dst))); }
public static ArrayList<x64> xgen(String o, String src, String dst) {
    return l64(new x64(o,loc(src),loc(dst))); }
public static ArrayList<x64> xgen(String o, String opnd) {
    return l64(new x64(o, loc(opnd))); }
```

In the preceding code snippet, the l64() method just creates a single `ArrayList` element containing an x64 object. The rest are just many implementations of xgen() that take different parameter types. Now, finally, it's time to present the x64 code generator method.

Implementing the x64 code generator method

The Unicon implementation of the x64code() method in the j0 class looks like this. The implementation must fill in one case branch for each opcode in the three-address instruction set. There will be a lot of cases, so we present each one separately in the sections to follow, starting with the first one shown here:

```
method x64code(icode)
    every i := 1 to *\icode do {
        instr := icode[i]
        case instr.op of {
            "ADD": { ... append translation of ADD to xcode }
            "SUB": { ... append translation of SUB to xcode }

            . . .

        }
    }
end
```

The Java implementation of x64code() is shown here:

```
public static void x64code(ArrayList<tac> icode) {
    int parmCount = -1;
    for(int i=0; i<icode.size(); i++) {
        tac instr = icode.get(i);
        switch(instr.op) {
        case "ADD": { ... append translation of ADD to xcode}
        case "SUB": { ... append translation of SUB to xcode}

            ...

        }
    }
}
```

Within the framework of this x64code() method, we now provide translations for each of the three-address instructions. We will start with simple expressions.

Generating x64 code for simple expressions

The cases for three-address opcode have many elements in common. The code for addition shows many common elements. Note that this is for integer addition. Floating point or other kinds of addition would be represented by different intermediate code instructions and result in the selection of different x64 native instructions. In Unicon, the x64 code for integer addition is created like this:

```
"ADD": { xcode |||:= xgen("movq", instr.op2, "%rax") |||
                 xgen("addq", instr.op3, "%rax") |||
                 xgen("movq", "%rax", instr.op1) }
```

In this code, operand 2 is read from memory into register rax. The x64 ADD instruction has many variations; this instance of ADD reads operand 3 from memory and adds it to what is already in rax, treating the register as an accumulator register. The result is then placed into operand 3. In Java, the implementation of addition consists of the following code:

```
case "ADD": { xcode.addAll(xgen("movq", instr.op2, "%rax"));
              xcode.addAll(xgen("addq", instr.op3, "%rax"));
              xcode.addAll(xgen("movq", "%rax", instr.op1));
              break; }
```

There are 19 or so three-address instructions. The final code generation pattern illustrated by the preceding ADD instruction is used for the other arithmetic instructions. For a unary operator such as NEG, the pattern is slightly simplified, as we can see here:

```
"NEG": { xcode |||:= xgen("movq", instr.op2, "%rax") |||
                 xgen("negq", "%rax") |||
                 xgen("movq", "%rax", instr.op1) }
```

In Java, the implementation of negation consists of the following code:

```
case "NEG": { xcode.addAll(xgen("movq", instr.op2, "%rax"));
              xcode.addAll(xgen("negq", "%rax"));
              xcode.addAll(xgen("movq", "%rax", instr.op1));
              break; }
```

An even simpler instruction such as ASN may be worth treating as a special-case, since x64 code features direct memory-to-memory move instructions, but one option is to stick with the same script and simplify the preceding pattern further, like this:

```
"ASN": { xcode |||:= xgen("movq", instr.op2, "%rax") |||
                 xgen("movq", "%rax", instr.op1) }
```

In Java, the implementation of an assignment might look like this:

```
case "ASN": { xcode.addAll(xgen("movq", instr.op2, "%rax"));
              xcode.addAll(xgen("movq", "%rax", instr.op1));
              break; }
```

Expressions are the most common elements in code. The next category is pointers.

Generating code for pointer manipulation

Three of the three-address instructions pertain to the use of pointers: ADDR, LCON, and SCON. The ADDR instruction turns an address in memory into a piece of data that can be manipulated to perform operations, such as pointer arithmetic. It pushes its operand, an address reference in one of the memory regions, as if it were an immediate mode value. The code is illustrated in the following snippet:

```
"ADDR": { xcode |||:= xgen("leaq", instr.op2, "%rax")
          xcode |||:= xgen("%rax", instr.op1) }
```

In Java, the implementation of the ADDR instruction consists of the following code:

```
case "ADDR": { xcode.addAll(xgen("leaq", instr.op2, "%rax"));
               xcode.addAll(xgen("%rax", instr.op1));
               break; }
```

The LCON instruction reads from memory pointed at by other memory, as follows:

```
"LCON": { xcode |||:= xgen("movq", instr.op2, "%rax") |||
                      xgen("movq", "(%rax)", "%rax") |||
                      xgen("movq", "%rax", instr.op1) }
```

In Java, the implementation of the LCON instruction consists of the following code:

```
case "LCON": { xcode.addAll(xgen("movq", instr.op2, "%rax"));
               xcode.addAll(xgen("movq", "(%rax)", "%rax"));
               xcode.addAll(xgen("movq", "%rax", instr.op1));
               break; }
```

The SCON instruction writes to memory pointed at by other memory, as follows:

```
"SCON": { xcode |||:= xgen("movq", instr.op2, "%rbx") |||
                      xgen("movq", instr.op1, "%rax")
                      xgen("movq", "%rbx", "(%rax)") }
```

In Java, the implementation of the SCON instruction consists of the following code:

```
case "SCON": { xcode.addAll(xgen("movq", instr.op2, "%rbx"));
               xcode.addAll(xgen("movq", instr.op1, "%rax"));
               xcode.addAll(xgen("movq", "%rbx", "(%rax)"));
               break; }
```

These instructions are important to support structured data types such as arrays. Now, let's consider bytecode code generation for control flow, starting with the GOTO instruction.

Generating native code for branches and conditional branches

Seven intermediate code instructions pertain to branch instructions. The simplest of these is the unconditional branch or GOTO instruction. The GOTO instruction assigns a new value to the instruction pointer register. It should be no surprise that the GOTO bytecode is the implementation of the three-address GOTO instruction, as illustrated in the following code snippet:

```
"GOTO": {   xcode |||:= xgen("goto", instr.op1) }
```

In Java, the implementation of the GOTO instruction consists of the following code:

```
case "GOTO": { xcode.addAll(xgen("goto", instr.op1));
               break; }
```

The conditional branch instructions in the three-address code are translated down into simpler final code instructions. For the x64 instruction set, this means executing a compare instruction that sets condition codes prior to one of the x64 conditional branch instructions. The Unicon implementation of the BLT instruction looks like this:

```
"BLT": { xcode |||:= xgen("movq", instr.op2, "%rax") |||
                 xgen("cmpq", instr.op3, "%rax") |||
                 xgen("jle", instr.op1) }
```

In Java, the implementation to generate bytecode for the BLT instruction consists of the following code:

```
case "BLT": { xcode.addAll(xgen("movq", instr.op2, "%rax"));
              xcode.addAll(xgen("cmpq", instr.op3, "%rax"));
              xcode.addAll(xgen("jle", instr.op1));
              break; }
```

This pattern is employed for several of the three-address instructions. Now, let's consider the more challenging forms of control flow transfer that relate to method calls and returns.

Generating code for method calls and returns

Three of the intermediate code instructions handle the very important topic of function and method calls and returns. A sequence of zero or more PARM instructions pushes values onto the stack, after which the CALL instruction performs a method call. From inside the called method, the RET instruction returns from a method to the caller.

This three-address code calling convention must be mapped down onto the underlying x64 instruction set, preferably with the standard calling conventions on that architecture, which requires the first six parameters to be passed into specific registers.

To pass parameters into correct registers, the PARM instruction must track which parameter number it is. The Unicon code for the PARM instruction consists of the following:

```
"PARM": { if /parmCount then {
            parmCount := 1
            every j := i+1 to *icode do
                if icode[j].op == "CALL" then break
                parmCount +:= 1
          }
          else parmCount -:= 1
          genParm(parmCount, instr.op1) }
```

For the first parameter, the every loop counts the number of parameters before the CALL instruction. The genParm() method is called with the current parameter number and the operand. In Java, the implementation of the PARM instruction is similar, as we can see here:

```
case "PARM": { if (parmCount == -1) {
                  for(int j = i+1; j<icode.size(); j++) {
                      tac callinstr = icode.get(j);
                      if (callinstr.op.equals("CALL"))
                          break;
                      parmCount++;
                  }
              }
              else parmCount--;
              genParm(parmCount, instr.op1);
              break; }
```

The preceding cases for parameters depend on a genParm( ) method that generates code, depending on the parameter number. Before loading registers for a new function call, register values that have been modified must be saved to their main memory locations, as follows. Since save( ) clears the dirty flag when it generates a register save instruction, multiple calls to genParm( ) will only save registers once for a given call:

```
method genParm(n, addr)
   every (!regs).save()
   if n > 6 then xcode |||:= xgen("pushq", addr)
   else xcode |||:= xgen("movq", addr, case n of {
      1: "%rdi"; 2: "%rsi"; 3: "%rdx";
      4: "%rcx"; 5: "%r8";    6: "%r9"
   })
end
```

The corresponding Java implementation of genParm( ) looks like this:

```
public static void genParm(int n, address addr) {
   for (RegUse x : regs) x.save();
   if (n > 6) xcode.addAll(xgen("pushq", addr));
   else {
      String s = "error:" + String.valueOf(n);
      switch (n) {
      case 1: s = "%rdi"; break; case 2: s = "%rsi"; break;
      case 3: s ="%rdx"; break; case 4: s = "%rcx"; break;
      case 5: s = "%r8"; break; case 6: s = "%r9"; break;
      }
      xcode.addAll(xgen("movq", addr, s));
   }
}
```

The CALL instruction is next. After the call, the op1 destination in the three-address code is saved from the rax register. The op2 source field is the method address that was used prior to the first PARM instruction. The op3 source field gives the number of parameters, which is not used on x64. The code is illustrated in the following snippet:

```
"CALL": { xcode |||:= xgen("call", instr.op3)
          xcode |||:= xgen("movq", "%rax", instr.op1)
          parmCount := &null }
```

In Java, the implementation of the CALL instruction looks like this:

```
case "CALL": { xcode.addAll(xgen("call", instr.op3));
               xcode.addAll(xgen("movq", "%rax", instr.op1));
               parmCount = -1;
               break; }
```

The Unicon implementation of the RETURN instruction looks like this:

```
"RETURN": { xcode |||:= xgen("movq", instr.op1, "%rax") |||
                 xgen("leave") ||| xgen("ret", instr.op1) }
```

In Java, the implementation of the RETURN instruction looks like this:

```
case "RETURN":{ xcode.addAll(xgen("movq", instr.op1, "%rax"));
                xcode.addAll(xgen("leave"));
                xcode.addAll(xgen("ret", instr.op1));
                break; }
```

Generating code for method calls and returns is not too difficult. Now, let's consider how to handle the pseudo-instructions in the three-address code.

Handling labels and pseudo-instructions

Pseudo-instructions such as labels do not translate into code, but they are present in the linked list of three-address instructions and require consideration in the final code. The most common and obvious pseudo-instruction is a `label`. If the final code is generated in a human-readable assembler format, labels can be generated almost as-is, modulo any format differences such as name mangling that might be necessary to make them legal in the assembler file. If we were generating the final code in a binary format, labels would require precise calculation at this point and would be entirely replaced by actual byte offsets in the generated machine code. The code is illustrated here:

```
"LAB": { every (!regs).save()
         xcode |||:= xgen("lab", instr.op1) }
```

In Java, the equivalent implementation is shown here:

```
case "LAB": { for (RegUse ru : regs) ru.save();
              xcode.addAll(xgen("lab", instr.op1)); break; }
```

As a representative of other types of pseudo-instructions, consider which x64 code to output for the beginnings and ends of methods.

At the beginning of a method in intermediate code, all you've got is the pseudo-instruction proc x,n1,n2. The Unicon code for this pseudo-instruction is shown here. Several of the assembler directives beginning with the prefix .seh are beyond the scope of this book and will not be discussed, other than to mention that they support structured exception handling:

```
"proc": {
    n := ((\(instr.op2)).integr() + (\(instr.op3)).integr()) * 8
    xcode |||:= xgen(".text") |||
                xgen(".globl", instr.op1.region) |||
                xgen(".def\t" || instr.op1.region ||
                    ";\t.scl\t2;\t.type\t32;\t.enddef") |||
                xgen(".seh_proc\t" || instr.op1.region) |||
                xgen("lab", instr.op1.region) |||
                xgen("pushq", "%rbp") |||
                xgen(".seh_pushreg\t%rbp") |||
                xgen("movq", "%rsp", "%rbp") |||
                xgen(".seh_setframe\t%rbp, 0") |||
                xgen("subq", "$" || \n, "%rsp") |||
                xgen(".seh_stackalloc\t" || n) |||
                xgen(".seh_endprologue")
    if instr.op1.region === "main" then
        xcode |||:= xgen("call", "__main")
    every i := !((\(instr.op2)).intgr()) do
        regs[i].loaded := regs[i].dirty := "true"
    /i := 0
    every j := i+1 to 11 do
        regs[i].loaded := regs[i].dirty := &null
}
```

Line by line in the preceding code, the assignment to n calculates the total number of local region bytes, including space for parameters passed in registers but copied into stack memory if the method calls another method. The .text directive tells the assembler to write to the code section. The .globl directive states that the method name should be linkable to other modules. The .def, .scl, .type, and .enddef directives introduce the symbol as a function. The lab directive declares the (mangled) function name as an assembler label, which is to say that the mangled name can be used as a reference to this function entry point in the code region. The pushq instruction saves the previous base pointer on the stack. The movq instruction establishes the base pointer for the new function at the current stack top.

The subq instruction allocates memory by moving the stack pointer further down by that amount in the stack. The if statement inserts a call to __main() to initialize the runtime system when program execution starts from main(). The two loops mark used parameters while noting that the other registers are clear. In Java, the corresponding code for a method header looks like this:

```java
case "proc": {
    int n = 0;
    if (instr.op2 != null) n += instr.op2.intgr();
    if (instr.op3 != null) n += instr.op3.intgr();
    n *= 8;
    xcode.addAll(xgen(".text"));
    xcode.addAll(xgen(".globl", instr.op1.region));
    xcode.addAll(xgen(".def\t" + instr.op1.region +
                      ";\t.scl\t2;\t.type\t32;\t.endef"));
    xcode.addAll(xgen(".seh_proc\t" + instr.op1.region));
    xcode.addAll(xgen("lab", instr.op1.region));
    xcode.addAll(xgen("pushq", "%rbp"));
    xcode.addAll(xgen(".seh_pushreg\t%rbp"));
    xcode.addAll(xgen("movq", "%rsp", "%rbp"));
    xcode.addAll(xgen(".seh_setframe\t%rbp, 0"));
    xcode.addAll(xgen("subq", "$"+n, "%rsp"));
    xcode.addAll(xgen(".seh_stackalloc\t"+n));
    xcode.addAll(xgen(".seh_endprologue"));
    if (instr.op1.region.equals("main"))
        xcode.addAll(xgen("call","__main"));
    int j = 0;
    if (instr.op2 != null)
        for ( ; j < instr.op2.offset; j++)
            regs[j].loaded = regs[j].dirty = true;
    for (; j < 11; j++)
        regs[j].loaded = regs[j].dirty = false;
    break;
}
```

The end pseudo-instruction is somewhat simpler, as we can see here. We do not want to fall off the end of a method, so if the function does not end with a return statement, we emit instructions to restore the old frame pointer and return, along with assembler directives for the end of a function:

```
"end": {
```

```
        if xcode[-1].op ~== "ret" then
            Xcode |||:= xgen("leave") ||| xgen("ret")
        Xcode |||:= xgen(".seh_endproc")
    }
```

The matching Java implementation of the end pseudo-instruction is shown here:

```
    case "end": {
        if (! Xcode.get(xcode.size()-1).op.equals("ret")) {
            xcode.addAll(xgen("leave"));
            xcode.addAll(xgen("ret"));
            }
        xcode.addAll(xgen(".seh_endproc"));
        break;
    }
```

The last few sections produced a data structure containing a representation of the bytecode and then showed how various three-address instructions are translated. Now, let's move on to producing the output native x64 code from a list of x64 objects.

Generating x64 output

As with many traditional compilers, the native code for Jzero will be produced by carrying out the following steps. First, we will write out a linked list of x64 objects in a human-readable assembler language with the .s extension. We then invoke the GNU assembler to turn that into a binary object file format with the .o extension. An executable is constructed by invoking a linker, which combines a set of .o files specified by the user with a set of .o files containing runtime library code, and data referenced from the generated code. This section presents each of these steps, starting with producing the assembler code.

Writing the x64 code in assembly language format

This section provides a brief description of the x64 assembler format as supported by the GNU assembler, which uses AT&T syntax. Instructions and pseudo-instructions occur on a line by themselves with a tab (or eight spaces) of indentation on the left. Labels are an exception to this rule, as they contain no leading spaces of indentation and consist of an identifier followed by a colon. Pseudo-instructions begin with a period. After the mnemonic for the instruction or pseudo-instruction, there may be a tab or spaces followed by zero, one, or two comma-separated operands, depending on the requirements of the instruction.

As an example of all this, here is a simple x64 assembler file containing a function that does nothing and returns a value of 42. In the assembler, this is how it might look:

```
        .text
        .globl  two
        .type   two, @function
two:
.LFB0:
        pushq   %rbp
        movq    %rsp, %rbp
        movl    $42, -4(%rbp)
        movl    -4(%rbp), %eax
        popq    %rbp
        ret
.LFE0:
        .size   two, .-two
```

The j0 class has a method named x64print() that outputs a list of x64 objects into a text file in this format. As you can see in the Unicon code from j0.icn shown next, it calls the print() method on each of the x64 objects in the xcode list:

```
method x64print()
    every (!xcode).print()
end
```

The Java implementation of x64print() in the j0.java file is shown here:

```
public static void x64print() {
    for(x64 x : xcode) x.print();
}
```

Having shown how the assembler code is written, it's time to look at how to invoke the GNU assembler to produce an object file.

Going from native assembler to an object file

Object files are binary files containing actual machine code. An assembler file written out in the preceding section is assembled using the as command, as shown here:

```
as --gstabs+ -o two.o two.s
```

In this command line, `--gstabs+` is a recommended option that includes debugging information. `-o two.o` is an option that specifies the output filename.

The resulting `two.o` binary file is not readily understandable by humans as-is but can be viewed using various tools. Just for fun, the first 102 bytes of ones and zeros from `two.o` are shown in the following screenshot; each row shows six bytes, with the **American Standard Code for Information Interchange (ASCII)** interpretation shown on the right. The screenshot shows you the ones and zeros in text form, thanks to a tool named `xxd` that prints the bits out literally in textual form. Of course, a computer usually processes them from 8 to 64 bits at a time, without first transliterating them into text form:

```
01111111 01000101 01001100 01000110 00000010 00000001  .ELF..
00000001 00000000 00000000 00000000 00000000 00000000  ......
00000000 00000000 00000000 00000000 00000001 00000000  ......
00111110 00000000 00000001 00000000 00000000 00000000  >.....
00000000 00000000 00000000 00000000 00000000 00000000  ......
00000000 00000000 00000000 00000000 00000000 00000000  ......
00000000 00000000 00000000 00000000 00010000 00000010  ......
00000000 00000000 00000000 00000000 00000000 00000000  ......
00000000 00000000 00000000 00000000 01000000 00000000  ....@.
00000000 00000000 00000000 00000000 01000000 00000000  ....@.
00001011 00000000 00001010 00000000 01010101 01001000  ....UH
10001001 11100101 11000111 01000101 11111100 00000100  ...E..
00000000 00000000 00000000 10001011 01000101 11111100  ....E.
01011101 11000011 00000000 01000111 01000011 01000011  ]..GCC
00111010 00100000 00101000 01010101 01100010 01110101  : (Ubu
01101110 01110100 01110101 00100000 00110111 00101110  ntu 7.
00110101 00101110 00110000 00101101 00110011 01110101  5.0-3u
```

Figure 14.2: Binary representations are not human-friendly, but computers prefer them

It is not a coincidence that bytes 2–4 of the file say `ELF`. **Executable and Linkable Format (ELF)** is one of the more popular multiplatform object file formats, and the first four bytes identify the file format. Suffice it to say that such binary file formats are important to machines but difficult for humans. Now, let's consider how object files are combined to form executable programs.

Linking, loading, and including the runtime system

The task of combining a set of binary files to produce an executable is called `linking`. This is another subject about which an entire book could be written. For Jzero, it is a very good thing that under either Linux or by using the Mingw64 Windows version of gcc, we can let the `ld` GNU linker program do the work. It takes a `-o` file option to specify its output filename, and then any number of `.o` object files.

The object files for a working executable include a startup file that will initialize and call main( ), often called crt1.o, followed by the application files, and then zero or more runtime library files. If we build a Jzero runtime library named libjzero.o, the ld command line might look like this:

```
ld -o hello /usr/lib64/crt1.o hello.o -ljzero
```

If your runtime library calls functions in a real C library, you will have to include them as well. A full ld-based link of a runtime system built on top of the **GNU Compiler Collection's (GCC's)** glibc looks like this:

```
ld -dynamic-linker /lib64/ld-linux-x86-64.so.2 \
    /usr/lib/x86_64-linux-gnu/crt1.o \
    /usr/lib/x86_64-linux-gnu/crti.o \
    /usr/lib/gcc/x86_64-linux-gnu/7/crtbegin.o \
    hello.o -ljzero \
    -lc /usr/lib/gcc/x86_64-linux-gnu/7/crtend.o \
    /usr/lib/x86_64-linux-gnu/crtn.o
```

Your users would not often have to type this command line itself, since it would be boiled into your compiler's linker invocation code. However, it has the fatal flaws of being non-portable and version-dependent. To use an existing GCC C library from within your runtime system, you might prefer to let an existing GCC installation perform your linking for you, by baking something like this into your compiler's linker invocation code:

```
gcc -o hello hello.o
```

The linker must assemble one big piece of binary code from several binary object code inputs. In addition to bringing together all the instructions in the object files, the linker's primary job is to determine the addresses of all the functions and global variables in the executable. The linker must also provide a mechanism for each object file to find the addresses of functions and variables from other object files. Whether you use ld or gcc to invoke the linker, your compiler may need a mechanism by which it can find your runtime library, as denoted by the -ljzero library references in the ld example above.

For functions and variables that are not defined in user code but are instead part of the language runtime system, the linker must have a mechanism to search the runtime system and incorporate as much of it as is needed. The runtime system includes startup code that will initialize the runtime system and set things up to call main( ). It may include one or more object files that are always linked to any executable for that language. Most importantly, the linker provides a way to find and link in only those portions of the runtime system that are explicitly called from the user code.

In modern systems, things have gotten more complicated over time. It is standard to defer various aspects of linking and loading to runtime, particularly to allow processes to share library code that has already been loaded for use by other processes.

Summary

This chapter showed you how to generate native code for x64 processors. Among the skills you learned, the main task was to traverse a linked list of intermediate code and convert it into instructions in the x64 instruction set. In addition, you learned how to write out x64 code in the GNU assembler format. Lastly, you learned how to invoke the assembler and linker to turn the native code into an ELF object and executable file format.

The next chapter looks in more detail at the task of implementing new high-level operators and built-in functions in your language's runtime system.

Questions

1. What are the main new concepts that become necessary to generate x64 native code, compared with bytecode?

2. What are the advantages and disadvantages of supplying the addresses of global variables as offsets relative to the %rip instruction pointer register?

3. One of the big issues affecting the performance of modern computers is the speed of performing function calls and returns. Why is function call speed important? In which circumstances is the x64 architecture able to perform fast function calls and returns? Are there aspects of the x64 architecture that seem likely to slow down function calling?

Join our community on Discord

Join our community's Discord space for discussions with the authors and other readers:

```
https://discord.com/invite/zGVbWaxqbw
```

15
Implementing Operators and Built-In Functions

New programming languages are invented because, occasionally, new ideas and new computational capabilities are needed to solve problems in new application domains. Libraries of functions or classes are the most common means of extending mainstream languages with additional computational capabilities, but adding a library is not always sufficient.

This chapter describes how to support very high-level and domain-specific language features by adding operators and functions that are built into the language. The following chapter will discuss adding control structures.

Adding operators and built-in functions may shorten and reduce what programmers must write to solve certain problems in your language, improve its performance, or enable language semantics that would otherwise be difficult. This chapter illustrates the ideas within the context of Jzero, emphasizing the string and array types. By way of comparison, the later sections describe how operators and functions are implemented in Unicon.

This chapter will cover the following main topics:

- Implementing operators
- Writing built-in functions
- Integrating built-ins with control structures
- Developing operators and functions for Unicon

In this chapter, you will learn how to write parts of the runtime system that are too complex to be instructions in the instruction set. You will also learn how to add domain-specific capabilities to your language. Let's start with how to implement high-level operators!

Implementing operators

Operators are expressions that compute a value. Simple operators that compute their results via a few instructions on the underlying machine were covered in the preceding chapters. This section describes how to implement an operator that takes many steps. You can call these operators **composite operators.** In this case, the underlying generated code may perform calls to functions that run natively on the underlying machine.

It may be useful to compare implementing composite operators that are built into to the language, as we are discussing in this chapter, with the feature of operator overloading in some languages such as C++. Operator overloading allows composite operators to be implemented, usually for new user-defined types, as part of a program's source code. The purpose of operator overloading is usually to enable arbitrary new user-defined types to use the same concise arithmetic notation enjoyed by primitive atomic types in a language. However, since they are implemented in the source language, they can only do things the source language knows how to do. Also, operator overloading systems usually only allow new definitions of existing operators with prescribed precedence and association rules.

When you build a new programming language, there are three languages involved. The language you are writing is called the source language. The language that you generate as output is called the target language. The language that you are writing your language processing tool in is called the implementation language.

This chapter is about functions called from generated code to perform composite built-in operators that are written in the implementation language rather than the source language. For example, let's consider Jzero as our source language. The implementation language may be Unicon if we are running our Jzero programs on a bytecode interpreter written in Unicon. But the implementation language might be C if we are running our Jzero programs by generating native code and linking that native code to a runtime system written in C.

Operators written in the implementation language may be lower level and do things that are impossible in the source language. For example, parameter-passing rules might be different in the implementation language than they are in the programming language that you are creating. As new built-ins in your language, the operators being discussed here are not limited to new definitions of existing operators for new types; you can add completely new operators if you are willing to add new tokens and new grammar rules to specify their precedence and association.

If you are wondering when you should make a new computation into an operator, you can refer to *Chapter 2, Programming Language Design*. Rather than repeat that material, we will note that operators are generally constrained to operate on, at most, three operands, and that most operators use one or two operands. Perhaps a note of caution is also needed: adding new definitions for operators that may be completely unrelated to their customary uses may reduce the semantic transparency of your language, making it harder to read and maintain programs that are written in it, since a new developer may misinterpret the code that they are looking at.

If you can leverage analogies to arithmetic that will let programmers in your new language reuse appropriate familiar operators for your new computations, great. Otherwise, you are expecting programmers to learn and memorize new patterns, which is asking a lot. You can add hundreds of operators to your language, but human brains will not memorize that many. If you try to introduce more operators than we have keyboard keys, for example, your language may fail to achieve widespread popularity due to the excessive cognitive load. Now, let's consider to what extent the act of adding new operators to a language relates to, or follows because of, adding new hardware capabilities.

Comparing adding operators to adding new hardware

In the same way that you may discover that a common computation of yours might deserve to be an operator in your language, hardware designers might realize that computers should support a common computation with native instructions. When language designers realize that a computation should be an operator in their language, that makes that computation a candidate for hardware implementation. Similarly, when hardware designers implement a common computation in their hardware, language designers should ask whether that computation should be supported directly with operators or some other specialized syntax. Here is an example.

Before the 80486 in 1994, most PCs did not come with hardware capable of directly performing floating-point calculations; on some platforms, a floating-point co-processor was an expensive add-on needed only for scientific computing. If you were implementing a compiler, you probably implemented a floating-point data type in software as a set of functions. These runtime system functions were called from generated code but were transparent to the programmer. A program that declared two float variables, f1 and f2, and executed the f1 + f2 expression would compute the floating-point sum without noting that the generated code included function calls that might be slower by 10x, 100x, or more compared with adding two integers.

Here is another example that may be a sore point for some readers. After a program called Doom created enormous demand for 3D graphics in the 1990s, GPUs were developed. GPUs now support computations far beyond their original scope of games and similar 3D programs. However, GPU programming is not supported directly in most mainstream languages, and the steep learning curve and difficult programming for GPUs have lessened and slowed their enormous impact. To summarize: there is a rich juicy gray area in between those operators that should be built into the programming language to make programming simple, as well as operators that should be built into the hardware. Now, let's learn how to add compound operators by adding one to Jzero: string concatenation.

Implementing string concatenation in intermediate code

For Jzero, the string type is essential but was not implemented in the preceding chapters on code generation or bytecode interpretation, which focused on integer computation. The String class has a concatenation operator that we must implement. Some computers support concatenation in hardware for some string representations. For Jzero, String is a class and concatenation is comparable to a method – either a factory method or a constructor since it returns a new string rather than modifying its arguments.

In any case, it is time to implement s1+s2, where s1 and s2 are strings. For intermediate code, we can add a new instruction called SADD. If you don't want to add a new instruction, you can generate code that calls a method for string concatenation, but we are going to run with an intermediate code instruction for this example. The code generation rule for the plus operator will generate different code depending on the type. Before we can implement that, we must modify the check_types() method in the tree class so that the s1 string plus the s2 string is legal and computes a string. In the Unicon implementation, change the lines in tree.icn where addition is type-checked to allow the String type, as follows:

```
if op1.str() === op2.str() === ("int"|"double"|"String")
   then return op1
```

In the Java implementation, add the following OR in tree.java:

```
if (op1.str().equals(op2.str()) &&
    (op1.str().equals("int") ||
     op1.str().equals("double") ||
     op1.str().equals("String")))
   return op1;
```

Having modified the type checker to allow string concatenation, the intermediate code generation method, genAddExpr(), is similarly extended. The Unicon modifications in tree.icn are
highlighted in the method body shown here:

```
method genAddExpr()
   addr := genlocal()
   icode := kids[1].icode ||| kids[2].icode
   if typ.str() == "String" then {
     if rule ~= 1320 then
       j0.semErr("subtraction on strings is not defined")
     icode |||:= gen("SADD", addr,
         kids[1].addr, kids[2].addr)
   }
   else icode |||:= gen(if rule=1320 then "ADD" else "SUB",
       addr, kids[1].addr, kids[2].addr)
end
```

The check for production rule 1320 is because the String type does not support subtraction. The
corresponding Java modifications in tree.java are as follows:

```
void genAddExpr() {
  addr = genlocal();
  icode = new ArrayList<tac>();
  icode.addAll(kids[0].icode); icode.addAll(kids[1].icode);
  if (typ.str().equals("String")) {
    if (rule != 1320)
      j0.semErr("subtraction on strings is not defined");
    icode.addAll(gen("SADD", addr,
                    kids[0].addr,kids[1].addr);
  }
  else icode.addAll(gen(((rule==1320)?"ADD":"SUB"),
                    addr, kids[0].addr, kids[1].addr));
}
```

At this point, we have added an intermediate code instruction for string concatenation. Now, it
is time to implement it in the runtime system. First, we will consider the bytecode interpreter.

Adding String concatenation to the bytecode interpreter

Since the bytecode interpreter is software, we can simply add bytecode instructions for string operations, as we did for intermediate code. An opcode for the SADD instruction must be added to Op.icn and Op.java, along with SPUSH and SPOP instructions to push and pop strings. We must modify the bytecode generator to generate a bytecode SADD instruction for an intermediate code SADD instruction. In the bytecode() method in j0.icn, the Unicon implementation looks as follows:

```
"SADD": {
   bcode |||:= j0.bgen(Op.SPUSH, instr.op2) |||
                j0.bgen(Op.SPUSH, instr.op3) |||
                j0.bgen(Op.SADD) |||
                j0.bgen(Op.SPOP, instr.op1)
   }
```

If this looks like the code for the ADD instruction, that is the point. As with the ADD instruction, the final code consists mainly of converting a three-address instruction into a sequence of one-address instructions. The Java implementation in j0.java is shown here:

```
case "SADD": {
   rv.addAll(j0.bgen(Op.SPUSH, instr.op2));
   rv.addAll(j0.bgen(Op.SPUSH, instr.op3));
   rv.addAll(j0.bgen(Op.SADD, null));
   rv.addAll(j0.bgen(Op.SPOP, instr.op1));
   break;
   }
```

We must also implement the bytecode instruction for SADD, which means we must add it to the bytecode interpreter. Since the Unicon and Java implementation languages both have high-level string types with semantics similar to Jzero, we can hope that implementation will be simple. If the Jzero representation of a String in the j0x bytecode interpreter is an underlying implementation language string, then the implementation of the SADD instruction will just perform string concatenation. However, in most languages, the source language semantics differ from the implementation language, so it is usually necessary to implement a representation of the source language type that models source language semantics in the underlying implementation language.

Having issued that warning, let's see if we can implement Jzero strings as plain Unicon and Java strings. To implement string concatenation in bytecode, we push two strings onto the stack with SPUSH, concatenate them with SADD, and pop the result with SPOP. The SPUSH instruction is the same as PUSH on Unicon but will be different on Java:

```
Op.SPUSH: {
    val := deref(opr, opnd)
    push(stack, val)
}
```

The corresponding Java implementation of SPUSH in j0machine.java looks like the following. The stackbuf variable is a ByteBuffer that was sized for holding a good number of 64-bit integer values, but now, we must decide how to use it to also hold strings. If we store the actual string contents in stackbuf, we are not implementing a stack anymore – we are implementing a heap and it will be a can of worms. Instead, we store an integer code in stackbuf that we can use to obtain the string by looking it up in a **string pool**:

```
case Op.SPUSH: {
    String val = sderef(opr, opnd);
    long l = stringpool.put(val);
    stackbuf.putLong(sp++, l);
    break;
}
```

The SADD instruction in the interp() method in j0machine.icn is almost the same as that of the ADD integer:

```
Op.SADD: {
    val1 := pop(stack); val2 := pop(stack)
    push(stack, val1 || val2)
}
```

This Unicon implementation relies on the fact that the Unicon value stack does not care if you sometimes push integers and sometimes push strings. Unicon has an underlying string region where memory for the strings' underlying contents is stored, and the bytecode interpreter uses that implicitly.

The corresponding Java implementation in `j0machine.java` is shown below. Two strings are fetched from the string pool by popping their keys off the stack, and the new string constructed from their concatenation is placed in the string pool. Its key is pushed on the stack:

```
case Op.SADD: {
    String val1 = stringpool.get(stackbuf.getLong(sp--));
    String val1 = stringpool.get(stackbuf.getLong(sp--));
    long val3 = stringpool.put(val1 + val2);
    stackbuf.putLong(sp++, val3);
}
```

This code depends on the `stringpool` class, which uses unique integers to store and retrieve strings. These unique integers are references to the string data that can be conveniently stored on stackbuf, but now, the Java implementation requires the `stringpool` class, so here it is in the `stringpool.java` file. For any string, the way to retrieve its unique integer is to look it up in the pool. Once it's been issued like this, a unique integer can be used to retrieve the string on demand:

```
public class stringpool {
    static HashMap<String,Long> si;
    static HashMap<Long,String> is;
    static long serial;
    static { si = new HashMap<>(); is = new HashMap<>(); }
    public static long put(String s) { … }
    public static String get(long L) { … }
}
```

This class requires the following pair of methods. The put ( ) method inserts strings into the pool. If the string is already in the pool, its existing integer key is returned. If the string is not already in the pool, the serial number is incremented and that number is associated with the string:

```
public static long put(String s) {
    if (si.containsKey(s)) return si.get(s);
    serial++;
    si.put(s, serial);
    is.put(serial, s);
    return serial;
}
```

The get ( ) method retrieves a `String` from `stringpool`:

```
public static String get(long L) {
    return is.get(L);
}
```

Now, it is time to look at how to implement this operator for native code.

Adding String concatenation to the native runtime system

The Jzero native code is much lower level than the bytecode interpreter. Implementing the Jzero `String` class semantics from scratch in C is a big job. Jzero uses an extremely simplified subset of the Java `String` class, for which we only have room to describe the highlights. Here is an underlying C representation of a `String` class for use in Jzero:

```
struct String {
    struct Class *cls;
    long len;
    char *buf;
};
```

Within this struct, `cls` is a pointer to an as-yet-undefined structure for class information, `len` is the length of the string, and `buf` is a pointer to data. The Jzero string concatenation might be defined as follows:

```
struct String *j0concat(struct String *s1, struct String *s2){
    struct string *s3 = alloc(sizeof struct String);
    s3->buf = allocstring(s1->len + s2->len);
    strncpy(s3->buf, s1->buf, s1->len);
    strncpy(s3->buf + s1->len, s2->buf, s2->len);
    return s3;
}
```

This code raises as many questions as it answers, such as what the difference is between `alloc()` and `allocstring()`; we will get to those shortly. But it is a function that we can call from the generated native code via this addition in `j0.icn`:

```
"SADD": {
    bcode |||:= xgen("movq", instr.op2, "%rdi") |||
               xgen("movq", instr.op3, "%rsi") |||
               xgen("call", "j0concat") |||
               xgen("movq", "%rax", instr.op1)
    }
```

The corresponding Java implementation in j0.java is shown here:

```
case "SADD": {
  rv.addAll(xgen("movq", instr.op2, "%rdi"));
  rv.addAll(xgen("movq", instr.op3, "%rsi"));
  rv.addAll(xgen("call", "j0concat"));
  rv.addAll(xgen("movq", "%rax", instr.op1));
  break;
  }
```

Here, you can see that substituting a function call to implement an immediate code instruction is straightforward. Let's compare this with the code that is generated for built-in functions, which we will present next.

Writing built-in functions

Low-level languages such as C have no built-in functions; they have standard libraries that contain functions available to all programs. Linking a function to your program and calling it is conceptually the same action, whether it is a library function or a user-defined function. The higher the language level, the more conspicuous the difference between what is written for its runtime system in a lower-level implementation language and what is written by end users in the language itself. Let's consider how to implement built-ins in the bytecode interpreter.

Adding built-in functions to the bytecode interpreter

Let's implement System.out.println() in the bytecode interpreter. One of our design options is to implement a new bytecode machine instruction for each built-in function, including println(). This doesn't scale well to thousands of built-in functions. We could implement a callnative instruction, providing us with a way to identify which built-in function we want to call. Some languages implement an elaborate interface for calling native code functions and implement println() (or some lower-level building block function) as a wrapper function written in Jzero that uses the native calling interface.

For Jzero, as described in the *Running a Jzero program* section of *Chapter 12, Bytecode Interpreters*, we chose to use the existing call instruction, with special function values to denote built-in functions. The special values we chose were small negative integers where a function entry point address would normally go. So, the function call mechanism must be built to look for small negative integers to distinguish between method types and do the correct thing for user-defined and built-in methods.

Let's look at the do_println() method, which we suggested in *Chapter 12, Bytecode Interpreters*. For Jzero, this runtime system method is hardwired to write to standard output, much like puts() in C. The string to be written is on the stack; it's no longer on the top since the call instruction pushed a function return address. In Unicon, do_println() might be implemented as follows:

```
method do_println()
    write(stack[2])
end
```

In Java, the do_println() method would look something like this:

```
public static do_println() {
    long l = stackbuf.getLong(sp-1)
    String s = stringpool.get(l);
    System.out.println(s);
}
```

Built-in functions in bytecode are simple. Now, let's look at writing built-in functions for native code.

Writing built-in functions for use with the native code implementation

Now, it is time to implement System.out.println() for the native code Jzero implementation. In a Java compiler, it would be a method of the System.out object, but for Jzero, we can do whatever is expedient. We can write a native function named System_out_println() in assembler, or if our generated native code adheres carefully to the calling conventions of a C compiler on the same platform, we can write it in C, put it in our Jzero runtime library, and link it to the generated assembler modules to form our executable. The function takes one string argument, struct String *, as shown in the previous section. Here is an implementation; you can put it in the System_out_println.c file:

```
#include <stdio.h>
void System_out_println(struct String *s) {
    for(int i = 0; i < s->len; i++) putchar(s->buf[i]);
    putchar('\n');
}
```

The more interesting part of all this is how the generated code gets access to this and other built-in native functions. You can compile it via the following command line for gcc:

```
gcc -c System_out_println.c
```

You can add the System_out_println.o output file to an archive library named libjzero.a with the following command line:

```
ar cr libjzero.a System_out_println.o
```

The preceding two command lines are not executed within your compiler at each compile or link time; instead, they are run when the Jzero compiler itself is being built, alongside potentially many other bits of the operator or built-in function library code. They create a library archive file called libjzero.a. This archive file can be linked to Jzero's generated code using the ld or gcc command, as described in *Chapter 14, Native Code Generation*, in the *Linking, loading, and including the runtime system* section.

The -lsomefile command-line option expands to match libsomefile.a so that our runtime can be invoked as -ljzero. Now, how does the Jzero compiler find the runtime library, which may be installed anywhere? The answer will vary by operating system, and some of the convenient options require administrative privileges. If you can copy libjzero.a into the same directory that's used by your linker for other system libraries such as C:\Mingw\lib on Windows or /usr/lib64 on Linux, you may find that everything works great. If that is not an option, you may resort to environment variables or command-line options, either to inform the linker where the library is or to inform the Jzero compiler itself on the Jzero command line where the runtime library can be found. Adding built-in functions like this is important because not every language addition can be made in the form of an operator. Similarly, not every language addition is best formulated as a function. Sometimes, such operators and built-in functions are more effective when they're part of new control structures that support some new problem domain. Let's consider how these operators and functions might profit from being integrated with syntactic additions in the form of control structures.

Integrating built-ins with control structures

Control structures are usually bigger things than expressions, such as loops. They are often associated with novel programming language semantics or new scopes in which specialized computations occur. Control structures provide a context in which a statement (often, this is a compound statement consisting of a whole block of code) is executed. This can be whether (or how many times) it is executed, on what associated data the code is computing, or even what semantics to apply during the evaluation of the operators.

Sometimes, these control structures are explicitly and solely used for your new operators or built-in functions, but often, the interactions are implicit byproducts of the problem-solving that your language enables.

Whether a given block of code is executed, selecting which of several blocks to execute or executing code repeatedly are the most traditional control structures, such as `if` statements and loops. The most likely opportunities for operators or functions to interact with these constructs include a special iterator syntax to control loops using your domain values, and a special switch syntax to select block(s) of code to execute.

The Pascal `WITH` statement is an old but good example of associating some data with a chunk of code that uses that data. The syntax is `WITH r DO statement`. A `WITH` statement attaches some record, `r`, to a statement – usually, this is a compound statement – within which the record's fields are in scope, and a field named x need not be prefixed by an accessor expression, such as `r.x`. This is a low-level building block that object orientation (and associated `self` or `this` references) is based on, but Pascal allows such object attachments for individual statements, a finer granularity than method calls. Pascal also allows multiple objects to be associated with the same block of code via nested `WITH` statements.

We can illustrate some of the considerations of interacting with control structures by considering the implementation of a `for` loop, which iterates over the characters in strings. Because Java is not perfect, you cannot write the syntax – that is, `for(char c:s) statement` – to execute `statement` once for each element of s, but you can write `for(char c:s.toCharArray()) statement`.

So, Java arrays interact nicely with the `for` control structure, but the Java `String` class is not as nice. There is an `Iterable` interface, but strings do not work with it without jumping through extra hoops. When you design your language, try to make common tasks straightforward. A similar comment would apply to accessing `String` elements. Nobody wants to write `s.charAt(i)` when they could be writing `s[i]`; this is a good argument for operator support. An example of integrating a built-in function with a control structure by providing parameter defaults will be provided in the next section. But first, let's look at how operators and built-in functions are implemented for Unicon.

Developing operators and functions for Unicon

Unicon is a very high-level language with many built-in features. For such languages, it will make sense to do some engineering work to simplify creating its runtime system. The purpose of this section is to share a bit about how this was done for Unicon, for comparison purposes. Unicon's operators and built-in functions are implemented using **RTL**, which stands for **Runtime Language**.

RTL is a superset of C developed by Ken Walker to facilitate type inference in the Icon runtime system; Unicon inherits it from Icon. RTL writes out C code, so it is almost a very specialized form of C preprocessor that maintains a database in support of type inferencing.

Operators and functions in RTL look like C code, with many pieces of special syntax. There is syntax support for associating different pieces of C code, depending on the data type of the operands. To allow for type inferencing, the Unicon result type that's produced by each chunk of C code is declared. The RTL language also has syntax support, which makes it easy to specify when an operand type conversion needs to take place. In addition, each chunk of C code is marked with syntax to specify whether to inline it in the generated code or execute the specified code via a C function call. First, we will describe how to write operators in RTL, along with their special considerations. After that, we will learn how to write Unicon built-in functions in RTL, which are coded much like operators.

Writing operators in Unicon

After various clever macro expansions and omitting #ifdefs, the addition operator in Unicon looks as follows. The following code shows three different forms of addition for C (long) integers, arbitrary precision integers, and floating-point. In the actual implementation, there is a fourth form of addition, for array-at-a-time data-parallel addition:

```
operator{1} + add(x, y)
    declare { C_integer irslt; }
    arith_case (x, y) of {
        C_integer: { abstract { return integer }
            inline { … }
        }
        integer: { abstract { return integer }
            inline { … }
        }
        C_double: { abstract { return real }
            inline { … }
        }
    }
end
```

In the preceding code, the special RTL case statement for arithmetic operators, called arith_case, is performed at compile time by the Unicon optimizing compiler, while in the bytecode interpreter, it is an actual switch statement that executes at runtime.

Hidden within `arith_case`, a set of language-wide standard automatic type conversion rules is applied; for example, strings are converted into their corresponding numbers if possible.

The case for regular C integer addition checks the validity of its result and triggers arbitrary precision addition, as per the middle case on integer overflow. The outline of this case body looks like this; some #ifdefs have been omitted for the sake of readability:

```
irslt = add(x,y, &over_flow);
if (over_flow) {
    MakeInt(x,&lx);
    MakeInt(y,&ly);
    if (bigadd(&lx, &ly, &result) == RunError)
        runerr(0);
    return result;
    }
else return C_integer irslt;
```

The `add()` function is called to perform regular integer addition. If there is no overflow, the integer result that is returned by `add()` is valid and is returned. By default, RTL returns from Unicon operators and functions using a generic Unicon value that can hold any Unicon type. If a C primitive type is returned instead, it must be specified. In the preceding code, `return` at the end is annotated in RTL to indicate that a C integer is being returned.

If the call to `add()` overflows, the C integers x and y are placed in Unicon descriptor structures using the macro `MakeInt()`, and then passed as parameters into the function `bigadd()`, which performs arbitrary precision addition and stores the answer in the descriptor `result`. Here is the Unicon runtime's implementation of the `add()` function, which performs integer addition and checks for overflow. There is no more RTL extended syntax going on here, just references to macros for the 2^63-1 and -2^63 values. Someone was probably careful when they wrote this code:

```
word add(word a, word b, int *over_flowp)
{
    if ((a ^ b) >= 0 &&
        (a >= 0 ? b > MaxLong - a : b < MinLong - a)) {
        *over_flowp = 1;
        return 0;
        }
    else {
```

```
        *over_flowp = 0;
        return a + b;
        }
    }
```

This is pretty straightforward C code, except for (a^b), the exclusive OR operator, which is a way of asking if the values are both positive or both negative. In addition to computing the sum, this function writes a Boolean value to the address given in its third parameter to report whether an integer overflow has occurred.

Because it does not have to check for overflow, the floating-point real number addition branch of arith_case, denoted by C_double in RTL, is much simpler. Instead of calling a helper function, the real number addition is done inline using the regular C + operator:

```
    return C_double (x + y);
```

We have omitted the corresponding implementation of the arbitrary precision addition function, bigadd(), which is called in this operator, which is many pages long. If you want to add arbitrary precision arithmetic to your language, you should read about the **GNU Multiple Precision (GMP)** library, which lives at https://gmplib.org/. Now, let's consider a few of the issues that come up when writing built-in functions for Unicon.

Developing Unicon's built-in functions

Unicon's built-in functions are also written in RTL (and C) and, as in the case of operators, the code for each function can be designated to be inlined or placed in a function that may be called. Built-in functions are longer than operators, on average, but perhaps in most cases, the RTL function syntax exists as an advanced form of wrapper that enables a C function to be called from Unicon, with conversions between the type representations of Unicon values and C values as needed. Unlike operators, many functions have multiple parameters for which designated default values may be specified via special syntax. As an example, here is the code for Unicon's string analysis function, any(), which succeeds if the character at the current position within a string is a member of a set of characters specified in its first parameter. The RTL reserved word, function, declares a Unicon built-in function instead of a regular C function. The {0,1} syntax indicates how many results this function can produce. In the case of any(), the function may produce either zero results (it fails) or one result; any() is fallible but it is not a generator. The if-then statement specifies that the first parameter must be convertible into a cset. If not, a runtime error occurs.

The body reserved word specifies that the generated code should call a function here, rather than inline the code. Since the code block is small, choosing body instead of inline is a debatable choice here. If the Testb() macro expands to be quite long, the decision to use body might be justified:

```
function{0,1} any(c,s,i,j)
    str_anal( s, i, j )
    if !cnv:tmp_cset(c) then
        runerr(104,c)
    body {
        if (cnv_i == cnv_j)
            fail;
        if (!Testb(StrLoc(s)[cnv_i-1], c))
            fail;
        return C_integer cnv_i+1;
        }
end
```

In addition to the bits of RTL syntax, macros play a huge role. str_anal is a macro that sets up a string for analysis, defaulting parameters 2–4 to the current string scanning environment. str_anal also ensures that s is a string and that i and j are integers, converting them into those types if necessary, and issuing a runtime error if an incompatible value is passed in. String scanning environments are created by the string scanning control structure; the location under study within the string can be moved around by other string scanning functions. Adding domain-specific control structures such as string scanning will be presented in the next chapter. This example serves to motivate them. One reason to use new control structures is to make operators and built-in functions more powerful and concise.

In this section, we presented a few highlights that show how Unicon's operators and built-in functions are implemented. A lot of the issues in the runtime system of a very high-level language were found to revolve around the big semantic difference between the source language (Unicon) and the implementation language (in Unicon's case, C). Depending on the level of the language you are creating and the language you write its implementation in, you may find it useful to resort to similar techniques.

Summary

This chapter showed you how to write high-level operators and built-in functions for the runtime system of your language. One of the main points that you are to take away is that the implementation of operators and functions can range from completely different to almost entirely the same, depending on the language you are inventing.

The examples in this chapter taught you how to write code in your runtime system that will be called from generated code. You also learned how to decide when to make something a runtime function instead of just generating the code for it using instructions.

The next chapter will continue the topic of implementing built-in features by exploring domain control structures.

Questions

1. It is mathematically provable that every computation that you could implement as an operator or built-in function can be implemented instead as a library method, so why bother implementing high-level operators and built-in functions?

2. What factors must you consider if you are deciding between making a new operator or a new built-in function?

3. There were probably some good reasons why Java decided to give strings only partial operator and control structure support, despite strings being important and supported better in languages such as Icon and Unicon (and Python, which was influenced by Icon). Can you suggest some of the reasons for this?

Join our community on Discord

Join our community's Discord space for discussions with the authors and other readers:

https://discord.com/invite/zGVbWaxqbw

16

Domain Control Structures

The code generation that was presented in the previous chapters covered basic conditional and loop control structures, but domain-specific languages often have unique or customized semantics that merit introducing novel control structures. Adding a new control structure is usually substantially more difficult than adding a new function or operator. However, when they are effective, the addition of domain control structures is a large part of what makes domain-specific languages worth developing instead of just writing class libraries. The examples in this chapter should support this assertion, but it is based on the Sapir-Whorf hypothesis, which claims that language influences and constrains what we are able to think. Programming languages that are Turing complete can compute anything, but that does not mean they are equally practical for all jobs. Adding domain control structures can make your language the most practical choice for some new application domains that are not well-served by existing languages.

This chapter covers the following main topics:

- Knowing when a new control structure is needed
- Processing text using string scanning
- Rendering graphics regions

The first section will help you learn how to determine when a domain control structure is needed. The second and third sections will present two example domain control structures.

This chapter will give you a better idea of when and how to implement new control structures as needed in your language design. More importantly, you will learn how to walk the thin line that balances the need to stick with generating familiar code for familiar structures and the need to reduce programmers' effort in new application domains by introducing novel semantics.

Java and its Jzero subset do not have comparable domain control structures, so the examples in this chapter come from Unicon and its predecessor, Icon. While this chapter outlines their implementation, at times using code examples, you are reading this chapter for the ideas rather than reading so you can type the code in and see it run. First, let's re-examine when a new control structure is justified.

Knowing when a new control structure is needed

If you google the definition of a **control structure**, it will say something like *"control structures determine the order in which one or more chunks of code execute."* This definition is fine for traditional mainstream languages. It focuses on control flow, and it addresses two kinds of control structures: choosing which (or whether) to execute and code (loops) that can repeat under some conditions. The if statements and while loops that we implemented for Jzero earlier in this book are good examples.

Higher-level languages tend to have a more nuanced view of control structures. For example, in a language with built-in backtracking, the order in which chunks of code may execute becomes more complicated. This book will paraphrase Ralph Griswold's definition of a control structure in the Icon programming language: a control structure is an expression containing two or more subexpressions in which one subexpression is used to control the execution of another subexpression. That definition is more general and more powerful than the traditional mainstream definition provided in the previous paragraph. A control structure can be not just about the order in which chunks of code execute but also the context in which chunks of code execute.

The phrase *"control the execution"* in Griswold's definition can be interpreted as broadly and as loosely as you want. Instead of just whether a chunk of code executes, or which chunk of code, or how many times, a control structure can determine how the code executes. This could mean introducing new scopes where names are interpreted differently, adding new operators, changing the behavior of existing operators, or a myriad of other possibilities.

A new control structure is needed when there are one or more major programming **pain points**. Often, pain points arise when people start writing software in support of a new class of computer hardware or for a new application domain. Often, the first support for new hardware or a new application domain comes in the form of a function or class library written for a mainstream language. The pain might be because the library design is suboptimal, in which case, maybe it should just be fixed if that's possible. Alternatively, the pain might be because the mainstream language is ill-suited to the task, or because the new hardware or application domain is inherently complex.

Awareness or knowledge of an application domain's pain points may or may not exist at language design time, but when writing for a new application domain in a language that predates that domain by decades, the language certainly couldn't anticipate domain programmers' needs. More often, the awareness of pain points is generated from early substantial experiences as developers attempt to write software for that domain.

Pain points are often due to complexity, frequent and pernicious bugs, code duplication, or several other famous bad smells or **anti-patterns**. Some **code smells** are described in *Refactoring: Improving the Design of Existing Code*, by *Martin Fowler*. Anti-patterns are described at `antipatterns.com` and in several books referenced on that site.

Individual programmers or programming projects may be able to reduce their code smells or avoid anti-patterns by performing code **refactoring**, but when the use of application domain libraries entails that most or all applications in that domain face such problems, an opportunity for one or more domain control structures becomes conspicuous. We will see interesting examples of control structures affecting how code executes later in this chapter. Let's start with a simple one.

Many general-purpose libraries have an API with the same parameters repeated across tens or even hundreds of related functions. Applications that use these APIs may feature many calls where the same sequence of parameters is provided to the library over and over. The classic Microsoft Windows Graphics API is a good example of this. Things such as windows, device contexts, colors, line styles, and brush patterns are provided repeatedly to many drawing calls. You can write any code you want, but when you call GetDC() to acquire a device context, there had better be exactly one corresponding call to ReleaseDC(). A lot of the code in between those two points will pass that device context as a parameter over and over.

For the sake of reducing the network traffic involved, Win32's open-source counterpart known as Xlib, the C library for writing applications under the X Window System, placed several common graphic drawing elements into a graphics context object that reduces the number of redundant parameters that need to be passed in each graphics call. Even with this clever feature, the Xlib API remains quite complex and contains a lot of parameter redundancy.

The designers of libraries are, in some cases, brilliant, but their APIs still might be relatively hostile to ordinary developers, with steep learning curves. Before the advent of graphical user interface builders that generated this code for us, creating graphical user interfaces disproportionately slowed down development and increased the cost of developing many graphic applications, and it lured many coders into poor practices such as block copying and modifying vast swaths of user interface code.

For a language where a new control structure is not an option, the problem of excessive redundant parameters may be unavoidable. If you build a language, a control structure is a real option for addressing this issue.

Pain points become a target for a new control structure when they can be solved within the domain that you are supporting. In that case, the new control structures can be seen as a by-product of traditional languages' lack of support for that domain. If your application domain has a corpus of existing libraries and applications written in a mainstream language, you can study that code to look for its pain points and craft control structures that ameliorate them in your programming language. If your application domain is quite new and no mainstream language APIs and application base are available, you might resort to guessing or writing example programs in your new language to look for pain points. Let's look at a novel control structure where these principles were applied successfully: string scanning in the Icon and Unicon languages.

Scanning strings in Icon and Unicon

Unicon inherits this domain control structure from its immediate predecessor, Icon. In Icon and Unicon, a control structure called string scanning is invoked by the s ? expr syntax. Within expr, the string s is the scanning *subject* and it is referenced by a global keyword called &subject. Within the subject string, a current analysis *position*, which is stored in the &pos keyword, denotes the index location in the subject string where that string is being scanned. The position starts at the beginning of the string and can be moved back and forth, typically working its way toward the end of the string. For example, in the following program, s contains "For example, suppose string s contains":

```
procedure main()
    s := "For example, suppose string s contains"
    s ? {
        tab(find("suppose"))
        write("after tab()")
    }
end
```

Now, let's say we were adding the above-mentioned scanning control structure:

```
s ? { … }
```

The &subject and &pos keywords would initially be in the following state:

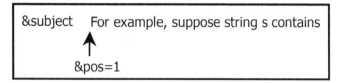

Figure 16.1: Subject and position at the start of a scan

The tab() function moves scanning forward by setting the string's scanning position to 14, the result produced by the find() function. The subsequent analysis would commence from the word *suppose*, as shown in the following screenshot:

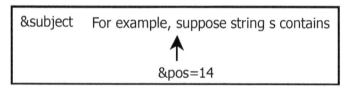

Figure 16.2: Subject and position after advancing to the start of the string "suppose"

This mechanism is very general and allows for a variety of pattern-matching algorithms. Now, it is time to dive into the details of how this control structure is utilized through its operations.

Scanning environments and their primitive operations

A (subject, position) pair is called a **scanning environment**. Within the string scanning control structure, there's one operator, two built-in position-moving functions, and six built-in functions that analyze the subject string. The six built-in string analysis functions are summarized in the following table. They are described in more detail in *Appendix, Unicon Essentials*:

Function	Purpose
any(C)	Is the character at the current position a member of character set c?
many(C)	Are 1+ characters at the current position members of character set c?
match(s)	Do the characters at the current position match a search string s?
find(s)	At what positions do the characters match a search string s?
upto(C)	At what positions are characters found that are members of c?
bal()	At what positions are characters balanced with respect to delimiters?

Figure 16.3: Built-in analysis functions of the string scanning control structure

The two built-in functions that move the position are named move() and tab(). The move(n) function slides the position index over by n letters relative to the current position, where positive n moves from left to right and negative n moves backward from right to left. The tab(n) function is an absolute move, setting the position to an index, n, within the subject. The position-moving built-in functions are commonly used in combination with the string analysis functions. For example, find("suppose") returns the index at which the "suppose" string may be found, and tab(find("suppose")) sets the position to that location. In the example shown in *Figure 16.1*, executing tab(find("suppose")) would be one of many ways that the scanning environment might be set to the state shown in *Figure 16.2*. Another way to get there would be to execute the following code:

```
tab(upto(',')) & move(1) & tab(match(" "))
```

It is typical to combine the string analysis primitives in such a fashion to form larger and more complex patterns. The language's built-in backtracking process, called goal-directed evaluation, means that earlier partial matches will undo themselves if a latter part of a pattern match fails, leaving the position unchanged.

The tab(match(s)) combination is deemed so useful that a unary prefix operator, =, is defined for it. This is not to be confused with the binary = operator, which performs a numeric comparison. In any case, the =s expression is equivalent to tab(match(s)). This set of primitives was invented for Icon and preserved in Unicon. Unicon adds a complementary mechanism here (a SNOBOL-style pattern type, featuring regular expressions as its literals). You may be wondering whether additional operators for other common combinations of string analysis and position-moving functions would add expressive power to string scanning.

Icon and Unicon's string scanning control structure contrasts strongly with the monolithic pattern-matching operations found in other string processing languages. String scanning is a more general mechanism than regular expressions, in which ordinary code can be mixed into the middle of the pattern match. The following string scanning example extracts proper nouns from the S string and stores them in a list, L:

```
S ? { L := []
    while tab(upto(&ucase)) do
        put(L, tab(many(&letters)))
}
```

The preceding while loop discards characters until an uppercase letter is found. It treats each such uppercase letter as the start of a proper name and places the name in a list.

This is not as concise as I might hope for, but it's extremely general and flexible. Let's look at how this control structure reduces the excess redundant parameters for string analysis functions.

Eliminating excessive parameters via a control structure

String scanning provides a standard set of default parameter values for the string analysis functions that are built into the language. These functions all end with three parameters: *the subject string, the start position, and the end position*. These parameters default to the current scanning environment, which consists of the subject string, &subject, the current position, &pos, and the end of the subject string. As a trivial example, the call find("suppose") searches for the string "suppose", but within what string? Starting from where? And searching until when? Outside a string scanning environment, the call requires up to four parameters and might look like find("suppose", subject, startpos, endpos). By itself, the control structure enables calls to find() to be shorter and more readable. Within a larger string analysis task, these extra parameters recur over and over in many string analysis functions. The string scanning control structure addresses one of the pain points in string analysis. However, parameter simplification is not the entirety of the impact and purpose of string scanning.

The current string scanning environment is visible within called functions and has a dynamic scope. It is common, and simple, to write helper functions that perform a part of a string analysis task, without having to pass the scanning environment around as parameters.

Scanning environments may be nested. As part of a scanning expression or helper function, when a substring requires further analysis, this can be performed by introducing another string scanning expression. When a new scanning environment is entered, the enclosing scanning environment must be saved and restored when the nested sub-environment is exited. This nesting behavior is preserved in Icon and Unicon's novel goal-directed expression semantics, in which expressions can be suspended and later resumed implicitly. The scanning environment is saved and restored on the stack. These operations are finer-grained but also depend on the procedure activity on the stack, such as procedure calls, suspends, resumes, returns, and failures.

For those who want more details, string scanning has been described extensively in other venues, such as Griswold and Griswold's *The Icon Programming Language, 3rd edition*. The implementation is described in *Section 9.6* of *The Implementation of Icon and Unicon*. In addition to saving and restoring scanning environments on the stack, two bytecode machine instructions are used to simplify code generation for this control structure. When you add a control structure, it is fair game to add intermediate and/or bytecode instructions to support it. Now, let's look at a second and completely unrelated domain control structure that we introduced into Unicon as part of its 3D graphics facilities: **rendering regions**.

Rendering regions in Unicon

This section describes a control structure called rendering regions, which was added to Unicon while writing this book. Since this feature is new, we will look at it in some detail. The rendering region control structure has been on Unicon's to-do list for a long time but adding a control structure can be a bit difficult, especially if the semantics are non-trivial, so it took writing this chapter to get around to it. First, though, we need to set the scene.

Rendering 3D graphics from a display list

Unicon's 3D graphics facilities specify what is to be drawn via a series of calls to a set of built-in functions, and an underlying runtime system renders code written in C and OpenGL that draws the scene as many times per second as possible. The Unicon functions and C render code communicate using a **display list**. Mainly, the Unicon functions place primitives on the end of the display list, and the rendering code traverses the display list and draws these primitives as quickly as possible.

In OpenGL's C API, there is a similar-sounding display list mechanism that serves to pre-package and accelerate sets of primitives by placing them on the GPU in advance, reducing the CPU-GPU bottleneck. However, Unicon is a dynamic language that prioritizes flexibility over performance. To manipulate the display list at the Unicon application code level, the Unicon display list is a regular Unicon list managed by the CPU, rather than a C OpenGL display list managed on the GPU.

Even managed on the CPU, the Unicon display list is a great improvement over just executing all the Unicon code to render a scene in each frame. This is because traversing the display list and drawing all the primitives from it can be performed entirely by C code in the Unicon runtime system and can be heavily optimized. In contrast, executing the Unicon code to render a scene incurs the overhead of the bytecode virtual machine interpretation... repeatedly, at whatever frame rate the code can run – hopefully, 60 frames per second or faster.

When Unicon's 3D facilities were first created, every graphic primitive in the display list was rendered every frame. This worked well for small scenes. For scenes with many primitives, it becomes impractical to reconstruct the display list from scratch on each frame. New capabilities were needed to enable applications to make rapid changes and be selective about which primitives on the display list will be rendered. Those capabilities' final form was non-obvious when we started. Now, let's look at how rendering regions started as a function API.

Specifying rendering regions using built-in functions

Selective rendering was introduced in Unicon initially using a function named WSection(). The W character in this function name stands for window and is a common prefix for many of Unicon's built-in functions about graphics and window systems, so this is the (window) section function. A call to WSection() with a string parameter s defines the beginning of a section named s, and a second call to WSection() with no parameter marks the end of the most recent open section, defining the bounds of a **rendering region**. Rendering regions make it easy to turn on or off whole collections of 3D primitives on the display list between each frame, without having to reconstruct the display list or insert or delete elements. It is common to nest rendering regions inside each other, for example, to introduce additional levels of detail that can be enabled as an object and viewer draw nearer to each other.

The call to WSection(s) that begins a new rendering region introduces a display list record with a skip field that can be turned on and off; the second call to WSection() is an end marker that helps determine how many display list primitives are to be skipped. The following example draws a yellow torus above a character's head:

```
    questR := WSection("Joe's halo")
        Fg("diffuse translucent yellow")
        PushMatrix()
        npchaloT := Translate(0, h.headx+h.headr*3, 0)
        ROThalo := Rotate(0.0, 0, 1.0, 0)
        DrawTorus(0, 0, 0, 0.1, h.headr*0.3)
        PopMatrix()
    WSection()
```

You can't run this example as a standalone since it has been taken from the middle of a 3D application that is busy rendering a 3D scene. The missing context includes an open 3D window that these functions all operate on, and a current object within which the h class variable denotes the character's head. But hopefully, this example illustrates how WSection() calls are used in pairs that define the beginning and end of a set of 3D operations.

Most Unicon 3D functions return the display list entry that they have added as their return value. The return value of WSection() is a record on the display list that affects display list behavior for however many primitives comprise that section.

In the preceding code example, once drawn, the halo remains present on the display list until it is explicitly removed or the screen is cleared. The halo can be made visible or invisible by clearing or setting the skip flag; assigning questR.skip := 1 causes the halo to disappear. Effectively, a rendering region introduces a conditional branch to the display list data structure.

Rendering regions also support 3D object selection. The parameter of a starting WSection() specifies a string value that is returned when the user touches or mouse clicks on one of the 3D primitives within that section.

Varying levels of detail using nested rendering regions

Rendering regions support nesting. In 3D scenes, complex objects may be rendered by traversing a hierarchical data structure where the largest or most important graphical elements are at the root. Nested rendering regions support levels of detail, where secondary and tertiary graphic details can be rendered within subsections and turned on and off, depending on how near or far the object is from the camera. Levels of detail can be important for performance, allowing details to be proportional to the approximate distance between the viewer and the objects being observed. There are fancy data structures that can be used to implement this level of detail, but rendering regions work well for it.

The code for rendering a chair, for example, might be organized into three levels of detail using three nested sections. The Chair class's lod1, lod2, and lod3 variables would be associated with the three nested sections within the code to fully render the chair:

```
method full_render()
    lod1 := WSection("chair level 1")
        ... render the big chair primitives
        lod2 := WSection("chair level 2")
            ... render smaller chair primitives
            lod3 := WSection("chair level 3")
                ... render tiny details in the chair
            WSection()
        WSection()
    WSection()
end
```

After the initial full_render() enters the primitives into the display list, each time the chair render level changes, the render() method in the Chair class updates how much should be rendered and how much should be skipped by setting the skip flags.

The following code can be read as follows: if the chair hasn't been rendered yet, perform a
full_render(). If it has been rendered, set some skip flags to indicate how much detail to render
based on the render_level parameter, ranging from 0 (invisible) to 3 (full detail):

```
method render(render_level)
    if /rendered then return full_render()
    case render_level of {
        0: lod1.skip := 1
        1: { lod1.skip := 0; lod2.skip := 1 }
        2: { lod1.skip := lod2.skip := 0; lod3.skip := 1 }
        3: { lod1.skip := lod2.skip := lod3.skip := 0 }
        }
end
```

This mechanism works marvelously, but some painful bug hunts identified a problem. As con-
ceived, the section mechanism was fragile and error-prone. Whenever a WSection() is accidentally
placed in the wrong spot or not nested properly, the program misbehaves or visual aberrations
ensue. Introducing a control structure simplifies the use of rendering regions and reduces the
frequency of errors related to the section boundary markers in the display list.

Creating a rendering region control structure

This subsection will describe an implementation of a rendering region control structure in Unicon,
to show some of the work involved in introducing novel control structures to support application
domains. The general syntax introduced is:

```
wsection expr1 do expr2
```

The idea of this control structure is to place a rendering region around expr2 in such a way that
the end of the section cannot be accidentally moved around or misplaced. For example, the earlier
"Joe's Halo" example would now look like the following:

```
questR := wsection "Joe's halo" do {
        Fg("diffuse translucent yellow")
        PushMatrix()
        npchaloT := Translate(0, h.headx+h.headr*3, 0)
        ROThalo := Rotate(0.0, 0, 1.0, 0)
        DrawTorus(0, 0, 0, 0.1, h.headr*0.3)
        PopMatrix()
        }
```

This book does not describe the full details of the Unicon implementation of rendering regions; instead, it presents the minimum of what is involved while keeping things readable. For further details on the Unicon implementation, you can consult *The Implementation of Icon and Unicon*. The source files in the implementation that are being modified here live in the uni/unicon sub-directory within the Unicon language distribution.

To add a control structure, you must define its requirements, syntax, and semantics. Then, you will have to add any new elements to the lexical analyzer, grammar, trees, and symbol tables. Compile-time semantic checks may be required. The main work of implementing a control structure will then proceed, which consists of adding rules to the code generator to handle whatever new shapes appear in the syntax tree for your control structure.

Adding a reserved word for rendering regions

Before you can add a new control structure to your grammar, new lexical elements (in this case, a new reserved word, wsection) must be added to Unicon's lexical analyzer. You learned how to add reserved words to Jzero in *Chapter 3, Scanning Source Code*. Adding one to Unicon is similar, in that the lexical analyzer and parser will both have to agree on a new integer code for the new reserved word, which is defined by the parser.

Unicon was developed before the uflex tool was created, which was presented in *Chapter 3, Scanning Source Code*. In the future, Unicon may be modified to use uflex, but this section describes how to add a reserved word to Unicon's current, hand-written lexical analyzer, which is called unilex.icn in the Unicon source code. Reserved words are stored in a table that contains, for each reserved word, two integers. One integer contains a pair of Boolean flags for semi-colon insertion rules, stating whether the reserved word is legal at the beginning (a Beginner) and/or the end (an Ender) of an expression. The other integer contains the integer terminal symbol category. The new reserved word, wsection, will be a Beginner of expressions, so semi-colons may be inserted on new lines that immediately precede it. The table entry for wsection in unilex.icn looks like this:

```
    t["wsection"] := [Beginner, WSECTION]
```

The reason this lexical analyzer addition is so small is that the pattern and code that are needed to recognize wsection are the same as for other reserved words and identifiers. For this lexical analyzer code to work, WSECTION must have been declared in the grammar, as described in the following section, and the ytab_h.icn file containing #define rules for the terminal symbols must be regenerated using the -d option to iyacc.

Now, it is time to use this new reserved word in a grammar rule.

Adding a grammar rule

The addition of the wsection control structure is intended to feel consistent with the rest of the Icon and Unicon syntax. The do reserved word almost makes it sound too much like a loop; a precedent is the Pascal language's with statement, which uses do and is not a loop. The addition of this grammar rule in unigram.y consists of two bits. In the terminal symbol declarations, the following is added:

```
%token WSECTION    /* wsection  */
```

In the main grammar section, the grammar rules to add this control structure to unigram.y are as follows:

```
expr11 : wsection ;
wsection : WSECTION expr DO expr {
        $$ := node("wsection", $2, $4)
    };
```

Many, or most, control structures will have semantic requirements, such as the fact that the first expression in the preceding rule – the section identifier – must be a string. Since Unicon is a dynamically typed language, the only way that we could enforce such a rule at compile time would be if we restricted section identifiers to string literals. We elect not to do that and instead enforce the string requirement for the first expression in the generated code, but if your language is typed at compile time, you would add that check to the appropriate point in your tree traversals where other type checks are performed. Now, let's consider the other semantic checks that are needed.

Checking wsection for semantic errors

The purpose of the wsection control structure is to make rendering regions less prone to errors. In addition to the wsection construct, which makes it impossible to omit a closing call to WSection() or accidentally write two rendering regions that overlap, under what other circumstances might rendering regions get messed up? Statements that transfer the control flow out of the rendering region in an unstructured way are problematic. In Unicon, these include return, fail, suspend, break, and next. However, if the rendering region has loops inside it, a break or next expression inside of such a loop is perfectly reasonable.

So, the Unicon compiler's task is to decide what to do in the event of an abnormal control flow departure from within a rendering region. For the string scanning control structure, the correct thing to do was implement saving and restoring scanning environments on the stack, but rendering regions are different.

A **rendering region** is used at display list construction time to ensure that the display list entries are well-formed. The display list is then used later – repeatedly – in the runtime system whenever the screen is to be redrawn. The original control flow that was used when the display list was constructed has nothing to do with it. For this reason, in a wsection, attempting to exit prematurely without reaching the end of the render region results in an error. If a programmer wants to code a render region in an unstructured manner, they can call WSection() explicitly in pairs at their peril.

Enforcing these semantic rules requires some logic to be in a (sub)tree traversal whenever a wsection is encountered in the syntax tree. Tree traversals will look a bit different in the Unicon translator than they do in Jzero, but overall, they resemble the Unicon implementation of Jzero. If you were adding a comparable control structure to Jzero, the best place to introduce this check is in the j0 class's semantic() method, right after the root.check_types() method call, which performs type checking. The new check at the end of the semantic() method would look like this:

```
root.check_wsections();
```

The following check_wsections() method has been added to Unicon's tree.icn:

```
method check_wsections()
    if label == "wsection" then check_wsection()
    else every n := !children do
            n.check_wsections()
end
```

The helper method called to check that each wsection construct is called check_wsection(). It is a subtree traversal that looks for tree nodes that could exit a wsection abnormally and reports a semantic error if the code attempts this. However, it would be possible to generate code that performs these checks at runtime, providing lazy enforcement. The check_wsection() method takes an optional parameter, which tracks nested loops contained within the wsection construct so that any break or next expressions nested within a wsection are allowed, so long as they do not break out of wsection:

```
method check_wsection(loops:0)
    case label of {
        "return"| "Suspend0"| "Suspend1":
            yyerror(label || " inside wsection")
        "While0"|"While1"|"until"|"until1"|
        "every"|"every1"|"repeat":
            loops +:= 1
```

```
        "Next"|"Break":
           if loops = 0 then
               yyerror(label || " inside wsection")
           else loops -:= 1
        "wsection": loops := 0
        }
    every n := !children do {
       if type(n) == "treenode" then
          n.check_wsection(loops)
       else if type(n) == "token" then {
          if n.tok = FAIL then
             yyerror("fail inside wsection")
       }
    }
end
```

The preceding code performs semantic checks so that the wsection control structure can enforce
its requirement that every opening WSection(id) has a closing WSection(). Now, let's look at
generating the code for wsection.

Generating code for a wsection control structure

Code generation for the wsection control structure can be modeled using the equivalent calls to
the WSection() function.

To understand the code generation for wsection, we need a semantic rule for the wsection syntax that solves the problem in the general case. The following table shows such a semantic rule.
Instead of intermediate code generation instructions, the code is expressed as a source-to-source
transformation. A wsection control structure is implemented with some semi-fancy Icon code
that executes a matching pair of WSection() calls, producing the result of the opening call to
WSection() as the result of the entire expression. Because of this, the display list record can be
assigned to a variable by a surrounding expression if desired:

Production	Semantic Rule
wsection : WSECTION expr$_1$ DO expr$_2$	wsection.code = "1(WSection(" || expr$_1$ || "),{" || expr$_2$|| ";WSection();1})"

Figure 16.4: Semantic rule for generating code for the wsection control structure

The Icon code in the preceding semantic rule requires some explanation. The expression

```
{expr2; WSection(); 1}
```

executes expr2, followed by a closing WSection(). The 1 character after the second semi-colon ensures that the whole expression succeeds when evaluated by the surrounding expression; it precludes backtracking. The surrounding expression

```
1(WSection(…), {…})
```

evaluates the opening WSection(…) first and then executes the body, but produces the return value of the opening WSection() as the result of the entire expression. In full disclosure, this semantic rule is incomplete in that further semantic restrictions must be enforced in order to guarantee that matching calls to WSection() occur. For example, expr2 had better not return or suspend out of the function in which this control structure executes, or the required matching-pairs semantics of WSection() will not be enforced. Adding these semantic checks, similar to those shown earlier for precluding break or next in a wsection body, are left as an exercise for you.

To implement the semantic rule shown in the preceding table and make the actual output of the code happen, the Unicon code yyprint() generator procedure must be modified. yyprint(n) generates code for syntax tree node n as string output to a file named yyout. It has a lot of different code branches – pretty much one for each kind of tree node – and these branches call many helper functions as needed. For a wsection, the yyprint() function should utilize the following code, which can be added to the treenode case clause:

```
else if node.label == "wsection" then {
    writes(yyout, "1(WSection(")
    yyprint(node.children[1])
    writes(yyout,"),{")
    yyprint(node.children[2])
    write(yyout, ";WSection();1})")
    fail
}
```

The reason this works, where the domain control structure is simply being written out as an artful arrangement of some underlying function calls, is because the main Unicon compiler is a semi-transpiler that writes out an intermediate form that looks almost like source code. Specifically, Unicon's intermediate form is almost Icon source code. A great many languages can be invented very quickly if the underlying representation is another very high-level language such as Icon or Python.

All this extending of the Unicon language has probably made you excited to try adding your domain control structures. Hopefully, as we head into the summary, you have an idea of how to go about doing that.

Summary

This chapter explored the topic of domain control structures. Domain control structures go way beyond libraries, or even built-in functions and operators, in terms of supporting programmers' abilities to solve problems in new application domains. Most of the time, domain control structures simplify code and reduce the occurrence of bugs in programming that would be prevalent when programmers develop their code using generic mainstream language features.

The next chapter will present the challenging topic of garbage collection. Garbage collection is a major language feature that often distinguishes low-level system programming languages from higher-level application languages and domain-specific languages.

Questions

Answer the following questions to test your knowledge of this chapter:

1. Control structures are just if statements and loops. What's the big deal?

2. All application domain-specific control structures let you do is provide some default values for some standard library functions. Why bother using them?

3. What additional primitives or semantics would make the string scanning control structure more useful to application domain programmers?

4. Would it be a good idea for the wsection control structure to generate code, including a PushMatrix() and a PopMatrix() that surround its code body? This would make the example shorter and higher-level.

Join our community on Discord

Join our community's Discord space for discussions with the authors and other readers:

https://discord.com/invite/zGVbWaxqbw

17

Garbage Collection

Memory management is one of the most important aspects of modern programming. Almost any language that you invent should provide automatic heap memory management via **garbage collection.** Garbage collection refers to any mechanism by which heap memory is automatically freed and made available for reuse when it is no longer needed for a given purpose. The heap, as you may already know, is the region in memory from which objects are allocated by some explicit means such as the reserved word new (in Java). In lower-level languages, such objects live until the program explicitly frees them, but in many modern languages, heap objects are retained in memory as long as they are needed. After a heap object is of no further use in the program, its memory is made available to the program for other purposes by a garbage collection algorithm that runs behind the scenes in the programming language runtime system.

This chapter presents a couple of methods with which you can implement garbage collection in your language. The first method, called **reference counting**, is not very difficult to implement and has the advantage of freeing memory incrementally as soon as the program is no longer using it. However, reference counting has a fatal flaw, which we will discuss in the section titled *The drawbacks and limitations of reference counting*. The second method, called **mark-and-sweep collection**, is a more robust mechanism, but it is much more challenging to implement. It has the downside that execution pauses periodically for however long the garbage collection process takes. These are just two of many possible approaches to memory management; for a more advanced and in-depth treatment of the subject, you may want to check out *The Garbage Collection Handbook*, by Jones, Hosking, and Moss, from CRC Press. There is no silver bullet. Implementing a garbage collector with neither a fatal flaw nor periodic pauses to collect free memory is liable to have other costs associated with it.

This chapter covers the following main topics:

- Grasping the importance of garbage collection
- Counting references to objects
- Marking live data and sweeping the rest

The goal of this chapter is to explain why garbage collection is important and show you how you can do it. The skills you'll learn include the following: making objects track how many references are pointing to them; identifying all the pointers to live data, including pointers located within other objects; and freeing memory and making it available for reuse. Let's start with a discussion of why you should bother with all this anyway.

Grasping the importance of garbage collection

In the beginning, programs were small, and the **static allocation** of memory was decided when a program was designed. The code was not that complicated, and programmers could lay out all the memory that they were going to use during the entire program as a set of global variables. Life was good. A lot of programmers would prefer to just stick with static allocation, and in certain niche application domains, that remains feasible.

For the rest of us, **Moore's Law** happened, and computers got bigger. Data got bigger. Customers started demanding that programs handle arbitrary-sized data instead of accepting the fixed upper limits inherent in static allocation. Programmers invented **structured programming** and used function calls to organize larger programs in which most memory allocation was on the **stack**.

A stack provides a form of **dynamic memory allocation**. Stacks are great because you can allocate a big chunk of memory when a function is called and deallocate memory automatically when a function returns. The lifetime of a local memory object is tied directly to the lifetime of the function call within which it exists.

Eventually, things got complicated enough that folks noticed that software was not keeping up with hardware advances and was becoming the bottleneck in development. We had a **software crisis** and attempted to wish **software engineering** into existence to try and address this crisis. There were occasional bugs where pointers to memory that had been freed on the stack were left hanging around, but those were rare and usually just a sign of novice programmers at work. Life was still relatively good, unless your software was complex, and lives depended on it. But then, Moore's Law happened some more.

Now, even programs running on our wristwatches are too large to understand, and we have a software environment where, at runtime, a program may have billions and billions of objects. Customers expect to be able to create as many objects as they want, and they expect such objects to live for as long as they are needed. The pre-eminent form of allocated memory is dynamic memory, and in most applications, most of the memory is allocated from the memory region called the **heap**. The correct and efficient use of the heap region is a primary concern in modern programming language design and implementation.

In the software engineering literature, it has long been common to see claims that 50% to 75% (or more) of total software development time is spent debugging. This translates into a lot of time and money. In my personal experience over several decades of helping student programmers, in languages where programmers manage their own memory, 75% or more of debugging time is spent on memory management errors.

This is especially true for novices and non-expert programmers, but memory problems happen to beginners and experts alike. C and C++, I am looking at you. Improved tools for memory debugging exist now that did not in the previous century, but memory management is still a central issue.

Now, pretend the concern is not just about how much time or money will be sunk into memory management. As program size and complexity increases, the probability of developers correctly manually managing a software project's memory decreases, resulting in a high probability that the project will fail outright during development or fail critically after deployment.

"What kinds of memory management errors?" you ask. You can start with these: not allocating enough memory; attempting to use memory beyond the amount you allocated; forgetting to allocate memory; not understanding when you need to allocate memory; forgetting to deallocate memory so that it can be reused; deallocating already-deallocated memory; and attempting to use memory for a given purpose after it has been deallocated or repurposed. These are just a few examples.

When programs are only a modest size and the computers involved are terribly expensive, it makes sense to maximize efficiency by throwing programmer time at manual memory management as much as necessary. But as programs grow ever longer and computers become cheaper with larger memory sizes, the practicality of managing memory by hand decreases. Automatic memory management is inevitable. Doing it all on the stack went by the wayside long ago, when structured programming gave way to the **object-oriented** paradigm.

Now, for many or most programs, most of the interesting memory is allocated out of the heap, where objects live an arbitrary length of time until they are explicitly freed, or unused memory is automatically reclaimed. This is why garbage collectors are of paramount importance and deserve your attention as a language implementer.

Having said all this, implementing garbage collection can be difficult, and making it perform well is even more difficult. Garbage collection does not magically solve all memory management problems. Since implementing garbage collection is difficult, if you are overwhelmed, you might get away with putting this off until the success of your language demands it. Sun's original Java implementation got away with a missing garbage collector for years. But, if you are serious about your language, you will eventually want a working garbage collector for it. Let's start with the simplest approach, which is called reference counting.

Counting references to objects

In **reference counting**, each object keeps a count of how many pointers refer to it. This number starts out as 1 when an object is first allocated and a reference to it is provided to a surrounding expression. The reference count is incremented when the reference is stored in a variable, including when it is passed as a parameter or stored in a data structure. The count is decremented whenever a reference is overwritten by assigning a variable to refer elsewhere, or when a reference no longer exists (such as when a local variable ceases to exist because a function returns). If the reference count reaches 0, the memory for that object is garbage because nothing points to it. At that point, the memory can be reused for another purpose. This all seems reasonable; look at what it would take to add reference counting to our example language in this book, Jzero.

Adding reference counting to Jzero

Jzero allocates two kinds of things from the heap that could be garbage collected: strings and arrays. For such heap-allocated memory entities, Jzero's in-memory representation includes the object's size in a word at the beginning. Under reference counting, a second word at the beginning can hold the number of references that point to that object. The representation for a string is shown in *Figure 17.1*:

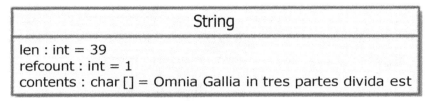

Figure 17.1: An in-memory representation of a string

In the example given, if `len` and `refcount` are 8 bytes each and there are 39 bytes of string data, `refcount` added 8 to a total size of 55 bytes (perhaps rounded to 56), so the addition of `refcount` is only a 14% overhead. But if the average string length were 3 bytes and you had billions of little strings to manage, adding a reference count represents a significant overhead that might limit the scalability of your language on big data. Given this representation, reference counting comes into play when objects get created in the first place, so let's look at example operations whose generated code involves heap allocation.

Reducing the number of heap allocations for strings

When an object such as `String` is created, memory must be allocated for it. In Java (and Jzero), all objects are allocated memory out of the heap. For strings, this results in an interesting situation in which the Java source code string constants are generally resolved at compile time to statically allocated addresses, but Java `String` heap objects are always allocated at runtime. Suppose the code is as follows:

```
String s = "hello";
```

On one hand, the memory contents of the `hello` string can be allocated in the static memory region. On the other hand, the Jzero `String` object that we assign to `String s` should be a class instance allocated from the heap that contains the length and reference count along with the reference to the character data. If we are naïve, the code we generate in this case might resemble the following:

```
String s = new String("hello");
```

If this code executes a billion times, we don't want to allocate a billion instances of the `String`, we only want one. In Java, the runtime system uses a string pool for string constants, so that it only needs to allocate one instance. Although our implementation uses them internally, Jzero does not provide programmers with user-level access to the full Java `String` or `Stringpool` classes, but we will put a static method named `pool()` in the Jzero `String` class that returns a reference to a String, allocating the instance if it is not already in the string pool. Given this method, the generated code can look more like the following:

```
String s = String.pool("hello");
```

This avoidance of allocating redundant `String` objects is facilitated by the fact that `String` objects are immutable. There are many ways that one might generate this code. I guess the easiest way that comes to mind is to distinguish the type of a `STRINGLIT` node from the type `String`, and add a type promotion rule that says `STRINGLIT` can be turned to `String`, inserting a tree node at each of these type promotion sites.

During a tree traversal, replace these type promotion nodes with the subtree that invokes the pool() method. Wherever a STRINGLIT node is being turned into a String, replace that subtree with the constructed set of nodes shown in *Figure 17.2*:

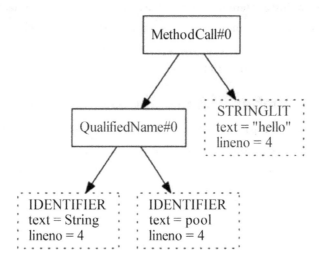

Figure 17.2: Substituting a STRINGLIT leaf for a call to the pool method

The code for a poolStrings() method that traverses the syntax tree and performs this substitution is shown below. The Unicon implementation in tree.icn is as follows:

```
method poolStrings()
    every i := 1 to *\kids do
        if type(\(\kids[i])) == "tree__state" then {
            if kids[i].nkids>0 then kids[i].poolStrings()
            else kids[i] := kids[i].internalize()
        }
    end
```

This method walks through the tree, calling an internalize() method to replace all leaves. The Java implementation of poolStrings() in tree.java is shown here:

```
public void poolStrings() {
    if (kids != null)
    for (int i = 0; i < kids.length; i++)
        if ((kids[i] != null) && kids[i] instanceof "tree") {
```

```
            if (kids[i].nkids>0) kids[i].poolStrings();
            else kids[i] = kids[i].internalize();
        }
    }
```

The tree method named `internalize()` in this traversal constructs and returns a subtree that invokes the `String.pool()` method if it is invoked on `STRINGLIT`. Otherwise, it just returns the node. In Unicon, the code looks as follows:

```
method internalize()
  if not (sym === "STRINGLIT") return self
  t4 := tree("token",parser.IDENTIFIER,
    token(parser.IDENTIFIER,"pool", tok.lineno, tok.colno))
  t3 := tree("token",parser.IDENTIFIER,
    token(parser.IDENTIFIER,"String", tok.lineno, tok.colno))
  t2 := j0.node("QualifiedName", 1040, t3, t4)
  t1 := j0.node("MethodCall",1290,t2,self)
  return t1
end
```

The corresponding code in Java looks like this:

```
public tree internalize() {
  if (!sym.equals("STRINGLIT")) return this;
  t4 = tree("token",parser.IDENTIFIER,
    token(parser.IDENTIFIER,"pool", tok.lineno, tok.colno));
  t3 = tree("token",parser.IDENTIFIER,
    token(parser.IDENTIFIER,"String", tok.lineno, tok.colno));
  t2 = j0.node("QualifiedName", 1040, t3, t4);
  t1 = j0.node("MethodCall",1290,t2,this);
  return t1;
}
```

This code in the compiler depends upon a runtime system function that implements the `String.pool()` method, using a hash table to avoid duplicates. Now, let's look at the code generation changes that are needed for the assignment operator.

Modifying the generated code for the assignment operator

Reference counting depends upon modifying the behavior of assignment to enable objects to track the references that point to them. In intermediate code for Jzero, there is an instruction named ASN that performs such an assignment. Our new reference counting semantics for the x = y assignment might consist of the following:

- If y points to an object, it gains a new reference. Increment its counter.
- If the old destination (x) points to an object, decrement its counter. If the counter is zero, free the old object.
- Perform assignment.

It is an interesting question whether this sequence of operations should be implemented by generating many three-address instructions for an assignment, or whether the semantics of the ASN instruction should be modified to do the bulleted items automatically when an ASN instruction executes. This is a design decision and there is no one answer that is always the best. The answer in some languages might be to add a new opcode such as OASN for object assignment.

Modifying the generated code for method call and return

Assignment is not the only point at which reference counts change. Every time objects are passed as parameters, under reference counting, their counts must be incremented to reflect the new object reference received in the called method. Worse yet, when a method returns, all the reference counts of parameters and local variables must be decremented. In the case of parameters, some of the same considerations for generated code apply as in the assignment discussion in the previous section; the start of each method might include a series of OASN instructions, one for each parameter. To handle method return, one implementation would be to use OASN to assign null values to each parameter and local variable prior to the RET instruction. Performance requirements might dictate optimization of this approach, since the speed of method call and return is important to overall performance. Adding one or two more specialty opcodes to streamline reference count updates may be well worth it. Having looked at how to implement this simplest of garbage collection methods, let's consider why reference counting might not be sufficient in most languages.

The drawbacks and limitations of reference counting

Reference counting has several downsides and one fatal flaw. One downside is that the assignment operator and method call and return are both made slower. For example, assignments must decrement counts of objects held prior to assignment and increment counts of objects being assigned. Suppose that triples the cost of an assignment.

This is a serious drawback, as assignment is a very frequent operation. Another downside is that the size of objects becomes larger to hold reference counts, which is unfortunate, especially for multitudinous small objects for which an extra counter is a significant overhead. Drawbacks are one thing; we can live with them if we must. But depending on your language requirements, some problems with reference counting may be deal breakers.

A more serious, perhaps fatal flaw is that even when it works, reference counting has no strategy for avoiding **fragmentation**; it may leave you with many tiny chunks of free memory, none of which are big enough to satisfy a new request. Another fatal flaw occurs if a chain of object references can have a cycle. This is a common practice in data structures. Many real-world problems are represented by graphs that can have cycles. In the case of a cycle, objects that point at each other never reach a reference count of 0, even after they are unreachable from the rest of the program. The diagram in *Figure 17.3* illustrates a circular linked list after it has become garbage. No outside pointer can ever reach this structure, but according to reference counting, the memory used by these objects is not reclaimable:

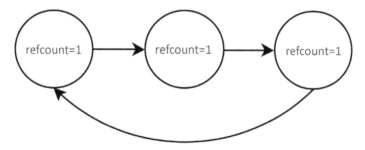

Figure 17.3: A circular linked list cannot be collected under reference counting

Despite these flaws, reference counting is relatively simple and easy, and it works well enough that it has served as the primary garbage collection method for some early Lisp implementations, for the Python language, and likely others. The Lisp community went on to pioneer many more robust garbage collection algorithms, and Python eventually implemented a more robust garbage collector in addition to continuing to use reference counting. Once you implement a more robust garbage collector, reference counting might be seen as unnecessary and wasteful of time and space. In any case, due to its fatal flaws, most general-purpose languages will not find reference counting sufficient, so let's look at an example of a more robust garbage collector, namely, the mark-and-sweep garbage collector used by the Unicon programming language.

Marking live data and sweeping the rest

This section gives an overview of the Unicon garbage collector, which is a mark-and-sweep style of garbage collector that was developed for the Icon language and then extended. It is written in (an extended dialect of) C, like the rest of the Icon and Unicon runtime system. Since Unicon inherited this garbage collector from Icon, much of what you see here is due to the many folks who implemented that language. All I did to it was add its "multiple regions" support, which seemed like a good idea at the time of 64KB heap limits on 640KB MS-DOS computers. Other aspects of this garbage collector are described in the book *The Implementation of Icon and Unicon: a Compendium.*

In almost all garbage collectors other than reference counting, the approach to collection is to find all the live pointers that are reachable from all the variables in the program; everything else in the heap is garbage. In a mark-and-sweep collector, live data is marked when it is found, and then all the live data is moved up to the top of the heap, leaving a big, beautiful pool of newly available memory at the bottom. The collect() C function from Unicon's runtime system is presented in outline form as follows:

```
int collect() {
   grow_C_stack_if_needed();
   markprogram();
   markthreads();
   reclaim();
}
```

Interestingly, the act of garbage collecting the heap begins with making sure we have enough C stack region memory to perform this task. Unicon has two stacks: the VM interpreter stack and the stack used by the C implementation of the VM. The necessity of growing the C stack was discovered the hard way. The reason for this is that the garbage collection algorithm is recursive, especially the operation of traversing live data and marking everything it points at. On some C compilers and operating systems, the C stack might grow automatically as needed, but on others, its size can be explicitly set. The garbage collector code does so by using an operating system function from the POSIX standard called setrlimit(). The code for growing the C stack looks like the following:

```
void grow_C_stack_if_needed() {
   struct rlimit rl;
   getrlimit(RLIMIT_STACK , &rl);
   if (rl.rlim_cur < curblock->size) {
```

```
        rl.rlim_cur = curblock->size;
        if (setrlimit(RLIMIT_STACK , &rl) == -1) {
           fprintf(stderr,"iconx setrlimit(%lu) failed %d\n",
                            (unsigned long)(rl.rlim_cur),errno);
           fflush(stderr);
           }
        }
   }
```

The preceding code checks how big the C stack is, and if the current block region is larger, it requests that the C stack be increased proportionally. This is overkill for most programs but corresponds roughly to the worst-case requirements. Fortunately, memory is cheap.

The main premise of the Unicon garbage collector is that frequent operations must be fast, even when that is at the expense of infrequent operations. In my presence, the famed computer scientist Ralph Griswold repeatedly remarked that most programs never garbage collect; they complete execution before they ever collect. This is only true from a certain point of view. It is true in a variety of application domains with short program executions, such as text processing utilities, and untrue in other application domains, such as servers and any other application that runs for an extended period.

Under the fast frequent operations doctrine, assignments are extremely frequent and must be kept as fast as possible – reference counting is a very bad idea for this reason. Similarly, memory allocations are quite frequent and must be as fast as possible. Garbage collection is infrequent, and it is OK for its cost to be proportional to the work involved.

To make matters more interesting, Icon and Unicon are specialty string- and text-processing languages, and the string data type is heavily special-cased in the implementation. Optimum efficiency for the string type might make some programs that are string-heavy perform extra well in this language, while other programs that are less string-centric do not benefit from these design decisions. Let's look at how Unicon manages memory differently for strings than for non-string data.

Organizing heap memory regions

Due to the important special case of strings, Unicon has two kinds of heaps. A general heap called the **block region** allows any data type other than strings to be allocated. A separate heap called the **string region** is maintained for string data.

Blocks are self-describing for garbage collection purposes; the layout of the block region is a sequence of blocks. Each block starts with a title word that identifies its type. Many block types are of fixed size; block types that are of varying size have a `size` field after the title word. *Figure 17.4* illustrates a block region. The rectangle on the left is a struct region that manages the block region (shown as the rectangle on the right). The block region being managed may be many megabytes in size, containing thousands or millions of blocks:

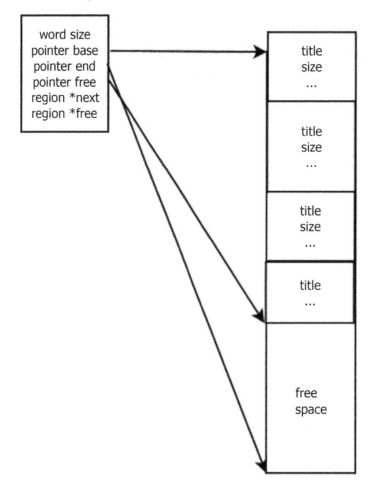

Figure 17.4: A block region

Within the block region, allocating is very fast. To allocate a block of size n for a class instance or other structure such as a list or table, just verify that n is less than the remaining free space between the pointers named `free` and `end`. In that case, the new block is located at the `free` pointer, and the region is updated to account for it by adding n to the `free` pointer.

In contrast to the block region, the string region is raw unstructured string data. String regions are organized as shown in *Figure 17.4*, except that the actual string data has no titles, sizes, or other structure – it is raw text. By not allocating strings as blocks like everything else, some common operations on strings, such as slices, are no-ops. Similarly, the string region can be byte-aligned with no wasted space when many small strings are allocated, unlike the block region, which is word-aligned. Also, data in the string region never contains any references to other live memory, so separating strings out from the block region reduces the total amount of memory within which references must be found.

At any given time, there is one current block region and one current string region from which memory may be allocated. Each program, and each thread, has its current block and string regions, which are the active regions within a bidirectional linked list of all heap regions allocated for that program or thread.

When the region is full and more memory is requested, a garbage collection of the current heaps is triggered. Older regions on the linked list are tenured regions and are only collected when a garbage collection on the current region fails to free enough memory for a request. When no region on the list can satisfy a request after all have been collected, a new current region of the same type is created and added to the linked list, generally twice as large as the previous one. Having discussed how Unicon's memory regions are organized, it is time to dig deeper into the details of how garbage collection determines what pieces of data are still in use, by traversing all the data that is reachable from the program's variables.

Traversing the basis to mark live data

In the first pass of garbage collection, live data is marked. All pointers to heap memory in the program must be found. This starts from a **basis set** of variables, consisting of global and static memory, and includes all local variables on the stack, which must be traversed. The heap objects pointed to by all these global and local variables are marked.

The task of marking live data in Unicon's runtime system is presented in an outline of its form in the following code example. The first two elements of the basis set consist of variables allocated on a per-program and per-thread basis. In Icon, there was only one program and one thread, so these were originally global variables. The Unicon virtual machine evolved to support multiple programs and, eventually, multiple threads. The current code uses a struct named progstate to hold all the variables that are maintained on a per-program basis. Under this organization, many global variables in Icon's runtime system became struct progstate fields, and finding all the basis variables in these categories became a series of data structure traversals to reach them all:

```
static void markprogram(struct progstate *pstate) {
```

```
        struct descrip *dp;
        PostDescrip(pstate->K_main);
        PostDescrip(pstate->parentdesc);
        PostDescrip(pstate->eventmask);
        PostDescrip(pstate->valuemask);
        PostDescrip(pstate->eventcode);
        PostDescrip(pstate->eventval);
        PostDescrip(pstate->eventsource);
        PostDescrip(pstate->AmperPick);
        PostDescrip(pstate->LastEventWin);/* last Event() win */
        PostDescrip(pstate->Kywd_xwin[XKey_Window]);/*&window*/
        postqual(&(pstate->Kywd_prog));
        for (dp = pstate->Globals; dp < pstate->Eglobals; dp++)
            if (Qual(*dp)) postqual(dp);
            else if (Pointer(*dp)) markblock(dp);
        for (dp = pstate->Statics; dp < pstate->Estatics; dp++)
            if (Qual(*dp)) postqual(dp);
            else if (Pointer(*dp)) markblock(dp);
    }
```

The task of marking all the global variables in a program is straightforward:

```
    static void markthreads() {
        struct threadstate *t;
        markthread(&roottstate);
        for (t = roottstate.next; t != NULL; t = t->next)
            if (t->c && (IS_TS_THREAD(t->c->status))) {
                markthread(t);
            }
    }
```

Each thread is marked by a call to markthread() as follows. Some of the pieces of the thread state contain things that do not contain references to heap variables, but those fields that might contain heap pointers must be marked:

```
    static void markthread(struct threadstate *tcp) {
        PostDescrip(tcp->Value_tmp);
        PostDescrip(tcp->Kywd_pos);
        PostDescrip(tcp->ksub);
```

```
      PostDescrip(tcp->Kywd_ran);
      PostDescrip(tcp->K_current);
      PostDescrip(tcp->K_errortext);
      PostDescrip(tcp->K_errorvalue);
      PostDescrip(tcp->T_errorvalue);
      PostDescrip(tcp->Eret_tmp);
   }
```

The actual marking process is different for strings and objects. Since Unicon variables can hold any type of value, a macro named PostDescrip() is used to determine whether a value is a string, another sort of pointer, or neither. References to strings are called qualifiers and they are marked using a function called postqual(). Other types of pointers are marked using the markblock() function:

```
#define PostDescrip(d) \
   if (Qual(d)) postqual(&(d)); \
   else if (Pointer(d)) markblock(&(d));
```

In order to interpret this macro, you need more than the postqual() and markblock() helper functions; you also need to know what the Qual() and Pointer() test macros are doing. A short answer would be that they perform a bitwise AND to check the value of a single bit within the descriptor word of a Unicon value. The value is a string if the descriptor word's topmost (sign) bit called F_Nqual is 0, but if that bit is 1, it is not a string and the other flag bits can be used to check other properties, of which the F_Ptr pointer flag would indicate that the value word contains a pointer – possibly a pointer to a value in the heap:

```
#define Qual(d) (!((d).dword & F_Nqual))
#define Pointer(d) ((d).dword & F_Ptr)
```

These tests are fast, but they are performed many times during garbage collection. If we came up with a faster way than shown in the PostDescrip() macro to identify values for the potential marking of live strings and blocks, it might affect the garbage collection performance significantly.

Marking the block region

For blocks, the mark overwrites part of the object with a pointer back to the variable that points at the object. If more than one variable points at the object, a linked list of those live references is constructed as they are found. This linked list is needed so that all those pointers can be updated to point at the new location when the object is moved.

The `markblock()` function is over 200 lines of code. It is presented in a summarized form in the following code example:

```
void markblock(dptr dp) {
    dptr dp;
    char *block, *endblock;
    word type;
    union block **ptr, **lastptr;
    block = (char *)BlkLoc(*dp);
    if (InRange(blkbase, block, blkfree)) {
        type = BlkType(block);
        if ((uword)type<=MaxType)
            endblock = block+BlkSize(block);
        BlkLoc(*dp) = (union block *)type;
        BlkType(block) = (uword)&BlkLoc(*dp);
        if ((uword)type <= MaxType) {
            ...traverse any pointers in the block
        }
    else ... handle other types of blocks that will not move
}
```

Traversing pointers within a block depends on how blocks are organized in the language. Pointers within a block are always a contiguous array. A global table within the garbage collector named `firstp` tells at what byte offset for each type of block its nested pointers can be found. A second global table named `firstd` tells at what byte offset for each block type its descriptors (nested values, which can be anything, not just a block pointer) are found. These are traversed by the following code within the `markblock()` function shown above:

```
            ptr = (union block **)(block + fdesc);
            numptr = ptrno[type];
            if (numptr > 0) lastptr = ptr + numptr;
            else
                lastptr = (union block **)endblock;
            for (; ptr < lastptr; ptr++)
                if (*ptr != NULL)
                    markptr(ptr);
            }
        if ((fdesc = firstd[type]) > 0)
```

```
        for (dp1 = (dptr)(block + fdesc);
            (char *)dp1 < endblock; dp1++) {
        if (Qual(*dp1)) postqual(dp1);
        else if (Pointer(*dp1)) markblock(dp1);
        }
```

The nested block pointers are visited by walking through the ptr variable and calling markptr() on each one. The markptr() function is similar to markblock() but may visit other types of pointers besides blocks. The nested descriptors are visited by walking through the dp1 variable and calling postqual() for strings and markblock() for blocks.

For strings, an array named quallist is constructed of all the live string pointers (including their lengths) that point into the current string region. The function named postqual() adds a string to the quallist array:

```
void postqual(dptr dp) {
    if (InRange(strbase, StrLoc(*dp), strfree)) {
        if (qualfree >= equallist) {
            newqual = (char *)realloc((char *)quallist,
                (msize)(2 * qualsize));
            if (newqual) {
                quallist = (dptr *)newqual;
                qualfree = (dptr *)(newqual + qualsize);
                qualsize *= 2;
                equallist = (dptr *)(newqual + qualsize);
            }
            else {
                qualfail = 1;
                return;
            }
        }
        *qualfree++ = dp;
    }
}
```

Most of the preceding code consists of expanding the size of the array if needed. The array size is doubled each time additional space is needed.

Furthermore, if the object contains any other pointers, they must be visited, and what they point at must be marked and traversed recursively following all pointers to everything that can be reached.

Reclaiming live memory and placing it into contiguous chunks

In the second pass of a garbage collection process, the heaps are traversed from top to bottom, and all live data is moved to the top. The overall reclamation strategy is shown in the following code. Note that garbage collection is complicated by concurrent threads – we do not consider concurrency here:

```
static void reclaim()
   {
   cofree();
   if (!qualfail)
      scollect((word)0);
   adjust(blkbase,blkbase);
   compact(blkbase);
   }
```

Reclaiming memory consists of freeing up unreferenced static memory consisting of co-expressions that have become garbage in a cofree() function, then moving all the live string data up in the scollect() function, and then moving the block data up within the block region by calling adjust() followed by compact().

The cofree() function walks through each co-expression block. These blocks cannot be allocated in the block region because they contain variables that cannot be moved. For this reason, co-expression blocks are allocated using C's malloc() function and must be freed explicitly using free(). The cofree() function consists of the following code:

```
void cofree() {
   register struct b_coexpr **ep, *xep;
   register struct astkblk *abp, *xabp;
   ep = &stklist;
   while (*ep != NULL) {
      if ((BlkType(*ep) == T_Coexpr)) {
         xep = *ep;
         *ep = (*ep)->nextstk;
         for (abp = xep->es_actstk; abp; ) {
            xabp = abp;
            abp = abp->astk_nxt;
            if ( xabp->nactivators == 0 )
```

```
            free((pointer)xabp);
        }
        free((pointer)xep);
        }
    else {
        BlkType(*ep) = T_Coexpr;
        ep = &(*ep)->nextstk;
        }
    }
}
```

The preceding code walks through a linked list of co-expression blocks. When the code visits a co-expression block whose title still says T_Coexpr, that indicates that the block was not marked as live. In that case, the co-expression and its associated bookkeeping memory blocks are freed using the standard free() library function.

The scollect() function collects the string region using the list of all the live pointers into it. It sorts the quallist array using the standard qsort() library function. Then, it walks through the list and copies live string data up to the top of the region, updating the pointers into the string region as the new locations of the strings are determined. Care is taken for pointers to overlapping strings so that they remain contiguous:

```
static void scollect(word extra) {
    char *source, *dest, *cend;
    register dptr *qptr;
    if (qualfree <= quallist) { strfree = strbase; return; }
    qsort((char *)quallist,
        (int)(DiffPtrs((char *)qualfree,
            (char *)quallist)) / sizeof(dptr *),
            sizeof(dptr), (QSortFncCast)qlcmp);
    dest = strbase;
    source = cend = StrLoc(**quallist);
    for (qptr = quallist; qptr < qualfree; qptr++) {
        if (StrLoc(**qptr) > cend) {
            while (source < cend) *dest++ = *source++;
            source = cend = StrLoc(**qptr);
        }
        if ((StrLoc(**qptr) + StrLen(**qptr)) > cend)
```

```
            cend = StrLoc(**qptr) + StrLen(**qptr);
        StrLoc(**qptr) = StrLoc(**qptr) +
                            DiffPtrs(dest,source)+(uword)extra;
        }
    while (source < cend) *dest++ = *source++;
    strfree = dest;
    }
```

The `adjust()` function is the first part of collecting the block region. It walks through the block region, moving pointers into the block region up to where the blocks will point. During marking, a linked list of all pointers to each block was constructed; this is used to update all those pointers to the block's new location. The source code for `adjust()` is shown next:

```
void adjust(char *source, char *dest) {
    register union block **nxtptr, **tptr;
    while (source < blkfree) {
        if ((uword)(nxtptr = (union block **)BlkType(source))>
            MaxType) {
            while ((uword)nxtptr > MaxType) {
                tptr = nxtptr;
                nxtptr = (union block **) *nxtptr;
                *tptr = (union block *)dest;
                }
            BlkType(source) = (uword)nxtptr | F_Mark;
            dest += BlkSize(source);
            }
        source += BlkSize(source);
        }
    }
```

The `compact()` function is the final step in collecting the block region, as shown in the following code block. It consists of moving the blocks of memory themselves up into their new location. The title words of the live blocks are used as the head of a linked list of all live references to that block, and cleared when the block is moved to its new location and live references are updated:

```
void compact(char *source) {
    register char *dest;
    register word size;
    dest = source;
```

```
    while (source < blkfree) {
        size = BlkSize(source);
        if (BlkType(source) & F_Mark) {
            BlkType(source) &= ~F_Mark;
            if (source != dest)
                mvc((uword)size,source,dest);
            dest += size;
            }
        source += size;
        }
    blkfree = dest;
    }
```

From this section, you should be able to conclude that a mark-and-sweep garbage collector is a non-trivial and relatively low-level undertaking. If you need encouragement, consider this: the work you do in building a garbage collector is for a good cause – it will save countless efforts from the programmers who use your language, and they will thank you for it. Many language inventors before you have implemented garbage collection successfully, and so can you.

Summary

This chapter showed you a lot about garbage collection. You learned what garbage is, how it comes about, and saw two very different ways to deal with it. The easy way, popularized by some early Lisp systems and early versions of Python, is called reference counting. In reference counting, the allocated objects themselves are made responsible for their collection. This usually works.

The more difficult form of garbage collection involves finding all the live data in the program and usually moving it to avoid memory fragmentation. Finding the live data is generally recursive, requires traversing stacks to find references in parameters and local variables, and is usually an onerous and low-level task. Many variations on this general idea have been implemented. One of the primary observations, which some garbage collectors exploit, is that most allocated objects are used for only a short time and then become garbage almost immediately. Collecting recently-allocated objects aggressively, and collecting longer-lived objects more and more infrequently as they age, has improved garbage collection performance in some modern language implementations.

Any method you employ to save programmers from having to manage their own memory will be greatly appreciated. In the next chapter, we will conclude the book with some thoughts on what we have learned.

Questions

1. Suppose a specific Unicon value, such as the `null` value, was particularly common when marking live data. Under what circumstances would it make sense to modify the `PostDescrip()` macro to check for that value to see if the tests in the `Qual()` and `Pointer()` macros can be avoided?

2. What would be the advantages and disadvantages of creating a separate heap region for each class type?

3. The `reclaim()` operation of Unicon's mark-and-sweep collector moves all the live non-garbage memory up to the top of the region. Would it be beneficial to modify this collector so that live data did not move?

Join our community on Discord

Join our community's Discord space for discussions with the authors and other readers:

`https://discord.com/invite/zGVbWaxqbw`

18

Final Thoughts

After learning so much about building a programming language, you may want to reflect on what was learned and think about areas you might want to study in more depth. This chapter reviews the main subjects presented in the book and gives you some food for thought by covering the following topics:

- Reflecting on what was learned from writing this book
- Deciding where to go from here
- Exploring references for further reading

The heart of this chapter suggests a number of avenues for further study. Bibliography details or links are provided for all those resources at the end. Let's start with what extra bonuses could be learned from this book.

Reflecting on what was learned from writing this book

I learned some useful things from writing this book. As an old-school C and UNIX person, this was not a given for me, but Java is very suitable for writing compilers at this point. Sure, Andrew Appel might have published *Modern Compiler Implementation in Java* in 1997, and other compiler-writing books in Java exist. These might be great, but how many compiler writers won't consider using Java to this day because Java is not a systems programming language, or because its official compiler construction tools are non-standard? Using a standard lex/YACC toolchain for Java makes it more interoperable with compiler code bases created for other languages.

I want to express my appreciation to the Byacc/J maintainer *Tomas Hurka* for accepting and improving my `static import` patch to make Byacc/J play more nicely with Jflex and similar tools (including my Merr tool, as covered in *Chapter 4, Parsing*) that generate `yylex()` or `yyerror()` in separate files. Supporting `yylex()` and `yyerror()` in separate files obviates the need for stupid workarounds, such as writing a stub `yylex()` method inside the parser class that turns around and calls `yylex()` generated in another file. Also, various small improvements to Java after its initial release, such as being able to use `String` values for `switch` cases, make a difference to the compiler writer. At this point, Java's conveniences and advantages, compared to C, almost outweigh its disadvantages, which are many. Java proponents: C programmers will not agree with you if you claim that Java's rigid package-and-class directory and file structure and lack of `#include` or `#ifdef` mechanisms are no big deal.

I didn't write this book to decide whether Java was good for compilers. I wrote this book to make Unicon great for compilers. This book's small miracle was finding a way to use the same lexical and syntax specifications for both Unicon and Java. I am happy that I was able to use both languages in the same way I would traditionally write a compiler in C. After that vital bit of lex/YACC specification sharing, Unicon didn't provide as much added advantage as I had expected compared to Java. Unicon skips many of Java's pain points, is somewhat more concise, and has an easier time with heterogeneous structure types. Ultimately, both languages were great for writing the Jzero compiler, and I'll let you be the judge of which code was more readable and maintainable. Now, let's consider where you might go from here.

Deciding where to go from here

You may want to study more advanced work in any number of areas. These include programming language design, bytecode implementation, code optimization, monitoring and debugging program executions, programming environments such as IDEs, and GUI builders. In this section, we will explore just a few of these possibilities in more detail. This section reflects many of my personal biases and priorities.

Studying programming language design

It is probably more difficult to identify strong works in programming language design than most of the other technical topics mentioned in this section. Harold Abelson and Gerald Sussman wrote a book called *Structure and Interpretation of Computer Programs*, which is widely reputed to be useful. It is not a programming language design book, but its insights occasionally delve into that subject.

Browsing casually, you may find many general programming language books. Rafael Finkel's *Advanced Programming Language Design* is one, covering a range of advanced topics. For other sources, language design books and papers written by actual language inventors have the potential to be more real and useful than books written by third parties.

One of the luminaries of language design, Niklaus Wirth, wrote many influential books. *Algorithms and Data Structures*, as well as *Project Oberon*, provide valuable insights regarding language design, as well as implementation. As the designer of several successful languages, including Pascal and Modula-2, Niklaus Wirth has great authority in arguing for simplicity with language designs that protect programmers from themselves. He is a giant on whose shoulders a person would do well to stand.

The **Prolog** programming language has produced rich literature describing many of the design and implementation problems that have been addressed for that language and logic programming in general. Prolog is important because it features *extensive implicit backtracking*. One of the important works on Prolog is *The Art of Prolog*, by Leon Sterling and Ehud Shapiro. Another important contribution is the **Byrd box model** of functions, in which, instead of understanding a function's public interface as a call followed by a return, a function is seen as having a call, producing a result, and being resumed repeatedly, until an eventual failure.

The next great programming language family that deserves attention is **Smalltalk**. Smalltalk did not invent the object-oriented paradigm, but it purified it and popularized it. A summary of some of its design principles was published in Byte magazine in an article titled *Design Principles Behind Smalltalk*, by Dan Ingalls – `http://worrydream.com/refs/Ingalls%20-%20Design%20Principles%20Behind%20Smalltalk.pdf`. While considering object-oriented languages, it is also prudent to consider semi-object-oriented languages such as C++, for which the book *Design and Evolution of C++* by Bjarne Stroustrup is of value.

The dramatic rise in popularity of very high-level languages such as Python and Ruby is one of the most important developments in recent decades. It is depressing how poorly represented many extremely popular languages are overall in the programming language design literature. TCL's inventor, John Ousterhout, wrote two important works on topics related to the design of very high-level languages. *Scripting: Higher-Level Programming for the 21st Century* is a good paper, albeit one that reflects its author's biases. Ousterhout also gave an important invited talk, humorously titled *Why Threads Are a Bad Idea*, arguing for event-driven programming and synchronous coroutines in preference to threads for most parallel workloads.

The Icon and Unicon languages are two more heavily documented examples of very high-level languages. The design of the Icon language is described in Griswold's *History of the Icon Programming Language*. Having looked at some fine options for studying language design further, let's consider options for studying their implementation.

Learning about implementing interpreters and bytecode machines

Advanced programming language implementation topics should include implementing all types of interpreters and runtime systems for advanced programming languages with novel semantics. The first very high-level language was **Lisp**. Lisp inventor John McCarthy is credited with inventing a mathematical notation that could be executed on the computer. Lisp is one of the first interactive interpreters, and arguably the first just-in-time compiler. Other Lisp implementors have written notable books. One of special note is John Allen's *Anatomy of Lisp*.

Any description of bytecode machines would be remiss if it omitted the Pascal bytecode machines. Many of the seminal works on Pascal's implementation are collected in *PASCAL: The Language and Its Implementation*, edited by David Barron. The UCSD Pascal system that popularized bytecode machines was based on the work of Urs Ammann at ETH Zurich, which is well represented in Barron's book. Another significant work on Pascal is Steven Pemberton and Martin Daniels' *Pascal Implementation: Compiler and Interpreter, which has the virtue of being a publicly available resource* – https://homepages.cwi.nl/~steven/pascal/.

A collection of books authored by Adele Goldberg and her collaborators document Smalltalk. The documentation that Goldberg et al. produced for this pure object-oriented language defined a high bar for language reference materials. Their work includes *Smalltalk-80: The Language and its Implementation*.

In the logic programming world, the **Warren Abstract Machine (WAM)** is one of the premier means of reasoning about the underlying semantics of Prolog and how to implement it. It is described in *An Abstract PROLOG Instruction Set*.

The implementation of Unicon is described in *The Implementation of Icon and Unicon: a Compendium*. This book combines and updates several previous works on the implementation of the Icon language, plus descriptions of the implementation of various subsystems that have been added to Unicon.

Acquiring expertise in code optimization

Code optimization is generally a subject of advanced graduate-level textbooks on compilers. The classic *Compilers: Principles, Techniques, and Tools* contains substantial documentation on various optimizations. Cooper and Torczon's *Engineering a Compiler* is a more recent treatment.

Code optimization for higher-level languages often requires more novel techniques. Various works on optimizing compilers for very high-level languages seem to suggest some unknown law that I will formulate here: it takes around 20 years for people to figure out how to execute a new very high-level language efficiently. For hints of this, I refer to examples such as *T: a Dialect of Lisp* and *The Design and Implementation of the SELF Compiler*, which came out 20 years after the Lisp and Smalltalk languages. Of course, how long it takes depends on the size and complexity of the language. I am biased, but one of my favorite works for such languages is Ken Walker's dissertation *The implementation of an optimizing compiler for Icon*, which is included in the Icon and Unicon implementation compendium.

Monitoring and debugging program executions

There are lots of books about debugging end user code, but there are few books on how to write program monitors and debuggers. Part of the problem is that the implementation techniques are low-level and highly platform-dependent, so much of what is written about debugger implementation might only be true for one particular operating system and may not be true in 5 years.

Regarding the big picture, you may want to think about how to design your debugger and what interface it should provide to the end user. Besides imitating the interface of mainstream debuggers, you should consider the notion of query-based debugging, as described in Raimondas Lencevicius' *Advanced Debugging Methods*. You should also consider the notions of relative debugging and delta debugging, which were popularized by the works of David Abramson et al. and Andreas Zeller.

One of the things you may want to read up on if you want to know more about debugger implementation is executable file formats and their debugging symbol information. The Microsoft Windows portable executable format is documented on the Microsoft website.

One of the most prominent corresponding UNIX formats is the **Executable and Linkable Format (ELF)**, which stores debugging information in a format called **Debugging With Arbitrary Record Formats (DWARF)**. These formats are dated but still in use, with some newer formats also on the scene.

The GNU debugger, known as GDB, is prominent enough that it has a *GDB Internals* manual, and GDB has frequently been used as the basis upon which research debugging capabilities are developed. Building experimental debugger features atop an existing mainstream debugger is a very different task than building a new debugger from scratch, or building a debugging and monitoring framework. `https://aarzilli.github.io/debugger-bibliography/` lists a few other debugger implementation resources, mainly oriented toward the Go language.

For a substantial discussion of the classic (a.k.a. ancient history) program execution monitoring literature, you can consult *Monitoring Program Execution: A Survey,* or the related work chapter in the book *Program Monitoring and Visualization.*

Designing and implementing IDEs and GUI builders

One element in the success of programming languages is the extent to which their programming environments support writing and debugging code. This book only briefly touches on these topics, and you might want to explore more on how IDEs and their user interface builders are implemented.

There is good news and bad news here. The bad news is that no one has written a *build your own integrated development environment* book. If you were going to build one from scratch, you might start by teaching yourself how to write a text editor, and then add other features. In that case, you might wish to consult *The Craft of Text Editing* by Craig Finseth. That book was written by a person who studied how the Emacs text editor was implemented for his Bachelor's thesis. There is also an appendix to the GNU Emacs Manual titled *GNU Emacs Internals.* You might want to check out *Smalltalk-80: The Interactive Programming Environment,* by Adele Goldberg.

The good news is that almost no one need write the text editor portion of an integrated development environment anymore. Each of the major graphical computing platforms comes with a user interface library that includes a text editor as one of its widgets. You can assemble the interface of an integrated development environment using a *graphical interface builder tool.* Unfortunately, graphic user interface libraries are usually non-portable and short-lived, which means that work spent programming on them is almost doomed to be discarded within a decade or two. It takes extraordinary effort to write code that runs on all platforms and it is nigh impossible to write graphics code that will live forever in internet years.

You may be able to find more books about multi-platform portable graphical user interface libraries and how to use them to write integrated development environments and user interface builder tools. Java is one of the most portable languages, and despite some false starts, it is still likely that some of the best, most multiplatform portable, and long-lived user interface libraries today might be Java libraries. Unfortunately, I don't know of a good book on implementing IDEs in Java.

Exploring references for further reading

Here is a detailed bibliography of the works discussed in the previous sections. Within each sub-section, the works are listed alphabetically by author.

Studying programming language design

In the area of programming language design, you may find the following items to be of interest:

- Harold Abelson and Gerald Sussman, *Structure and Interpretation of Computer Programs*, Second edition, MIT Press, 1996.

- Rafael Finkel, *Advanced Programming Language Design*, Pearson 1995.

- Ralph Griswold, *History of the Icon Programming Language*, Proceedings of HOPL-II, ACM SIGPLAN Notices 28:3 March 1993, pages 53–68.

- Daniel H.H. Ingalls, *Design Principles Behind Smalltalk*, Byte Magazine August 1981, pages 286–298.

- John Ousterhout, *Scripting: Higher-Level Programming for the 21st Century*, IEEE Computer 31:3, March 1998, pages 23–30.

- John Ousterhout, *Why Threads Are a Bad Idea (for most purposes)*, Invited talk, USENIX Technical Conference, September 1995 (available at https://web.stanford.edu/~ouster/cgi-bin/papers/threads.pdf).

- Leon Sterling and Ehud Shapiro, The *Art of Prolog*, MIT Press, 1986.

- Bjarne Stroustrup, *The Design and Evolution of C++*, Addison-Wesley, 1994.

- Niklaus Wirth, *Algorithms and Data Structures*, Prentice Hall 1985.

- Niklaus Wirth, *Project Oberon: The Design of an Operating System and Compiler*, Addison Wesley/ACM Press 1992.

This is a tiny sample of the best works in a rich body of literature, and it will certainly contain many grievous omissions. Now, let's look at a similar list for implementation.

Learning about implementing interpreters and bytecode machines

In the area of interpreter and bytecode machine implementation, you may find the following items to be of interest:

- John Allen, *Anatomy of Lisp*, McGraw Hill, 1978.

- Urs Ammann, *On Code Generation in a PASCAL Compiler*, Software Practice and Experience 7(3), 1977, pages 391–423.

- David W. Barron, ed., *PASCAL-The Language and Its Implementation*, John Wiley, 1981.

- Adele Goldberg, David Robson, *SmallTalk-80: The Language and its Implementation*, Addison-Wesley, 1983.

- Clinton Jeffery and Don Ward, eds., *The Implementation of Icon and Unicon: a Compendium*, Unicon Project, 2020 (available at `http://unicon.org/book/ib.pdf`).

- A. B. Vijay Kumar, *Supercharge Your Applications with GraalVM*, Packt, 2021.

- Steven Pemberton and Martin Daniels, *Pascal Implementation: The P4 Compiler and Interpreter*, Ellis Horwood, 1982 (available at `https://homepages.cwi.nl/~steven/pascal/`).

- David Warren, *An Abstract PROLOG Instruction Set*, Technical Note 309, SRI International, 1983 (formerly available at `http://www.ai.sri.com/pubs/files/641.pdf`; now you have to get it from the Internet Archive's Wayback Machine).

Now, let's look at a similar list for native code and code optimization.

Acquiring expertise in native code and code optimization

Regarding native code and code optimization, you may find the following items to be of interest:

- Al Aho, Monica Lam, Ravi Sethi, and Jeffrey Ullman, *Compilers: Principles Techniques and Tools*, Second edition, Addison Wesley, 2006.

- Craig Chambers, *The Design and Implementation of the SELF Compiler, an Optimizing Compiler for Object-Oriented Programming Languages*, Stanford dissertation, 1992.

- Keith Cooper and Linda Torczon, *Engineering a Compiler*, Second edition, Morgan Kaufmann, 2011.

- Chris Lattner and Vikram Adve, LLVM: A Compilation Framework for Lifelong Program Analysis & Transformation, in *Proceedings of the 2004 International Symposium on Code Generation and Optimization (CGO'04)*, Palo Alto, California, March 2004. Available at `https://llvm.org/pubs/2004-01-30-CGO-LLVM.html`.

- Jonathan Rees and Norman Adams, *T: a dialect of Lisp*, Proceedings of the 1982 ACM symposium on LISP and functional programming, pages 114–122.

- Kenneth Walker, *The implementation of an optimizing compiler for Icon*, University of Arizona dissertation, 1991.

After optimization, you might want to look further at the highly specialized area of program execution monitoring and debugging.

Monitoring and debugging program executions

In the area of monitoring and debugging, you may find the following items to be of interest:

- David Abramson, Ian Foster, John Michalakes, and Roc Sosic, *Relative Debugging: A new methodology for debugging scientific applications*, Communications of the ACM 39(11), November 1996, pages 69–77.

- DWARF Debugging Information Format Committee, *DWARF Debugging Information Format Version 5* (`http://www.dwarfstd.org`), 2017.

- John Gilmore and Stan Shebs, *GDB Internals*, Cygnus Solutions, 1999. The most recent copy is in wiki format and available at `https://sourceware.org/gdb/wiki/Internals`.

- Clinton Jeffery, *Program Monitoring and Visualization*, Springer, 1999.

- Raimondas Lencevicius, *Advanced Debugging Methods*, Kluwer Academic Publishers, Boston/Dordrecht/London, 2000.

- Microsoft, PE Format, available at `https://docs.microsoft.com/en-us/windows/win32/debug/pe-format`.

- Bernd Plattner and J. Nievergelt, *Monitoring Program Execution: A Survey. IEEE Computer*, Vol. 14. November 1981, pages 76–93.

- Andreas Zeller, *Why Programs Fail: A Guide to Systematic Debugging*, Second edition, Morgan Kaufmann, 2009.

Along with monitoring and debugging, it would be useful to consider integrated programming tools for your language.

Designing and implementing IDEs and GUI builders

In the area of development environments and user interface builders, you may find the following items to be of interest:

- Craig Finseth, *The Craft of Text Editing: Emacs for the Modern World. Springer*, 1990.

- Adele Goldberg, Smalltalk-80: The Interactive Programming Environment, Addison-Wesley, 1983.

- Bil Lewis, Dan LaLiberte, Richard Stallman, the GNU Manual Group, et al., GNU Emacs Internals, *GNU Emacs Lisp Reference Manual, Appendix E*. GNU Project, 1990–2021.

Honestly, I wish I had more good reading to recommend on the subject of IDEs and GUI builders. If you know of good works on this subject, send me your suggestions. Similarly, we might want to study available books on integrating new languages into existing IDEs. At least one such book exists:

- Nadeeshaan Gunasinghe and Nipuna Marcus, *Language Server Protocol and Implementation: Supporting Language-Smart Editing and Programming Tools*, Apress, 2021.

Now, let's wrap things up with a summary.

Summary

This book showed you a thing or two about building programming languages. We did this by showing you an implementation of a toy language called Jzero. However, Jzero is not what is interesting; what is interesting is the tools and techniques used in its implementation. We even implemented it twice!

If you thought that maybe programming language design and implementation was a swimming pool to enjoy, your new conclusion might be that it is more like an ocean. If so, the tools that have been placed at your disposal in this book, including versions of flex and YACC for use with Unicon or Java, are a luxury cruise liner capable of sailing you about on that ocean to wherever you want to go.

The first high-level language compiler is said to have taken 18 person-years to create. Perhaps now it is a task of a few months, although it is still an open-ended task where you can spend as much time as you can spare making improvements to any compiler or interpreter that you care to write.

The holy grail of compilers has long been a high-level declarative specification of the code generation problem, to match the declarative specification of lexical and syntax rules. Despite the earnest work of many people far smarter than me, this hoped-for breakthrough has been resistant. In its place, several crutches have proliferated. The very notion of a bytecode machine implemented in a portable system language such as C has made many languages portable to a myriad of processors... once someone ports a C compiler to them. This has become part of the mainstream due to technologies such as the .NET CLR and the JVM and GraalVM Java bytecode machines. Similarly, transpilers that generate code in the form of source code to another high-level language (such as C) have become widespread.

The third form of increased portability that's available to programming language inventors is the proliferation of intermediate-level target instruction formats such as LLVM. All of these widely used means of making your programming language portable dodge the common bullet of generating code for a new CPU. Perhaps the fourth form of increased portability has come from the fact that new CPU instruction sets are generated infrequently at this point, as the industry has collectively invested so much in the small number of hardware instruction sets for which optimizing code generators are available.

Thanks for reading this book. I hope that despite its many shortcomings, you were able to enjoy my book and that you found it useful. I look forward to seeing what new programming languages you invent in the future!

Join our community on Discord

Join our community's Discord space for discussions with the authors and other readers:

```
https://discord.com/invite/zGVbWaxqbw
```

Section IV

Appendix

This section will include materials that will help the readers to understand the main text.

This section comprises the following chapters:

- *Appendix, Unicon Essentials*
- *Answers*

Appendix: Unicon Essentials

This appendix presents enough of the Unicon language to help you understand the Unicon code examples in this book. This appendix is intended for experienced programmers and does not spend time introducing basic programming concepts. Instead, it presents Unicon while focusing on its interesting or unusual features compared to mainstream languages.

If you know Java, then most of the Unicon code in this book can be understood by looking at the corresponding Java code to see what is going on. You can look up whatever is not self-evident and not explained by Java comparison here. This appendix is not a complete Unicon language reference; for that, see *Appendix A of Programming with Unicon*, which is available in standalone public domain form in *Unicon Technical Report #8*. Both *Programming with Unicon* and *Unicon Technical Report #8* are hosted at unicon.org.

Syntactic shorthand

The notation in this appendix uses square brackets, [], to denote optional features and asterisks, *, to denote features that can occur zero or more times. When square brackets or asterisks are highlighted, this means they are occurring in the Unicon code rather than as optional or repeated features.

This appendix covers the following topics:

- Running Unicon
- Using Unicon's declarations and data types
- Evaluating expressions
- Debugging and environmental issues
- Function mini-reference
- Selected keywords

To begin, let's provide an expanded discussion of how to run Unicon programs.

Running Unicon

Unicon utilizes compile and link steps to translate source code programs into executable code. In this way, it is more like C than Java. Unicon source files end in the four characters .icn; it shares this extension with its predecessor, Icon.

Unicon object files are called ucode and end with the two characters .u. Here are some example invocations of the Unicon translator. Only the most common command-line options are presented:

- `unicon mainname [ filename(s) ]`

  Compile and link `mainname.icn` and other filenames to form an executable named `mainname.exe` on Windows or just `mainname` on most other platforms. The other filenames may have the extension .icn or .u; if no extension is provided, .icn is automatically added.

- `unicon -o exename [ filename(s) ]`

  The -o option directs the translator to compile and link an executable named exename, or on Windows, `exename.exe`. It's the same as the previous example, except the output filename is supplied explicitly instead of being inferred from the first filename given on the command line.

- `unicon -c filename(s)`

  The -c option directs the Unicon compiler to compile .icn files into .u files but not link them. No executable is produced.

- `unicon -u filename(s)`

  The -u option directs the Unicon translator to warn about undeclared variables. By default, undeclared variables become local.

- `unicon -version`

  Print the Unicon version, including a build date. No compilation, linking, or execution is performed.

- `unicon -features`

  Print the features of this Unicon build. When Unicon is built, many language features are only present if the underlying platform has header files and libraries available. This includes file compression and image formats, graphics and other kinds of I/O, and concurrency-related features.

- `unicon foo -x`

  The -x option directs the Unicon translator to both compile and run `foo.icn` in a single step.

You can read a more detailed description of how to run Unicon on Windows at http://unicon. org/utr/utr7.html. The full list of command-line options can be found at http://unicon.org/ utr/utr11.html.

If you don't like working from the command line, you may want to try out the Unicon IDE called ui. The ui program has options to compile and execute programs from inside a graphical interface. The following screenshot shows an example of this:

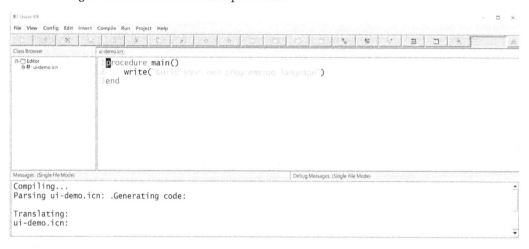

Figure A.1: A screenshot showing ui, the Unicon IDE

Developers in the Unicon community use many different programming environments, and the Unicon IDE is more of a technology demo than a production tool, but you may find it useful, if only for its beloved Help menu. It is written in about 10,000 lines of Unicon, not counting the GUI class libraries. Now, let's consider the kind of declarations that are allowed in Unicon and what data types it supports.

Using Unicon's declarations and data types

You can't write a Unicon program without declaring things. Declaring something is the act of associating a name, visible within some scope and lifetime, with some chunk of code or memory capable of holding a value. Next, let's learn how different program components can be declared.

Declaring program components

Unicon programs consist of one or more procedures beginning with main(). Program structure may also include classes. Unicon distinguishes user-defined procedures from functions that are built into the language. The following patterns show the syntax structure for the primary declarations of bodies of code in Unicon's procedures and methods:

- Declare procedure or method:

```
procedure X ( params ) [locals]* [initial] [exprs]* end
method X ( params ) [locals]* [initial] [exprs]* end
```

All procedures and methods have a name, zero or more parameters, and a body ending with the word end. The body may optionally start with local and static declarations and an initial section, followed by a sequence of expressions. Methods may only be declared inside classes, within which they have access to the names of class fields, the other methods of a class, and any superclasses. In Unicon, there are no static methods, and all methods are public.

- Declare field names or parameters:

```
[ var  [ , var  ]* ]
[ var [ : expr ]  [ , var [ : expr] ]* [ , var [] ] ]
```

Field names introduced in declaring classes or records are a list of names, separated by commas. Parameters for procedures and methods, including initially methods, are similar but have some extra options. Each of the zero or more parameters may optionally include a colon, followed by a default value or the name of a type coercion function (integer, real, string...). The final parameter may be followed by square brackets, indicating that a variable number of arguments will be passed in as a list.

- Declare globals, locals, and statics:

```
global variable [ , variable ]*
local variable [ := expr ] [ , variable [ := expr ]]*
static variable [ := expr ] [ , variable [ := expr ]]*
```

Variables are declared with a comma-separated list of names in one of the three scopes: global, local, or static. Local names can include an assignment to initialize the variable. Global variables live for the entirety of the program's execution. Local variables live for the duration of a single procedure or method call. Static variables live for the entirety of the program's execution, and one copy of each static variable is shared by all the calls to that procedure or method.

- Declare record or class type:

```
record R ( fields )
class C [ : super ]* ( fields ) [ methods ]* [ initially ] end
```

A record or class is declared by the corresponding reserved word followed by a name and a comma-separated list of field names, surrounded by parentheses. A record or class declaration declares a global variable that refers to the constructor function that creates instances of that record or class.

A class may also have a colon-separated list of superclass names prior to its field names. A class declaration contains zero or more methods and an optional `initially` section, followed by a reserved word end.

- Declare `initially` method:

```
initially [ ( params ) ] [locals]* [initial] [exprs]*
```

An `initially` section, also called an `initially` method, is a special, optional initialization method for a class that is called automatically by the class constructor. If an `initially` section is present, it must be after all other methods, immediately before the end of the class. It is not preceded by the word `method` and its parameter list is entirely optional, with no parentheses required when there are no parameters. The concept of an `initially` section in a class is related but distinct from the `initial` section that optionally may appear at the beginning of a procedure or method; an `initial` section identifies code that executes the first time that method is called.

- Reference library modules:

```
link module [ , module ]*
```

Unicon programs may include multiple files on the command line, but separately compiled modules that are used by a file may also be declared in the source code. Modules may either be string constant filenames or identifiers that are used as filenames. The extension .u is implied and generally omitted.

- Use package:

```
import package [ , package ]*
```

Unicon's global namespace may be composed of multiple named packages, which can be imported by supplying the package name(s). Now, let's look at Unicon's data types.

Using atomic data types

Unicon has a rich set of data types. Atomic types are immutable, while structure types are mutable. They appear directly in the source code as literal constant values or are computed and introduced into the program by operators or functions. When possible, Unicon coerces the various atomic types as needed to perform an operation. For example, `1+"2"` computes 3.

Numeric

Integers are signed arbitrary precision whole numbers. Integers are the most common type and work in an obvious way. There are literal formats in bases 2 through 36. The bases are expressed in the form baseRvalue, as in 16Rffff. There are also a set of suffixes such as K and M that multiply numbers by thousands or millions. For example, the integer literal 4G indicates a value of four billion. Integers in Unicon mostly just work without us paying much attention. All the usual arithmetic operators are provided, along with a handy exponentiation operator, x^y. The interesting unary operator, ?n, produces a random number between 1 and n. The unary operator, !n, generates integers from 1 to n.

The real data type provides floating-point approximations of real numbers. Real constants must contain either a decimal, an exponent, or both. The expressions 3.14, 2e9, and 5.3e24 are all examples of real constants. It is kind of amazing to think how much trouble real values used to cause programmers, and how they are now taken for granted: real values are the same size as 64-bit integers, although the binary format is different. One of the challenges you occasionally face is converting back and forth between integers and reals. Conversion is performed automatically as needed, but it does take time and potentially loses precision, especially if you do it repeatedly and unnecessarily.

Textual

Unicon has multiple built-in types for manipulating text, including strings, **character sets** (csets), and a rich **pattern type** borrowed from SNOBOL4. This book uses strings and csets but uses Flex and Yacc instead of patterns, so that the examples also compile and run under Java. For this reason, this appendix will not present the pattern type or its regular expression-based literal format.

Strings are ordered sequences of zero or more characters. A string literal is surrounded by double quotes and may include escape characters. The following table shows these escape sequences:

Code	Character	Code	Character	Code	Character	Code	Character
\b	backspace	\d	delete	\e	escape	\f	form feed
\l	line feed	\n	newline	\r	carriage return	\t	tab
\v	vertical tab	\'	quote	\"	double quote	\\	backslash
\ooo	octal	\xhh	hexadecimal	\^x	Control-x		

Table A.2: String and cset escape characters

Csets are unordered collections of zero or more non-duplicating characters. A cset literal is surrounded by single quotes and may include escape characters. Unicon has many cset keyword constants for predefined character sets that are sometimes found as macros or test functions in other languages. It turns out that having a full set data type for characters is useful when you're performing text processing, and considerably more general than having a few predefined macros or functions. The cset type supports the usual set operators, such as c1++c2, which computes a union, c1**c2, which computes an intersection, or c1--c2 for characters in c1 but not in c2. Now, let's move on and look at Unicon's structure types.

Aggregating multiple values using structure types

Structure types allow the construction of values that are composed of multiple values. In Unicon, structure types are generally mutable: the values within them can be modified or replaced. They are created at runtime by an action that allocates memory and initializes it from their component values. Many structure types are containers that allow values to be inserted or deleted. Structures can be constructed that contain other structures, including values of the same type; these are called recursive structures. The first structure type to consider is the class, which introduces a user-defined structure type.

Classes

If your data is not entirely numeric and not just textual, you probably want to write a class for it in Unicon. Each Unicon class is a new data type. Class types are used for values that represent entities or information from the application domain. They are usually used for things that contain several pieces of information, governed by complex behavior.

Classes are often defined as sets of additions or changes to other classes by means of inheritance. Unicon defines multiple inheritance semantics in an interesting way called **closure-based inheritance**, which allows cycles in the inheritance graph. The fact that Unicon classes are all public and all virtual keeps things simpler and focuses on expressive power rather than on protecting programmers from themselves. Now, let's look at Unicon's other structure types, which are often used to provide the associations between different class types. The first built-in structure type to consider is the list type.

Lists

This book presented classes before lists just to tease you a bit. In most programs, the **list** is the most common structure type. This book only shows a small inkling of what lists can do. Unicon lists are a wonderful cross between a linked list and an array that can grow and shrink and be used as a stack, queue, or deque. Internally and invisibly, the list type supports special representations for arrays of integers and arrays of real numbers that optimize their space representation compatibly with C.

In addition to being used as arrays or stacks and such, lists are commonly used as glue data structures within classes to implement aggregation and multiplicity relationships. Here is a warning for mainstream programmers about Unicon lists: their first subscript is 1, not 0. Like BASIC, Unicon and Icon are 1-based because that is more natural for humans, while 0-based indexing is arguably a concession that other languages make in order to be more natural for machines. Now let's compare the list type with the amazingly useful table data type.

Tables

A **table**, sometimes called a **dictionary** or an **associative array**, is an extremely flexible structure that corresponds to the mathematical concept of a function. A table maps keys from some domain onto values with some range. The Unicon table data type is named after its implementation, which is usually a hash table. A table feels like an array whose subscripts are not constrained to be contiguous integers starting from 1.

Any of Unicon's types may be used as keys, but in practice, strings and integers are almost the only types that are used as hash keys. Sure, you can use real numbers, but round-off errors make subsequent lookups tricky. And you can use csets as table keys; it is just very rare. If you use other structure values as keys in a table, everything works, but you don't compute their hash from their contents, because the contents are mutable. Without using the contents to hash and look up values, table performance does not scale well.

Sets

The **set** type is like a table in which the keys do not map onto anything. Keys may be inserted and deleted, and any value is either a member of the set or it is not. Operations such as union and intersection are supported. As with the table type, the keys in a set may be of arbitrary and mixed types.

Files

The **file** type is what you would expect. Files generally access persistent storage managed by the operating system. There are handy functions for processing lines at a time. Most forms of input and output are extensions of the file type, so most file functions can be applied to network connections, graphics windows, and so on.

Other types

Unicon has a host of other powerful built-in types for things such as windows, network connections, and threads. Unlike some languages, it does not have a global interpreter lock to slow its thread concurrency down. Given values in this rich set of data types, the bodies of Unicon programs are assembled into computations using various expressions.

Evaluating expressions

Unicon expressions are goal-directed. When they can, they compute a result, and this is called **success**. Expressions that have no result are said to **fail**. Failure will generally prevent a surrounding expression from being performed. **Generators** are a special category of Unicon expressions that are capable of computing more than one result; they are described in a section later in this chapter.

Goal-directed evaluation semantics with generators eliminates the need for a Boolean data type, which is usually found in other languages. It also dramatically increases the expressive power of the language, avoiding the need for a lot of tedious checking for sentinel values or writing explicit loops to search for things that can be found by goal-directed evaluation and backtracking. It takes time to get used to this feature, but once mastered, code is shorter and quicker to write.

Forming expressions using operators

Many of Unicon's operators will be familiar from other languages, but some operators are unique. Here is a summary of Unicon's operators. When chained together, the execution order of operators is determined by their precedence, which is generally as found in mainstream languages. Unary operators have higher precedence than binary, multiplication comes before addition, and so forth. When in doubt, force precedence using parentheses:

- Force precedence:

```
( exp )
```

Parentheses with no expression in front of them just force operator precedence and otherwise have no effect.

- Size:

  ```
  * x : int
  ```

  A unary asterisk is a size operator that returns how many elements are in a string, cset, queue, or structure, *x*.

- Is-null and is-nonnull:

  ```
  / x
  \ x
  ```

  These predicates just produce *x* if the test is true, but if the test is false, then they fail.

- Negate: and Unary plus:

  ```
  - num
  + num
  ```

  To negate a number is to flip its sign from positive to negative or vice versa. A unary plus operator coerces the operand into becoming a number but does not change its numeric value.

- Negate evaluation result:

  ```
  not exp
  ```

  A not converts an expression's, *exp*, success into a failure and vice versa. When it succeeds, the result that's produced is the null value.

- Tabmat:

  ```
  = str
  ```

  When the operand is a string, the unary equals is like `tab(match(s))`.

- Binary arithmetic (in three levels of decreasing precedence):

  ```
  num1 ^ num2
  num1 % num2        num1 * num2              num1 / num2
  num1 + num2        num1 - num2
  ```

  The usual binary numeric operators, along with the caret for exponentiation, may be followed immediately by a `:=` to perform an augmented assignment; for example, x `+:= 1` to add one to x. Almost all binary operators can be used with `:=` to perform augmented assignment.

- String and List concatenate:

```
str1 || str2
lst1 ||| lst2
```

To concatenate is to attach the first and second operand, in order, and produce the result.

- Assign a value:

```
variable := expr
```

In an assignment, the value on the right is stored in the variable on the left-hand side.

- Comparison:

```
num1 = num2              num1 ~= num2
str1 == str2             str1 ~== str2
num1 < num2              str1 << str2
num1 <= num2             str1 <<= str2
num1 > num2              str1 >> str2
num1 >= num2             str1 >>= str2
ex1 === ex2              ex1 ~=== ex2
```

The usual numeric comparison operators are provided, along with string versions that generally repeat the operator character. The tilde means NOT. The equivalent operator, ===, and not-equivalent operator, ~===, do not do any type conversion, while the others generally coerce operands to numeric or string types as needed. Comparison operators result in their second operand unless they fail.

- And:

```
ex1 & ex2
```

A binary ampersand operator tests the expression ex1 and if it succeeds, the result of the whole expression is the result of ex2. If ex1 fails, ex2 is not evaluated.

- Make an empty list:

```
[ ]
```

- Make an initialized list:

```
[ ex [ , ex ]* ]
```

- Make an expression results list:

  ```
  [: ex :]
  ```

- Make an initialized table:

  ```
  [ ex : ex [ ; ex : ex ]* ]
  ```

  When brackets enclose zero or more elements, lists and tables are created. Initialized lists contain elements separated by commas. Bracket-colons enclose an expression results list; the expression is fully evaluated (as per every) and all results are placed in the resulting list that is constructed. This notation should only be used on expressions that produce a finite sequence of results. For an initialized table, elements consist of **key-value pairs**, separated by semi-colons.

- Select subelement:

  ```
  ex1 [ ex2 [ , ex]* ]
  ```

- Slice:

  ```
  ex1 [ ex2 : ex3 ]
  ```

- Positive and negative relative slice:

  ```
  ex1 [ ex2 +: ex3 ]
  ex1 [ ex2 -: ex3 ]
  ```

  For lists and strings, when brackets have an expression to their left, an element or slice of that expression is taken. The expression L[1,2] is equivalent to L[1][2]. Regular element referencing picks out an element from a value, such as a string or a list. The element may be read and used in a surrounding expression or written into and replaced with an assignment. Subscripts normally start with a 1 for the first element. List and string indexes fail on out-of-range indices. Slicing is defined for both lists and strings. A string slice may be assigned if the original string is a variable. A list slice creates a list that contains a copy of the selected elements of the base list.

  The subscripts for tables are keys and may be of any type. Table indexes result in the table default value when an unknown key is looked up. Records accept both strings and integer subscripts as if they were both tables and lists.

- Access field:

  ```
  x . name
  ```

  The dot operator picks the name field out of a record or class instance, x.

Invoking procedures, functions, and methods

One of the most fundamental abstractions in all programming is the act of asking another piece of code, somewhere else, to compute a value that is needed in an expression. In Unicon, you can invoke, or call, a user-written procedure, a built-in function, or a class method by following its name or reference with parentheses while enclosing zero or more values.

- Call and Method Call:

  ```
  f ( [ expr1 [ , [ expri ] ]* ] )
  object . method ( [ expr1 [ , [ expri ] ]* ] )
  ```

  A procedure or function is called by following it with parentheses enclosing zero or more argument expressions, separated by commas. An omitted argument results in passing a null value in that position. Execution moves to that procedure or function and comes back when a result is produced, or no result is possible. A method is called by accessing the method name through an object.

- Finish call:

  ```
  return [ expr ]
  ```

  return produces expr as the result of a method or procedure. The call cannot resume. If the result, expr, is not provided, the null value is returned.

- Produce a result:

  ```
  suspend [ expr ]
  ```

  suspend produces expr as the result of a method or procedure. The call will be resumed to try for another result if the expression where the call was made fails. If the result, expr, is not provided, the expression produces the null value.

- End call without result:

  ```
  fail
  ```

  fail terminates a procedure or method call without a result. The call may not be resumed.

Iterating and selecting what and how to execute

Several Unicon control structures cover traditional control flow operations. These include sequencing, looping, and selecting pieces of code to execute:

- Execute in sequence:

  ```
  { expr1 [ ; expr2 ]* }
  ```

  Curly brackets denote expressions to be evaluated in sequence. Semi-colons terminate each expression in the sequence. Unicon features automatic semi-colon insertion, so semi-colons are rarely needed except for when two or more expressions are on the same line.

- If-then:

  ```
  if ex1 then ex2 [else ex3]
  ```

  if executes ex2 when ex1 succeeds; otherwise, it executes ex3.

- Evaluate until it fails:

  ```
  while ex1 [ do ex2 ]
  ```

  A while loop iterates executing ex1 followed by ex2, until ex1 fails.

- Consume a generator:

  ```
  every ex1 [ do ex2 ]
  ```

  An every loop executes a single evaluation of ex1 in which for each result produced by ex1, which is usually a generator, ex2 is executed. The loop terminates after all the results from ex1 have been produced.

- Loop body:

  ```
  do ex
  ```

  The do clause in loops is usually optional and provides a body expression, ex, to execute on each iteration of a loop. The loop body expression is often a compound expression enclosed in curly brackets.

- Evaluate forever:

  ```
  repeat ex
  ```

  The repeat expression is a loop that reevaluates ex over and over. Among other ways, ex may exit the loop using break, return, fail, or by halting program execution.

- Get out of loop:

```
break [ ex ]
```

break terminates a loop in the current procedure or method – always the nearest one. The ex expression is evaluated after the loop is terminated. You can write break break to get out of two loops, break break break to get out of three loops, and so on.

- Scan string:

```
str ? ex
```

The binary question mark looks like an operator but is actually a control structure that executes ex, setting &subject to str. The &pos keyword is started at 1. String scanning can be nested. It has a dynamic scope. The body expression, ex, is often a compound expression enclosed in curly brackets.

- Execute one branch:

```
case ex of { [ ex1 : ex2 ] * ; [default : exN ] }
```

case evaluates a key expression and compares the result against a sequence of case branches, tested in order. The comparison is performed without type conversions as per the === operator. If the key expression is equivalent to one of the expressions to the left of a colon, the expression on the right of that colon is executed as the result of the case expression.

- Run on first call:

```
initial ex
```

An initial clause is an optional part of a procedure or method body that evaluates an expression the first time that the procedure or method is called, but skipped on all subsequent calls. If present, the initial clause must be located at the front of a procedure or method body.

Generators

Some expressions in Unicon can produce multiple results; they are called generators. If a generator is resumed for a second or subsequent result, any new result triggers the re-execution of the surrounding expression, which may end up producing multiple results for its enclosing expression. For example, consider the call ord("="|"+"|"-"). The ord(s) function, which returns the ASCII code for s, is not a generator, but if its parameter expression is a generator, the whole call to ord() becomes a generator. In this case, "="|"+"|"-" is a generator that can produce three results.

If the enclosing expression needs all of them, ord( ) may get called three times and yield three results to an enclosing expression. As another example of this feature, consider the following expression:

```
\kids[1|2].first | genlabel()
```

This generator is capable of producing the .first field from either kids[1] or kids[2], provided that kids is not null and not empty, but if those did not occur or did not satisfy the surrounding expression, this expression would call genlabel( ) and produce its result(s) if it has any. Unicon's generator expressions include:

- Alternation:

  ```
  ex1 | ex2
  ```

  An alternation generates results from ex1 and then ex2. Although it can be read and used as an or expression, it is almost more of a concatenation of the sequences of results from the two subexpressions.

- Generate components:

  ```
  ! ex : any*
  ```

  A unary exclamation operator produces constituent pieces of a value in some defined order. Integers are generated by counting from 1 to the number. Strings, lists, records, and objects are generated by producing their values one at a time in a well-defined order. Tables and sets behave similarly but the order is undefined. Csets produce their element one-letter strings in ascending alphabetical order. Files generate their contents a line at a time.

- Finite numeric sequence:

  ```
  ex1 to ex2 [by ex3]
  ```

  A to expression generates numbers from ex1 to ex2. The default step is 1, but if by is provided, the sequence steps by that amount each time.

Debugging and environmental issues

This section contains information you may find useful when programming in Unicon. This includes a brief introduction to the Unicon debugger, some environment variables that you can set to modify Unicon runtime behavior, and a simple preprocessor that Unicon provides.

Learning the basics of the UDB debugger

Unicon's source-level debugger is named udb and is described in UTR 10, which can be read at `http://unicon.org/utr/utr10.html`. udb's command set is based on that of gdb, which lives at `https://www.gnu.org/software/gdb/`.

When you run udb, you provide the program to debug as a command-line argument. Alternatively, from within the debugger, you can run the load command to specify the program to debug. The debugger is normally exited using the quit (or q) command.

The udb prompt recognizes a lot of commands, often with an abbreviated form available. Perhaps after the quit command, the next most important command is help (or h).

The next most important command is the run (or r) command. It can be used to restart the program's execution from the beginning.

To set a breakpoint at a line number or procedure, you can use the break (or b) command, followed by the line number or procedure name. When execution hits that location, you will return to the udb command prompt. At that point, you can use step (or s) to execute one line at a time, next (or n) to run to the next line while skipping over any procedures or methods called, print (or p) to get the values of variables, or cont (or c) to continue execution at full speed.

Environment variables

Several environment variables control or modify the behavior of Unicon programs or the Unicon compiler. The most important of these are summarized here. By default, Unicon's block region heap and string region heap are sized proportional to physical memory, but you can set several of the runtime memory sizes explicitly:

Environment variable	Description
BLKSIZE	Bytes in the block heap
IPATH	List of directories to search for linking
LPATH	List of directories to search for includes
MSTKSIZE	Bytes on the main stack
STKSIZE	Bytes on co-expression stacks
STRSIZE	Bytes in the string heap
TRACE	Initial value of &trace

Table A.3: Environment variables and their descriptions

IPATH is also used to look for superclasses and package imports. Now, let's look at Unicon's pre-processor, which works like a simplified C preprocessor.

Preprocessor

The Unicon preprocessor performs file includes and replaces symbolic constants with values. The preprocessor also allows chunks of code to be enabled or disabled at compile time. This facilitates, for example, compiling different code for different operating systems.

Preprocessor commands

Lines that begin with a dollar sign are preprocessor directives:

```
$define sym text
```

The symbol, sym, is replaced with text. There are no macro parameters in this construct:

```
$include filenam
```

The file named filenam is incorporated into the source code where $include was found:

```
$ifdef sym
$ifndef sym
$else
$endif
```

Lines inside $ifdef are passed along to the compiler if sym was introduced by a previous $define. $ifndef passes along source code if a symbol was not defined. These two directives take an optional $else, followed by more code, and are terminated by $endif:

```
$line num [ filenam ]
```

The next line should be reported as starting at line num from the filenam file:

```
$undef sym
```

The definition, sym, is erased. Subsequent occurrences are not replaced by anything.

Built-in macro definitions

These symbols identify platforms and features that may be present and affect language capabilities. A leading underscore suggests their built-in status. The built-in macro definitions include the following:

Defined macro	Meaning	Defined macro	Meaning
CO_EXPRESSIONS	Synchronous threads	_MESSAGING	HTTP, SMTP, etc.
_CONSOLE_WINDOW	Emulated terminal	_MS_WINDOWS	Microsoft Windows
_DBM	DBM	_MULTITASKING	load(), etc.
_DYNAMIC_LOADING	Code can be loaded	_POSIX	POSIX
_EVENT_MONITOR	Code is instrumented	_PIPES	unidirectional pipes
_GRAPHICS	Graphics	_SYSTEM_FUNCTION	system()
_KEYBOARD_FUNCTIONS	kbhit(), getc(), etc.	_UNIX	UNIX, Linux, ...
_LARGE_INTEGERS	Arbitrary precision	_WIN32	Win32 graphics
_MACINTOSH	Macintosh	_X_WINDOW_SYSTEM	X Windows graphics

Table A.4: Built-in macros

These symbols, which you can check at compile time using $ifdef, have corresponding feature strings that can be checked at runtime using &features. For details, you can look at *Programming with Unicon*. Now, let's look at Unicon's built-in functions.

Function mini-reference

This section describes a subset of Unicon's built-in functions deemed most likely to be relevant to programming language implementers. For a full list, see *Appendix A of Programming with Unicon*. The parameters' required types in this section are given by their names. The names c or cs indicate a character set. The names s or str indicate a string. The names i or j indicate integers. A name such as x or any indicates that the parameter may be of any type. Such names may be suffixed with a number to distinguish them from other parameters of the same type. The colons and types after the parameters indicate return types, along with the number of returned values. Normally, a function will have exactly one return value. A question mark indicates that the function is a predicate that can fail with zero or one return value. An asterisk indicates that the function is a *generator* with zero or more return values.

Many functions also have default values for parameters, indicated in the reference using a colon and a value after their name. Functions with parameters ending in str, i1, and i2 are string analysis functions. String analysis functions' last three parameters default to &subject, &pos, and 0. The i1 and i2 parameters are swapped if i1 is greater than i2, so it does not matter in what order the indices are supplied. String analysis is always conducted from left to right:

- abs(n) : num

  abs(n) returns -n if n is negative. Otherwise, it returns n.

- `any(cs, str, i1, j2) : integer?`

  `any(cs, str, i1, i2)` produces i1+1 when `str[i1]` is a member of cset, `cs`, and fails otherwise.

- `bal(c1:&cset, c2:'(', c3:')', str, i1, i2) : integer*`

  `bal(c1, c2, c3, str, i1, i2)` produces indices in `str` where a member of c1 in `str[i1:i2]` is balanced as far as opener characters in c2 and closer characters in c3.

- `char(i) : str`

  `char(i)` returns the one-letter string encoding of i, which must be in the range 0-255.

- `close(f) : file`

  `close(f)` releases operating system resources associated with f and closes it.

- `copy(any) : any`

  `copy(y)` produces y. For structures, it returns a physical copy. For nested structures, the copy is one level deep.

- `delay(i) : null`

  `delay(i)` does nothing for at least i milliseconds, after which it returns, allowing execution to continue.

- `delete(y1, y2, …) : y1`

  `delete(y1, y2)` removes one or more values at the key location, y2, and any subsequent elements from the y1 structure.

- `exit(i:0) :`

  `exit(i)` quits the program execution and produces i as an exit status.

- `find(s, str, i1, i2) : int*`

  `find(s, str, i1, i2)` produces indices where s occurs in `str`, considering only indices between i1 and i2.

- `getenv(str) : str?`

  `getenv(str)` produces a value named `str` from the environment.

- `iand(i1, i2) : int`

  `iand(i1, i2)` returns i1 bitwise-ANDed with i2.

- `icom(i) : int`

  `icom(i)` flips the ones to zeros and the zeros to ones, producing the integer complement.

- `image(x) : str`

  `image(x)` produces a string that represents the contents of x.

- `insert(x1, x2, x3:&null) : x1`

  `insert(x1, x2, x3)` places x2 in the x1 structure. If x1 is a list, x2 is a position; otherwise, it is a key. If x1 is a table, the x2 key is associated with the x3 value. `insert()` produces the structure.

- `integer(x) : int?`

  `integer(x)` coerces x into the integer type. It fails when conversion is not possible.

- `ior(i1, i2) : int`

  `ior(i1, i2)` returns i1 bitwise-ORed with i2.

- `ishift(i1, i2) : int`

  `ishift(i1, i2)` shifts i2 bit positions over within i1 and returns the result. The shift goes right if i2<0 or left if i2>0. During the shift, i2 zero bits come in from the opposite direction of the shift.

- `ixor(i1, i2) : int`

  `ixor(i1, i2)` returns i1 bitwise-exclusive-ORed with i2.

- `kbhit() : ?`

  `kbhit()` returns whether a key on the keyboard has been pressed or not.

- `key(y) : any*`

  `key(y)` produces keys/indices with which a structure's y elements may be accessed.

- `list(i, x) : list`

  `list(i, x)` constructs a list with i elements that each contain x. x is not copied for each element of the list, so you have to allocate each element separately if you want a list of lists, for example.

- `many(cs, str, i1, i2) : int?`

  `many(cs, str, i1, i2)` produces the position in str that follows as many contiguous members of cs within str[i1:i2] as possible.

- `map(str1, str2, str3) : str`

  `map(str1, str2, str3)` returns `str1` transformed so that where `str1`'s characters may be found in `str2`, they are replaced with the corresponding characters in `str3`. `str2` and `str3` must be of the same length.

- `match(s, str, i1, i2) : int?`

  `match(s, str, i1, i2)` returns `i1+*s` when `s==str[i1+:*s]`. The function fails when there is no match.

- `max(num,…) : num`

  `max(…)` produces the numeric maximum of its parameters.

- `member(y,…) : y?`

  `member(y,…)` produces y when the other parameters are in structure y; otherwise, it fails.

- `min(num,…) : num`

  `min(…)` produces the numeric minimum of its parameters.

- `move(i) : str`

  `move(i)` increments or decrements `&pos` by i and returns a substring from the old to the new position within `&subject`. The position is reset if this function is resumed.

- `open(str1, str2, ...) : file?`

  `open(str1, str2,…)` asks the operating system to open the file named `str1` using mode `str2`. Subsequent arguments are attributes that may affect special files. The function recognizes the following modes, which are given in the `str2` argument:

Mode letter(s)	Description	Mode letter(s)	Description
a	add/append	nl	listen on a TCP port
b	open for both reading and writing	nu	connect to a UDP port
c	make a new file	m	connect to messaging server
d	GDBM database	o	ODBC (SQL) connection
g	2D graphics window	p	execute a command line and pipe it
gl	3D graphics window	r	read
n	TCP client	t	translate newlines
na	accept TCP connection	u	use a binary untranslated mode
nau	accept UDP datagrams	w	write

Table A.5: Modes and their descriptions

- `ord(s) : integer`

  `ord(s)` returns the ordinal (for example, ASCII code) of a one-letter string, s.

- `pop(L) : any?`

  `pop(L)` returns a value from the front of L and removes it from the list.

- `pos(i) : int?`

  `pos(i)` returns whether string scanning is at the location, i.

- `proc(x, i:1) : procedure?`

  `proc(str, i)` produces a procedure that is denoted `str` in the current scope. If i is 0, the built-in function named `str` is produced if there is one by that name. The official Unicon book defines additional behaviors of `proc()` when x is a thread, co-expression, or procedure, or when `str` denotes an operator such as `"-"` and i denotes its arity.

- `pull(L, i:1) : any?`

  `pull(L)` returns the last element of L and removes it. It can remove i elements.

- `push(L, y, ...) : list`

  `push(L, y1, …, yN)` pushes one or more elements onto the list, L, at the front. `push()` returns its first parameter, with new values added.

- `read(f:&input) : str?`

  `read(f)` inputs the next line of file f and returns it without the newline.

- `reads(f:&input, i:1) : str?`

  `reads(f, i)` inputs i bytes from the file, f, failing if no more bytes are there. `reads()` returns with available input, even if it is less than i bytes. When -1 bytes are requested, `reads()` returns a string that contains all the remaining bytes in the file.

- `ready(f:&input, i:0) : str?`

  `ready(f, i)` inputs i bytes from the file, f, usually a network connection. It returns without blocking and if that means less than i bytes are available, so be it. It fails when no input has arrived yet.

- `real(any) : real?`

  `real(x)` coerces x into its floating-point equivalent. It fails when no coercion is possible.

- `remove(str) : ?`

remove(str) deletes the file named str from the filesystem.

- rename(str1, str2) : ?

  rename(str1, str2) changes the str1 file's name to str2.

- repl(y, i) : x

  repl(x, i) produces i concatenated instances of x.

- reverse(y) : y

  reverse(y) produces a list or string that is in the opposite order of y.

- rmdir(str) : ?

  rmdir(str) deletes the folder with the name str or fails if it cannot be deleted.

- serial(y) : int?

  serial(y) produces an identifying integer for the structure, y. These numbers are as-signed when structures are allocated. Separate counters are used for each structure type. The identifying integer provides the chronological order in which instances of each type were allocated.

- set(y, …) : set

  set() allocates a set. Parameters are the initial values of the new set, except if they are lists; lists' elements are the initial values of the new set. Here, the parameters' contents are the initial values of the new set.

- sort(y, i:1) : list

  sort() allocates a list in which elements of y are sorted. When tables are sorted, keys are sorted when i is one or three, and values are sorted when i is two or four. When i is one or two, the return list's elements are two-element key-value sublists; when i is three or four, the return list's elements alternate between keys and values.

- stat(f) : record?

  stat(f) produces information about f. The argument may be a string filename or an open file. Three portable fields are size in bytes, mode access permissions, and the last modified time, mtime. The mode string resembles the long listing from ls(1). stat(f) fails when there is no filename or path, f.

- stop(s, ...) :

stop(args) writes its arguments to &errout, followed by a newline, and then quits the program.

- string(any) : str?

string(y) coerces y into a corresponding string. It fails when no conversion is possible.

- system(x, f:&input, f:&output, f:&errout, s) : int

system(x) runs a program given as a string command line or a list of command-line arguments. The program runs as a separate process. Optional arguments supply the standard I/O files. The process's exit status is returned. If the fifth parameter is "nowait", the function immediately returns with the new process ID instead of waiting for it to complete.

- tab(i:0) : str?

tab(i) assigns the location specified in i to &pos. It produces a substring between the new and former locations. The &pos keyword is reset to its former position if the function resumes.

- table(k, v, ..., x) : table

table(x) builds a table whose values default to x. table(k, v,…, x) initializes a table from alternating key and value arguments, ending (if the number of parameters is odd) with the table's default value.

- trim(str, cs:' ', i:-1) : str

trim(str, cs, i) produces a substring of str with members of cset, cs, deleted from the front (when i=1), the back (when i=-1), or both (when i=0). By default, it removes trailing spaces from the end.

- type(x) : str

type(x) produces the string name of the type x.

- upto(cs, str, i1, i2) : int*

upto(cs, str, i1, i2) generates the indices in str at which a member of cset, cs, may be found in str[i1:i2]. It fails otherwise.

- write(s|f, ...) : str|file

write(…) sends one or more string arguments appended by a newline to a file, defaulting to &output. write() produces the final parameter as its return value.

- writes(s|f, ...) : str|file

`writes(…)` sends one or more string arguments to a file, defaulting to &output. `writes()` produces the final parameter as its return value.

Selected keywords

Unicon has about 75 keywords. Keywords are global names beginning with an ampersand with a predefined meaning. Many keywords are constant or read-only values that are built into the language, while others are associated with built-in domain-specific language facilities such as string scanning or graphics. This section lists the most essential keywords, many of which appear in the examples in this book:

- `&clock : str`

  The `&clock` read-only keyword produces the current time of day.

- `&cset : cset`

  The `&cset` constant keyword denotes the cset containing every character.

- `&date : str`

  The `&date` read-only keyword produces the current date.

- `&digits : cset`

  The `&digits` constant keyword denotes the cset containing "0" through "9".

- `&errout : file`

  The `&errout` read-only keyword denotes the standard location for error output, often a terminal console window.

- `&fail :`

  The `&fail` keyword is an expression that fails to produce a result.

- `&features : str*`

  The `&features` read-only keyword produces a sequence of strings indicating any optional features and capabilities of this Unicon runtime system build. For example, if Unicon is built with graphics facilities, one or more strings indicate whether 2D or 3D are supported, whether various image file formats are available in this Unicon build, etc.

- `&input : file`

  The `&input` read-only keyword denotes the standard location from which to read input.

- `&lcase : cset`

  The `&lcase` constant keyword denotes the cset containing the letters "a" through "z".

- `&letters : cset`

  The `&letters` constant keyword denotes the cset containing the letters "A" through "Z" and "a" through "z".

- `&now : int`

  The `&now` read-only keyword produces the number of seconds since 1/1/1970 GMT.

- `&null : null`

  The `&null` constant keyword denotes a valueless value whose type is the null type. It is the default value in many language constructs, for things that haven't been initialized yet or have been omitted.

- `&output : file`

  The `&output` constant keyword denotes the standard default location where output is written, often a terminal console window.

- `&pos := int`

  The `&pos` keyword refers to the position within `&subject` where string analysis is performed. It starts at 1 in each string scanning environment and its value is constrained to always be a valid index within `&subject`.

- `&subject := str`

  The `&subject` keyword refers to the string under analysis in a string scanning control structure. Assigning a value to `&subject` has the side effect of setting `&pos` to 1.

- `&ucase : cset`

  The `&ucase` constant keyword denotes the cset containing letters "A" through "Z".

- `&version : str`

  The `&version` constant keyword reports the Unicon version as a string.

Join our community on Discord

Join our community's Discord space for discussions with the authors and other readers:

`https://discord.com/invite/zGVbWaxqbw`

Answers

The following answers sketch some possible solutions to the questions at the end of each chapter; these are provided for your reflection.

Chapter 1

1. It is much easier to generate C code than to generate machine code, but the resulting code may be larger or slower than native code, causing a performance cost. A transpiler depends on an underlying compiler that may be a bit of a moving target, but if the underlying compiler is highly portable, the transpiler will be far more portable than a compiler that generates native code.

2. Lexical, syntax, and semantic analysis, followed by intermediate and final code generation.

3. Classic pain points include input/output being overly difficult, especially on new kinds of hardware; concurrency; and making a program run across many different operating systems and CPUs. One feature that languages have used to simplify input/output has been to reduce the problem of communicating with new hardware via a set of strings in human-readable formats, for example, to play music or read touch input. Concurrency has been simplified in languages with built-in threads and monitors. Portability has been simplified in languages that provide their own high-level virtual machine implementation.

4. This depends on your application domain of interest, but here is one. The language will input programs written in a Java-like syntax stored in files with a .j0 extension, generating target code in the form of HTML5 and JavaScript that runs on websites. The language will support JDBC and socket communications via websockets, and 2D and 3D graphics by means of OpenGL. The language will support an intuitive square-bracket syntax for accessing string elements and HashMap keys. The language will support JSON syntax natively within the source code as a HashMap literal.

Chapter 2

1. Reserved words contribute both to human readability and ease of parsing for the language implementation, but they also sometimes preclude the most natural names for the variables in a program, and too many reserved words can make it more difficult to learn a programming language.

2. Integers in C or Java, for example, can be expressed as signed or unsigned in a decimal, octal, hexadecimal, or maybe even binary format, for small, medium, large, or super-sized words.

3. Several languages implement a semicolon insertion mechanism that makes semicolons optional. Sometimes, this involves using the newline character to replace the role of the semicolon as a statement terminator or separator. It is not usually a straightforward mapping of newline==semi-colon; there are often contextual rules involved. For example, Go adds semi-colons at newlines when the last token on a line is a member of a prescribed set of tokens that require semi-colons, which is a simple contextual rule. Icon and Unicon have additional context: in addition to looking at the last token on a line, Icon and Unicon also look at the first token of the next line, and they only insert a semi-colon if the first token on the next line can legally come after a semi-colon.

4. Although most Java programs do not make use of this capability, putting main() in several (or all) classes might be very useful in unit testing and integration testing.

5. While it is feasible to provide pre-opened input/output facilities, these can involve substantial resources and initialization costs that programs should not have to pay for unless a given input/output facility will be used in them. If you design a language that specifically targets a domain where one of these forms of input/output is guaranteed, it makes good sense to consider how to make access as simple as possible.

Chapter 3

1. A first approximation of the regular expression is [0-3][0-9]"/"[01][0-9]"/"[0-9]{4}. While it is possible to write a regular expression that matches only legal dates, such an expression is impractically long, especially when you consider leap years. In such cases, it makes sense to use the regular expression that provides the simplest close approximation of correctness, and then check the correctness in the semantic action or a subsequent semantic analysis phase.

2. yylex() returns an integer category for use in syntax analysis, while yytext is a string that contains the symbols matched, and yylval holds an object called a token that contains all the lexical attributes of that lexeme.

3. When a regular expression does not return a value, the characters that it matches are discarded and the yylex() function continues with a new match, starting with the next character in the input.

4. Flex matches the longest string that it can; it breaks ties among multiple regular expressions by selecting whichever one matches the longest string. When two regular expressions match the same length in a given point, Flex selects whichever regular expression occurs first in the lex specification file.

Chapter 4

1. A terminal symbol is not defined by a production rule in terms of other symbols. This is the opposite of a non-terminal symbol, which can be replaced by or constructed from the sequence of symbols on the right-hand side of a production rule that defines that non-terminal symbol.

2. A shift removes the current symbol from the input and pushes it onto the parse stack. A reduce pops zero or more symbols from the top of the parse stack that match the right-hand side of a production rule, pushing the corresponding non-terminal from the left side of the production rule in their place.

3. YACC gives you a chance to execute some semantic action code only when a reduce operation takes place.

4. The integer categories returned from yylex() in the previous chapter are exactly the sequence of terminal symbols that the parser sees and shifts during parsing. A successful parse shifts all the available input symbols and gradually reduces them back to the starting non-terminal of the grammar.

Chapter 5

1. The yylex() lexical analyzer allocates a leaf and stores it in yylval for each terminal symbol that it returns to yyparse().

2. When a production rule in the grammar is reduced, the semantic action code in the parser allocates an internal node, and it initializes its children to refer to the leaves and internal nodes corresponding to symbols on the right-hand side of that production rule.

3. yyparse() maintains a value stack that grows and shrinks in lock-step with the parse stack during parsing. Leaves and internal nodes are stored on the value stack until they are inserted as children into a containing internal node.

4. A value stack is fully generic and can contain whatever type(s) of values the compiler may require. In C, this is done using a union type, which is type-unsafe. In Java, it is done using a parserVal class that contains the tree nodes generically. In Unicon and other dynamic languages, no wrapping or unwrapping is needed.

Chapter 6

1. Symbol tables allow your semantic analysis and code generation phases to quickly look up symbols declared far away in the syntax tree, following the scoping rules of the language.

2. Synthesized attributes are computed using the information located immediately at a node, or using information obtained from its children. Inherited attributes are computed using information from elsewhere in the tree, such as parent or sibling nodes. Synthesized attributes are typically computed using a bottom-up, post-order traversal of the syntax tree, while inherited attributes are typically computed using a pre-order traversal. Both kinds of attributes are stored in syntax tree nodes in variables added to the node's data type.

3. The Jzero language calls for a global scope, a class scope, and one local scope for each member function. The symbol tables are typically organized in a tree structure corresponding to the scoping rules of the language, with child symbol tables attached or associated with the corresponding symbol table entries in the enclosing scope.

4. If Jzero allowed multiple classes in separate files, the symbol tables would need a mechanism to be aware of the classes. In Java, this may entail reading other source files at compile time while compiling a given file. This implies that classes must be easily found without reference to their filename, hence Java's requirement that classes be placed in files whose base name is the same as the class name.

Chapter 7

1. Type checking finds many errors that would prevent the program from running correctly, but it also helps determine how much memory will be needed to hold variables, and exactly what instructions will be needed to perform the various operations in the program.

2. A structure type is needed to represent arbitrarily deep composite structures, including recursive structures such as linked lists. Any given program only has a finite number of such types, so it would be possible to enumerate them and represent them using integer subscripts by placing them in a type table. However, references to structures provide a more direct representation.

3. If real compilers reported an OK line for every successful type check, non-toy programs would emit thousands of such checks on every compile, making it difficult to notice the occasional errors.

4. Picky type checkers may be a pain for programmers, but they help avoid unintended type conversions that hide logic errors, and they also reduce the tendency of a language to run slow due to silently and automatically converting types repeatedly at runtime.

Chapter 8

1. For any specific array access, the result of a subscript operator will be the array's element type. With a struct or class access, the name of the (member) field within the struct must be used to determine the resulting type, via a symbol table lookup or something equivalent.

2. A function's return type can be stored in the function's symbol table and looked up from anywhere within the function's body. One easy way to do this is to store the return type under a symbol that is not a legal variable name, such as return. An alternative would be to propagate the function's return type down into the function body as an inherited attribute. This might be relatively straightforward, but it seems like a waste of space in the parse tree nodes.

3. Generally, operators such as plus and minus have a fixed number of operands and a fixed number of types for which they are defined; this lends itself to storing the type checking rules in a table or a switch statement of some kind. Function calls have to perform type checking over an arbitrary number of arguments, which can be of an arbitrary type. The function's parameters and return type are stored in its symbol table entry. They are looked up and used to type check each site where that function is called.

4. Besides member access, type checking occurs when composite types are created, assigned, passed as parameters, and, in some languages, destroyed.

Chapter 9

1. It would be reasonable, and could be appropriate, to introduce three address instructions to do input and output. However, most languages' input and output operations are encapsulated by a function or method calling interface, as I/O tends to be encapsulated either by language runtime system calls or system calls.

2. Semantic rules are logical declarative statements of how to compute semantic attributes. Synthesized attributes can generally be computed by bottom-up, post-order tree traversals. Depending on their interdependences, inherited attributes can be computed by one or more top-down, pre-order tree traversals.

3. Computing labels during the same tree traversal that generates code may improve the compiler's performance. However, it is difficult to generate code with gotos whose labels point forward to code that has not been generated yet. A single-pass compiler may have to build auxiliary structures and backpatch code.

4. A naïve code generator that calls `genlocal()` for every new local variable may use far more space on the stack than is necessary. Excessive stack sizes may reduce performance due to poorer page caching, and in heavily recursive code, it may increase the possibility of a stack overflow or running out of memory.

Chapter 10

1. Colorblind individuals may be able to utilize a limited number of grayscales or textures in IDEs where color-seeing individuals use colors. For many users who have only partial color blindness, allowing users to customize the assignments of colors to various source code elements may be the best solution. If using textures to substitute for colors, one might use the background texture where text fonts are drawn. Other font styles such as bold, italics, underlining, or shadowing might also be used.

2. Reparsing does not have to depend on cursor motion. It could be based on the number of tokens inserted, deleted, or modified, the number of keystrokes, the amount of elapsed real time, or some combination of several of these factors.

3. There are many possible ways to indicate syntactic nesting. For example, nesting might be represented by indentation, a progressive darkening of the background color, or boundaries of scopes might be explicitly drawn with dashed lines.

Chapter 11

1. Preprocessors in C/C++ have been known to be a subtle source of bugs, particularly when programmers misuse them. Many of these bugs occur due to macros with parameters, whose bodies can call other macros with parameters. The output after all macros are expanded can be tricky, or surprising. Setting aside the issue of bugs, if one is not careful, macros can reduce readability instead of improving it, or they can give a false sense of security when a macro looks simple but what it expands to is complex.

2. Transpilers have a bootstrapping problem: they depend on some other high-level language being ported first. Transpilers leave you dependent on the underlying language working correctly, but the underlying language may change in ways that break the transpiler. Transpilers may also introduce problems with performance or the debugging of code.

Chapter 12

1. Complex instruction sets take more time and logic to decode and might make the implementation of the byte-code interpreter more difficult or less portable. On the other hand, the closer the final code comes to resembling intermediate code, the simpler the final code generation stage becomes.

2. Implementing bytecode addresses using hardware addresses provides the best performance that you might hope for, but it may leave an implementation more vulnerable to memory safety and security issues. A bytecode interpreter that implements addresses using offsets within an array of bytes may find it has fewer memory problems; performance may or may not be a problem.

3. Some bytecode interpreters may benefit from the ability to modify code at runtime. For example, bytecode that was linked using byte-offset information may be converted into code that uses pointers. Immutable code makes this type of self-modifying behavior more difficult or impossible.

Chapter 13

1. Operands from multi-operand instructions are pushed onto the stack by PUSH instructions. The actual operation computes a result. The result is stored in memory by a POP instruction.

2. A table that maps each of the labels to byte offset 120 is constructed. Uses of labels encountered after their table entry exists are simply replaced by the value 120. Uses of labels encountered before their table entry exists are forward references; the table must contain a linked list of forward references that are backpatched when the label is encountered.

3. On the Jzero bytecode stack machine, operands might already be on the stack and PARM instructions might be redundant, allowing for substantial optimization. Also, on the Jzero machine, the function call sequence calls for a reference/address to the method being called to be pushed before the operands; this is a very different calling convention from that used in the three-address intermediate code.

4. Static methods do not get invoked on an object instance. In the case of a static method with no parameters, you may need to push the procedure address within the CALL instruction, since it is preceded by no PARM instructions.

5. It is possible to guarantee no nesting of PARM...CALL sequences by re-arranging code and introducing additional temporary variables, but that can be cumbersome. If you determine that your three-address code for nested calls does in fact result in nested PARM...CALL sequences, you will need a stack of PARM instructions to manage it and will need to carefully search for the correct CALL instruction, skipping over any nested CALL instructions whose number of PARM instructions were placed on the stack after the PARM instruction that you are searching for. Have fun!

6. Whether portability trumps performance is a design decision, and there is no one right answer. It is possible to write a portable bytecode in which portable instructions rewrite themselves into native formats on the fly. In an extreme case, this might entail just-in-time compilation to pure native code.

Chapter 14

1. There are many new concepts in native code. These include many kinds and sizes of registers and main memory access modes. Choosing from many possible underlying instruction sequences is also important.

2. Even with the runtime addition required, addresses that are stored as offsets relative to the instruction pointer may be more compact and take advantage of instruction prefetching in the pipelined architecture, providing faster access to global variables than specifying them using absolute addresses. Disadvantages might include increased difficulty of debugging, reduced human readability of the assembler code, and/or higher complexity needed for alias analysis than what might be needed for optimization and reverse engineering tasks.

3. Function call speed is important because modern software is often organized into many frequently called tiny functions. The x64 architecture performs fast function calls if functions take advantage of passing the first six parameters in registers. Several aspects of x64 architecture seem to have the potential to reduce execution speed, such as a need to save and restore large numbers of registers to memory before and after a call.

Chapter 15

1. Although libraries are great, they have downsides. Libraries tend to have more version compatibility problems than the features that are built into the language. Libraries are unable to provide a notation that is concise and readable as built-ins. Lastly, libraries do not lend themselves to interactions with novel control structures to support new application domains.

2. If your new computation only needs one or two parameters, appears many times in typical applications in your domain, and computes a new value without side effects, it is a good candidate to be made into an operator. An operator is limited to two operands, or at the most, three; otherwise, it will not provide any readability advantage over a function. Unless new operators are familiar or can be used analogously to familiar operators, they may be less readable over time than using named functions.

3. Ultimately, we have to read the books written by the Java language inventors to hear their reasons, but one answer might be that Java designers wanted to use strings as a class and decided classes would not be free to implement operators, for the sake of referential transparency.

Chapter 16

1. Control structures in very high-level and domain-specific languages had better be a lot more expressive and powerful than just if statements and loops; otherwise, programmers would be better off just coding in a mainstream language. Often, a control structure can add power by changing the semantic interpretation of the code inside it, or by changing the data to which the code is applied.

2. We provided some examples in which control structures provided defaults for parameters or ensured an open resource was closed afterward. Domain-specific control structures can certainly provide additional high-level semantics, such as performing domain-specific input/output or accessing specialty hardware in a way that is difficult to accomplish within the context of a mainstream control flow.

3. The application domain is string analysis. Maybe some additional operators or built-in functions would improve Unicon's expressive power for string analysis. Can you think of any candidates you could add to the six-string analysis functions or the two position-moving functions? You could easily run some statistics on common Icon and/or Unicon applications and discover which combinations of tab() or move() and the six-string analysis functions occur most frequently in the code and are candidates for becoming operators, besides tab(match()). I doubt that tab(match()) is the most frequent. But beware: if you add too many primitives, it makes the control structure more difficult to learn and master. Also, the ideas from this control structure could be applied to the analysis of other sequential data, such as arrays/lists of numeric or object instance values.

4. It is tempting to bundle as much additional semantics into a domain control structure as possible so that you make the code more concise. However, if a good number of `wsection` constructs are not based on a hierarchical 3D model and would not make use of the built-in functionality of `PushMatrix()` and `PopMatrix()`, bundling that into `wsection` might slow down the construct's execution speed unnecessarily.

Chapter 17

1. You could modify the `PostDescrip()` macro to check for a null value before checking whether a value is a qualifier or a pointer. Whether such a check pays for itself depends on how costly the bitwise AND operator is, and the actual frequency of different types of data encountered during these checks, which can be measured but may well vary, depending on the application.

2. If each class type had its own heap region, it may become possible in the implementation that class instances might no longer need to track their size, potentially saving memory costs for classes that have many small instances. The freed garbage instances could be managed on a linked list and compared with a mark-and-sweep collector, and instances might never need to be moved or pointers updated, simplifying garbage collection. On the other hand, some program runs might only use a very few of the various classes, and allocating a dedicated heap region for such classes might be a waste.

3. While some time might be saved by not moving data during garbage collection, over time, a substantial amount of memory might be lost to fragmentation. Small chunks of free memory might go unused because later memory allocation requests were for larger amounts. The task of finding a free chunk of sufficient size might become more complex, and that cost might exceed the time saved by not moving data.

Join our community on Discord

Join our community's Discord space for discussions with the authors and other readers:

`https://discord.com/invite/zGVbWaxqbw`

packt.com

Subscribe to our online digital library for full access to over 7,000 books and videos, as well as industry leading tools to help you plan your personal development and advance your career. For more information, please visit our website.

Why subscribe?

- Spend less time learning and more time coding with practical eBooks and Videos from over 4,000 industry professionals
- Improve your learning with Skill Plans built especially for you
- Get a free eBook or video every month
- Fully searchable for easy access to vital information
- Copy and paste, print, and bookmark content

At www.packt.com, you can also read a collection of free technical articles, sign up for a range of free newsletters, and receive exclusive discounts and offers on Packt books and eBooks.

Other Books
You May Enjoy

If you enjoyed this book, you may be interested in these other books by Packt:

50 Algorithms Every Programmer Should Know – Second Edition

Imran Ahmad

ISBN: 9781803247762

- Design algorithms for solving complex problems
- Become familiar with neural networks and deep learning techniques
- Explore existing data structures and algorithms found in Python libraries
- Implement graph algorithms for fraud detection using network analysis
- Delve into state-of-the-art algorithms for proficient Natural Language Processing illustrated with real-world examples

- Create a recommendation engine that suggests relevant movies to subscribers
- Grasp the concepts of sequential machine learning models and their foundational role in the development of cutting-edge LLMs

Learn LLVM 17 – Second Edition

Kai Nacke

Amy Kwan

ISBN: 9781837631346

- Configure, compile, and install the LLVM framework
- Understand how the LLVM source is organized
- Discover what you need to do to use LLVM in your own projects
- Explore how a compiler is structured, and implement a tiny compiler
- Generate LLVM IR for common source language constructs
- Set up an optimization pipeline and tailor it for your own needs
- Extend LLVM with transformation passes and clang tooling
- Add new machine instructions and a complete backend

Packt is searching for authors like you

If you're interested in becoming an author for Packt, please visit `authors.packtpub.com` and apply today. We have worked with thousands of developers and tech professionals, just like you, to help them share their insight with the global tech community. You can make a general application, apply for a specific hot topic that we are recruiting an author for, or submit your own idea.

Share your thoughts

Now you've finished *Build Your Own Programming Language, Second Edition*, we'd love to hear your thoughts! Scan the QR code below to go straight to the Amazon review page for this book and share your feedback or leave a review on the site that you purchased it from.

`https://packt.link/r/1804618020`

Your review is important to us and the tech community and will help us make sure we're delivering excellent quality content.

Index

X

x64 instruction set 372, 373

  class, adding 373, 374

  memory regions, mapping 374

x64 output generation 393

  linking 395

  loading 395

  native assembler, invoking to
     object file 394, 395

  runtime system 396

  x64 code, writing in assembly
     language format 393, 394

Y

yacc (yet another compiler-compiler) 74

  advanced yacc declarations 75

  conflicts, fixing in parsers 79

  context-free grammar section 75, 76

  parsers 76-79

  specifications 74

  symbols, declaring in header section 75

  syntax error recovery 80

  toy example 80-85

  value stack, working with 112, 113

Download a free PDF copy of this book

Thanks for purchasing this book!

Do you like to read on the go but are unable to carry your print books everywhere?

Is your eBook purchase not compatible with the device of your choice?

Don't worry, now with every Packt book you get a DRM-free PDF version of that book at no cost.

Read anywhere, any place, on any device. Search, copy, and paste code from your favorite technical books directly into your application.

The perks don't stop there, you can get exclusive access to discounts, newsletters, and great free content in your inbox daily

Follow these simple steps to get the benefits:

1. Scan the QR code or visit the link below

https://packt.link/free-ebook/9781804618028

2. Submit your proof of purchase
3. That's it! We'll send your free PDF and other benefits to your email directly